Abrasive Methods Engineering

VOLUME 1

ABRASIVE METHODS ENGINEERING

VOLUME 1

by FRANCIS T. FARAGO, Ph.D

FORMERLY SENIOR RESEARCH ENGINEER,
GENERAL MOTORS CORPORATION

INDUSTRIAL PRESS
1980

Library of Congress Cataloging in Publication Data

Farago, Francis T
    Abrasive methods engineering.
    Includes index.
    1. Grinding and polishing—Handbooks, manuals, etc.

I. Title.
TJ1280.F3   621.9'2   76-14970
ISBN 0-8311-1134.8

Third Printing

CONTENTS

SECTION I - CYLINDRICAL GRINDING

CONTENTS

List of Tables

LIST OF TABLES

PREFACE

The use of abrasives, bonded into grinding wheels and, to a lesser degree, in other forms, as a means of machining and finishing, represents one of the most important areas of metalworking production. Because of the many and often unique advantages of machining with abrasives, the use of this type of metalworking is continually growing in most branches of modern machinery and related industries.

In spite of this it may seem surprising that the factual knowledge of the engineering community about grinding processes in general, does not reflect its important role in metalworking. This statement will be confirmed by most engineers associated with metalworking, and also is substantiated by the relative sparsity of technical books on grinding. In distinction to the extensive literature which has been published on most of the manufacturing methods which operate with metallic cutting tools, extremely few specialized books have appeared in the past, and possibly none in the last decade, on the technology of grinding.* Not counting the rather condensed chapters on grinding in general textbooks on machining practices, as well as various specialized publications on a limited number of subjects issued by manufacturers of abrasive equipment and tools, no up-to-date book on grinding technology is at present available to engineers engaged in different fields of metalworking.

One of the reasons for the lack of engineering books on grinding and other abrasive methods probably results from the fact that grinding, in general, is a process much more difficult to describe and specify in concrete terms than most other methods of metalworking. In the latter, well-defined and consistently repetitive actions take place between the cutting tool

(with the accurately designed shape of its elements), and with precisely controlled speeds and dimensions of penetration of the individual tool edges into the work material; while, in grinding, the elements of action are much less uniform.

In pointing out the relative lack of technical books on grinding as a family of specific metalworking processes, one should not infer a poor supply of published material in this field. As a matter of fact, prominent theoretical scientists and practical experts have published many valuable papers on particular aspects or applications of grinding. These publications embrace an extremely wide range of levels from the purely theoretical, through experimental, into empirical or observational. Many of these publications are important building blocks of a new branch of technological science, while others often supply useful hints or actual guidance to the practical engineer for resolving specific problems. However, even a substantial collection of technical papers, when serving different objectives with regard to the level of discussion and the area of coverage, cannot substitute the usefulness of a properly designed handbook of a particular field of technology.

A properly designed technological handbook should have, besides several others, the following properties:

—*Comprehensiveness* in covering most of the important aspects to a depth compatible with the practical size of the book.

—*Balance*, giving the proper weight and extent to treating the many different subjects in details commensurate with their relative significance.

—*Systematic structure* and method of presentation, to provide a logical survey of the subject material, discussing it in a manner which is easy to follow by those who wish to read the whole book or its main sections in their entirety.

---

*One of the recognized specialized books on grinding technology is also by the author. Published in 1952, in Budapest, Hungary, it is entitled (in translation): *Abrasive Materials and Grinding Processes.*

—*Specific information* on any particular subject which may be readily found by means of the Table of Contents or the Index, thus serving readers who may want to use the book occasionally when in need of guidance or indications for resolving particular problems.

In order to satisfy these and other similar requirements which are to be considered useful by a large array of readers with different educational and practical backgrounds, varying areas and degrees of interest, and who may wish to obtain information that is expected to be comprehensive, or specific only, a *method of presentation*, believed to be well-adapted to these diverse interests, has been applied.

Structurally the four individual *sections* of the book, each devoted to one of the four basic methods of grinding, are self-contained, almost as independent books bound into a common volume, with only a minimum of cross-references to the other sections.

Each section comprises several *chapters*, again substantially self-contained in their discussion of the subject, which is a distinct aspect or clearly distinguishable branch of the comprehensive field covered in the pertinent section.

The chapters are sub-divided into *titles*, each discussing a different territory of the subject matter, which are only functionally interrelated with those examined in other titles.

Finally, most titles comprise the discussion of several *topics*, such as specific mechanical or operational details, design or applicational varieties, etc.

Incidentally, this method of subject discussion should make this work also adaptable for *textbook applications*, should such need develop from the growing trend of teaching manufacturing technology as a part of engineering education.

In another book by the author (*Handbook of Dimensional Measurement*, Industrial Press Inc., New York, N.Y., 1966), a system of presentation, that of *synoptic tables* was used extensively and has been very well received by its readers. In comprehensive tables many aspects of the subject, such as basic and accessory equipment, methods of application, operational characteristics, etc., are listed individually; each item with a brief description, explanation or appraisal. These items are frequently accompanied by simple diagrams to assist in quick recognition and easy mental grasp of the essential characteristics and to facilitate a ready comparison with other alternatives. Synoptic survey tables of this kind are used where warranted throughout this book, resulting in a total of about 60 comprehensive tables, many of which extend over several book pages.

Grateful acknowledgement is expressed to the many manufacturers of equipment, including practically all the prominent grinding machine builders of this country, and several foreign manufacturers as well, for their assistance in providing comprehensive and up-to-date information on their products, and also supplying valuable illustration material. These latter are reproduced with appropriate credit notes, naming the respective manufacturers. Since the purpose of the illustrations is primarily to show typical examples, the selection of any particular make should not be considered as an expression of preference as against the available and comparable products of other reputable manufacturers.

While an extensive discussion of each subject in terms of theory and application has been the goal, the size of the book had to be kept within manageable limits, both as a publication and as a source of reference, but also to avoid obscuring the essential by an excessive discussion of the exceptional or the rarely used. Consequently, the subjects to be included, as well as their depth of examination, had to be judiciously appraised, with deliberate condensations and deletions made where needed in the interest of the whole.

Hopefully, the necessary lack of completeness and the occasional brevity of the discussion will not seriously detract from the value of this book, but will be accepted by the majority of readers as a means for achieving clarity and facilitating the retrieval of the information useful in their work.

INTRODUCTION

The Favorable Characteristics of Grinding

In most methods of machining metals, other than grinding, the workpiece is shaped by removing chips with cutting tools made of metals harder than the work material, It is characteristic of these machining processes employing cutting tools that: (a) the differential in the hardness of the tool and of the work is often quite limited, actually never of a magnitude which would exclude a significant degree of tool wear, and (b) in the process of detaching chips, heat is developed which, when exceeding a specific level, will harmfully affect the metallic tool material, a condition that always limits the applicable cutting speed.

These, and various other limitations of metal-cutting tools, do not exist at all, or to a much smaller degree, in the case of abrasives, as will be seen from the following brief review of the properties of *abrasive tools* in the machining of metallic and other work materials.

1. Abrasives are, in general, mineral crystals with far greater hardness than that of metallic workpieces, whether machined in their original state or after heat treatment.

2. The abrasive crystals are, for most practical purposes, not sensitive to the heat generated in metal-cutting operations, and can sustain much higher temperatures than can the best metallic tool materials—a property which permits operating abrasives at cutting speeds many times higher than that possible with any kind of metallic tool material.

3. When wear affects the edges of the individual grains in a grinding wheel, the self-sharpening properties of bonded abrasive tools become effective by releasing the dulled grains and exposing new sharp ones; a process which can be, and often is supported by occasional "dressing" or "truing" in the operating grinding machine. This property of grinding wheels practically eliminates the need for tool changes with the purpose of remote tool reconditioning. Under such circumstances continuous operation over a consider-able period of time is feasible, since the wheel dressing is considered an element of the process.

4. The simple reconditioning of the abrasive wheel, which is conveniently performed even when only a very small amount of face wear has occurred, avoids the significant effect of tool-edge dulling on size holding, which generally accompanies machining processes performed with cutting tools.

5. The process-integrated reconditioning of the abrasive wheel, by tying-in the truing with automatic position compensation, results in a degree of unattended dimensional accuracy of the work which exceeds by far the work size control attainable with conventional metalworking tools.

6. Work with particular, or even complex profiles, which would require very expensive, specially made form-cutting tools, can be produced accurately by grinding, primarily by means of relatively inexpensive truing templates.

7. The depth of abrasive grain penetration into the work material can be held to a very small amount and, when necessary, chips of microscopic size can easily be removed, thus avoiding the substantial machining allowances which the use of cutting tools may often require for the sake of satisfactory tool operation alone.

8. Cutting through the hard skin of certain raw materials or forgings, may require a minimum depth-of-cut for metallic tools, a condition which is not a factor to be considered in *abrasive* chip removal.

9. Finally, there are some work materials, both in their untreated condition and particularly after hardening, which cannot be worked with any tools other than abrasive tools.

The Principal Subject: The Basic Methods of Grinding

This book discusses the four basic methods of grinding, namely:

1. *Cylindrical grinding* - comprising operations comparable to turning on a lathe, with the workpiece

rotated around a fixed axis. However, a grinding wheel, which cuts by rotating at high speeds, is used instead of the firmly held metallic cutting tool.

2. *Centerless grinding* - a process carried out with a floating work axis, offers many particular advantages although it is practically limited to grinding, having no basic metal-cutting method as counterpart. This method is applicable primarily to the external round surfaces of parts in a wide range of dimensions and profiles, and also to the internal surfaces of short parts which have substantially annular shapes.

3. *Internal grinding* - a method serving the accurate machining of round surfaces inside the workpiece; in its system and scope of operation it is comparable to boring with the distinct difference, however, that internal grinding operates with rapidly rotating grinding wheels as its tools, and in the majority of cases, the workpiece is also rotated.

4. *Surface grinding* - for producing flat surfaces or those with parallel, straight line elements. Surface grinding can be carried out, depending on the applied machine system, with the periphery or with the face of the grinding wheel. In its scope of applications this method is comparable to planing or, even more closely, to plano-milling, although incorporating the many advantages offered by the self-sharpening and easily shaped grinding wheel, and the simple methods of work holding resulting from the much smaller force acting on the workpiece than in the comparable metal-cutting operations.

Each of these basic methods can be applied in any of several different systems, offering particular applicational potentials, degrees of adaptability, and levels of efficiency. A review of these and many other characteristics of the various systems belonging to each of the discussed basic methods, accompanied by an in-depth review of the generally available machinery and accessories, including their proven applications, represents one of the major objectives of this book.

Other abrasive methods. While the majority of mechanical parts have functional surfaces that are either round or straight, consequently well-suited to be machined by one of the basic grinding methods, there is a large variety of configurations, regular or special, which also require grinding, although the general shape and/or the surfaces to be ground on such parts, differ from those for which the basic methods are adapted. Examples of such part surfaces are the edges of cutting tools, screw threads, gear teeth, and many more.

Grinding wheels, although the most commonly used, do not represent the only kind of abrasive tools.

Abrasive grains may be bonded into other shapes than rotatable wheels, or they can be glued on a flexible backing, i.e., for producing coated abrasives. Finally, to remove work material, abrasive grains can be applied without being bonded or coated, as long as sufficient force is applied to cause these grains to penetrate into and to move along the surface of the workpiece in a properly controlled manner.

These many different applications of abrasives as the tools of machining and related operations are planned to be the subject of a complementary volume now being developed by the author.

Aspects Examined and Explained

The designation *grinding* comprises methods, equipment, and operations which are extremely diverse in their systems and purposes. This diversity of functions and objectives has a common denominator in the use of grinding wheels as the tools of machining the workpiece.

For discussing and explaining the essential aspects of such a wide variety of machining methods and their means, the following principles of presentation have been employed:

— Aspects which are considered essential are discussed in each section devoted to one of the major methods of grinding.
— Regularly and frequently used grinding methods and their means are examined to an extent and in detail commensurate with their relative importance.
— Special and uncommon methods of grinding are also mentioned and discussed either because they are required in particular, yet important, applications or when used as examples to indicate the potential capabilities or degree of adaptability of the basic grinding method.

Essential aspects which are extensively discussed in all sections are particularly the following:

— The general characteristics and the production purpose of the method.
— The functional elements and their motions which primarily determine the operation and results of the applied process.
— Commonly used machine types—their general dimensions and capabilities.
— Frequently selected accessories and supplemental equipment, describing briefly their individual purposes.
— Systems, with characteristics distinctly different from the common operation of the method, are examined and appraised in individual chapters.

Applications of each method and of several of their various systems are discussed extensively, substantially beyond the depth of the usual overall surveys. The ultimate purpose of selecting and employing any technological method is its application for accomplishing particular production objectives. Consequently, the aspects of where and when to use it, as well as the reasons for such choices, are given ample emphasis with regard to both the basic and supplemental equipment required for the implementation.

Process details and machining data—Setup instructions and values, also machining data such as speeds and feed rates, as well as operational and auxiliary process times, are discussed in conjunction with several grinding methods. Although ample data are presented, these mainly serve the purpose of examples for general information, however. No attempt has been made to supply comprehensive sets of instructions for actual process planning.

Grinding operations involve a very large number of variables, such as machinery and equipment, wheel composition, truing systems and devices, work material and design, grinding allowances, economic and quality objectives of the process, etc., the compounded effect of these variables excluding the practicality of generally valid operational values. As a matter of fact, unconditional reliance on general or average values, beyond their use as a starting level for developing the optimum factors, could be detrimental to deriving maximum benefits from any particular grinding method.

It is also well to keep in mind that notwithstanding the very extensive grinding research which is carried out for analyzing its mechanics, and for correlating tools, the work materials, process factors and results, as well as many other aspects, grinding still is often considered an art, relying on empirically determined elements. Without trying to define the mutual boundaries of science and art in this particular field of technology, it is obvious that experience plays an important role in determining the optimum conditions of specific grinding operations.

Gathering experience as a means of establishing empirically determined conditions can be carried out with a high degree of effectiveness when based on the knowledge or, at least, on the familiarity with the procedures and the equipment which are being successfully applied in solving problems similar to or even identical with those to be resolved by the manufacturing engineer. In order to contribute to the development of such a useful basis of empirically complemented knowledge, a large array of examples, comprising equipment, applications, and processes has been presented, considering all four basic grinding methods, as well as many of their different systems of execution.

The Potential Audience: The Professionals to Whom This Book is Addressed

The main purpose of a technical book is to supply useful information. In order to be useful, any technical or other kind of professional information, besides the essential requirements of its contents, must serve the interest and needs of a specific group of readers.

In the conception of a technical book, defining its potential readers is essential for several reasons, including the following:

(a) The proper determination of the degree of information to be presented, skipping the detailed discussion of generally known fundamentals, yet avoiding information gaps caused by assuming too much specific knowledge on any of the subjects.

(b) Adjusting the method of communication to the level of comprehension of the majority of readers, e.g., that of professionals of a specific branch of science.

(c) Use as a guide the type of work and professional responsibilities of the probable readers, when defining the material to be discussed in the framework of the general subject.

Considering the significant role of the potential reader for the proper structuring of a technical book which should prove really useful, it seems indicated to list those branches of engineering for whose members this book has been conceived and developed.

Manufacturing Management, with its wide scope of interests and extended range of responsibilities, cannot get deeply involved in any particular method of production. Nevertheless, a general understanding of the capabilities and potentials of a family of machining methods, particularly one of such importance as grinding, can have a great bearing on the correctness of many decisions made by manufacturing management. The attainable production rates and accuracy levels, the degree of possible automation and integration into automated processing lines, the flexibility of adaptation and the generally beneficial economic aspects or properly selected grinding methods are bits of information which may prove quite useful in decision-making processes. Even when perceived in outline only by glancing over the various chapters, or perusing only limited portions of the book, technological information of value for management purposes may be gathered with relatively little effort.

Manufacturing Engineering comprises the functions

responsible for determining the manner and means of making the product.

Methods engineering determines the most suitable way in which raw material or semi-finished parts can be converted into different stages of the final product.

Process engineering is responsible for establishing the technological details, including tooling and machining data, of the applicable process.

Facilities engineering covers the selection of the proper equipment, including both the basic machine and its supplemental accessories, for putting into effect the conceived methods and the projected processes.

While the preceding job designations are not uniformly applied nor always separated, the brief description of the responsibilities they involve, indicates how much the effectiveness of these functions is dependent on a thorough familiarity with equipment and processes, requiring a know-how such as this book is designed to provide on the broad subject of grinding.

Design Engineering, after having conceived the final product, carries out the design and establishes the dimensional specifications of its component parts. Many of these components can be produced, or actually must be produced, by grinding. Giving a shape to the component which, without interfering with its ultimate function, makes it best adapted to grinding, may often contribute to reduced manufacturing cost. Being acquainted with the accuracy capabilities of grinding, will assist the design engineer in specifying optimum product fits and tolerances without exceeding the common feasibility limits of particular grinding processes.

Value Engineering. The productivity of grinding in general, and of certain methods in particular, open cost-saving potentials which may occasionally even warrant the re-designing of a component part to make it better adapted for machining by grinding or, more specifically, by the grinding method offering the optimum balance between the required quality and the incurred production cost.

Cost Estimating. To assure the competitiveness of the product, the most efficient methods of machining may have to be selected. A basic understanding of grinding can be extended even by only glancing through the book or looking through the synoptic tables. By means of a detailed Table of Contents and Index, readily available and detailed information on feasible methods for particular grinding operations can provide hints, or even data for considering those methods with favorable cost-effectiveness. The proper selection of the applicable production method in this preliminary, yet decisive, phase of designing the process, can prove to be a meaningful factor in developing

favorable cost components.

Quality Control and Engineering. Dimensional tolerances which are generally superior to those attainable with any other metalworking process, and the easily controlled consistency of the product size, are some of the characteristics which make grinding a most desirable method with regard to quality control. Such aspects as process feasibility, automatic work sizing, and process-integrated inspection are examples of information available in this book. Means and methods of assuring uniform product size within very tight tolerance limits, reducing—or even entirely eliminating—the need for dimensional inspection as a separate operation, the various size-control devices, their applications and functions in grinding, are a few randomly listed subjects which are of potential interest to those in charge of quality control and of assuring the reliable accuracy of the product.

Tool Engineering. Toolmaking and tool maintenance are those areas of metalworking production where grinding fills a very important role for several reasons, since:
—Hardened steel or cemented carbides are used almost exclusively;
—A high degree of operational accuracy is needed in the majority of toolmaking operations;
—Irregular, often complex shapes are to be produced, generally in small lots or even single-piece production. Grinding wheels with easily trued shapes of their operating faces are by far the most economical and reliable means of such production tasks.

Abrasive Engineering, although the essence of the book, is named as the last group in this indicative listing of branches of engineering to which this book is addressed. More specifically, the book has as its subject the methods and the equipment by means of which abrasive materials are used for shaping the workpiece. This book is written for mechanical or metalworking engineers, in distinction to chemical engineers in whose domain the manufacture of abrasive materials and products really belongs. The reason for quoting abrasive engineering at the end of the list is the relatively infrequent appointment of one or a group of engineers to this particular function although, as the contents of this book indicates, this function requires a very broad field of specialized engineering know-how.

Many areas of this particular field of metalworking can, of course, be penetrated much more deeply than the necessarily limited extent of a book covering a comprehensive area allows. It is believed, however, that this book will contribute to the realization of the extent and depth of the know-how which the functional field of *abrasive methods engineering* involves, by pointing out the many aspects and details required for the effective execution of the pertinent responsibilities and last, but not least, the enormous product quality and production cost-benefits which can result from professionally engineered abrasive methods and processes. By substantiating the need for, as well as the potential benefits which may be expected from the functions of the specialized abrasive methods engineer, this book might prompt in many metalworking plants actual assignments to this field of responsibility. In such cases this book should prove a truly useful guide and reference source for both the fledgling abrasive methods engineer, and also to the more experienced practitioners of the field.

CYLINDRICAL GRINDING

The System, its Applications and Capabilities

Definition of the Category

The term cylindrical grinding, in general techno-logical usage, denotes a group of grinding processes whose common characteristic is the rotation of the workpiece around a fixed axis. Consequently, all surfaces machined by a cylindrical grinding process are in a definite relationship to a specific axis of rotation.

In cylindrical grinding, the workpiece receives a positive drive causing it to rotate around the selected axis. The drive torque can be transmitted by any of various means but it must never interfere with the retainment of the part's constant axis of rotation. The axis of rotation can be identical with the axis of the general part, this being the case for most work-pieces which are bodies of revolution, or the part can be made to rotate around the axis of a particular feature which is being ground.

The shapes produced in the process can be widely different in accordance with the requirement of the operation, as long as the relationship between the sur-face being ground and the part's axis of rotation is maintained. It follows then that processes designated as cylindrical grinding are not restricted to producing plain cylindrical surfaces alone. Cylindrical grinding can be applied efficiently to a wide variety of other configurations or to a combination of them, as long as the basic condition of a common axis of rotation is satisfied. Table 1-1.1 illustrates typical part configu-rations which are adaptable for grinding on cylindri-cal-type grinding machines.

The term cylindrical grinding, while generally used, is not dependably or fully descriptive. Thus, many shapes other than cylindrical are produced by pro-cesses of this category while, on the other hand, cy-lindrical surfaces can, and are extensively generated on grinding machines belonging to other major pro-cesses. Two examples of grinding machines assigned to other grinding process categories, yet still capable of producing cylindrical surfaces are *centerless* grin-ders and *internal* grinders, the latter being used main-ly for the grinding of cylindrical, or otherwise round, internal surfaces.

Capabilities of Cylindrical Grinding

In a majority of manufacturing processes grinding is used as a finishing method, for producing the final size, shape, and surface conditions of the manufac-tured part. Consequently, when establishing a pro-duction process in which the final machining will be grinding, the capabilities of the selected method for consistently producing the critical dimensional para-meters must be known.

Unless experience with similar workpieces, ground under comparable conditions, is available, capability studies, consisting of actual test runs, might be need-ed. From these test runs, dependable data can be obtained on the process and equipment capabilities with regard to specific workpieces, particularly those with tight dimensional tolerances.

When the setting up of such capability studies is planned, preliminary information on the potential performance of the selected method could prove to be of great help. For that purpose a few typical char-acteristic figures have been listed in Table 1-1.2. These represent close-to-top-performance values that can be obtained in cylindrical grinding under well-controlled conditions, using modern equipment and properly selected machining data.

3

Table 1-1.1 Examples of Typical Cylindrical Grinding Processes

DESIGNATION	DIAGRAM	DISCUSSION
Traverse grinding of cylindrical surfaces between dead centers		May be considered the basic method of cylindrical grinding, resulting in the highest degree of geometric accuracy. The need for a section on the surface of the part against which the driver can rest excludes the grinding of the part in its entire length in a single setup. Work rests (back rests) to provide additional support for long and/or slender parts are frequently needed.
Traverse grinding by holding the part clamped to the rotating headstock spindle		The accuracy of the operation is dependent on the running condition of the spindle and on the precision of the workholding device, which can be a chuck, collets, or one of the many other types of fixtures. Short and stubby parts need to be held on one end only. For longer parts the tailstock center or work rests can provide additional support.
Plunge grinding of cylindrical surfaces		Applicable to surface sections whose length does not exceed the width of the grinding wheel. Can considerably reduce the grinding time as compared to traverse grinding but requires more rigid parts and support. The obtained finish may be poorer than the best produced by traverse grinding. Possible wheel breakdown at the ends of the section may impede straightness unless frequent truing is applied.
Traverse grinding of slender tapers		Carried out by means of the swivel setting of the top table which supports the headstock and the tailstock. The included taper angle is limited to twice the amount of the max. table swivel. Part may be held between dead centers or in a holding device mounted on the rotating headstock spindle. Applicable to both short and long workpieces.

Table 1-1.1 (*Cont.*) **Examples of Typical Cylindrical Grinding Processes**

DESIGNATION	DIAGRAM	DISCUSSION
Plunge grinding of steep tapers		Either of two methods, or a combination of both can be used: (a) swiveling the headstock (for parts mounted at one end) or (b) by swiveling the wheel-head slide (on universal-type grinding machines which have wheel-head swivel; also applicable to parts mounted between centers). The length of the elements (generatrix) of the ground surface is limited by the width of the grinding wheel.
Traverse grinding of steep tapers		Can be carried out with manually operated traverse motion by applying either of two methods: on universal grinding machines with a double swivel type wheel head; see Fig. 1-2.2: (a) by grinding with the periphery of the straight wheel; (b) by dressing a bevel on the wheel and combining the swivel setting of the wheel slide with that of the table (see the diagram), thus reducing the depth of the bevel to be dressed on the wheel, and producing a good finish.
Plunge grinding of a cylindrical section with adjoining fillets or radii		The concurrent grinding of adjoining surface sections assures a stepless transition and excellent concentricity, thus improving both the appearance and the functional properties of the ground surface. Requires the application of a continuous radius and tangent type wheel-truing device.
Multiple diameter plunge grinding		Multiple diameters when extending over a length not in excess of the max. grinding wheel width which the machine spindle can accommodate, can be ground concurrently, thus improving the concentricity of the surfaces and the productivity of the process. The wheels are mounted on the spindle either side-by-side, or with spacers in-between. Special truing device is needed to consistently assure the required relationship between the wheel diameters.

(Continued on next page.)

Table 1-1.1 (*Cont.*) **Examples of Typical Cylindrical Grinding Processes**

DESIGNATION	DIAGRAM	DISCUSSION
Angular approach grinding; periphery and an adjoining face section are ground concurrently		By advancing the grinding-wheel head at an angle (frequently at 30 degrees), in relation to the radial direction, both the periphery and one face of the grinding wheel will cut effectively, thus permitting the simultaneous grinding of several work surface sections which are mutually perpendicular or even have intermediate contours. May be applied on universal grinders or, for continuous production, preferably on special angular grinding machines. Requires special truing device.
Traverse grinding combined with face grinding		Method applied when a longer cylindrical section which requires traverse grinding has an adjoining shoulder, to be ground in the same setup for assuring excellent squareness. First, the cylindrical section is ground to size, then the table is advanced, by hand, for grinding the shoulder. The rapid hand cross feed is used for traversing across the face of the shoulder. The wheel face should be recessed to assure a clean, square cut.
Form plunge grinding with profile-trued wheel		By truing the wheel to the inverse shape of the required part profile, form-contoured round parts can be produced either by radial or angular-approach plunge grinding. Profiles with a single convex or concave circular-arc profile can be shaped with a regular radius-truing device, while more complex or composite profiles require a form-truing attachment designed to duplicate the contour of a template.
Form plunge grinding with crush-shaped wheel		Wheel form truing by means of crushing rolls is usually applied on special cylindrical grinding machines expressly designed for such operations. Exceptionally, regular cylindrical grinding machines may also be adapted and equipped with a crush-forming device. This is a very fast and efficient method for the high-volume production of essentially round parts even with intricate profiles which must be ground to tight tolerances.

Table 1-1.1 (*Cont.*) **Examples of Typical Cylindrical Grinding Processes**

DESIGNATION	DIAGRAM	DISCUSSION
Internal grinding with special attachment		Internal grinding is listed here for indicating the adaptability of cylindrical grinding machines, particularly of the so-called universal type. Besides the opportunity for carrying out occasional internal work on an existing cylindrical grinding machine, significant benefits in product accuracy can result from grinding coaxially designed external and internal surfaces in the same setup.
Rotary surface grinding		Grinding the face of a rotating part with the periphery of the wheel is a method apt to produce a surface of excellent flatness and with grinding marks essentially concentric with the center of the part's rotation. While special rotary surface grinders are built for such operations, for occasional use it is convenient to carry out that operation on a cylindrical grinding machine whose headstock can be swiveled by 90 degrees. Using less than 90-degree headstock swivel, wide-angle taper surfaces can also be produced with reciprocating table and cross-slide feed.
Crankshaft grinding		While, in principle, most types of cylindrical grinding machines of sufficient swing can be tooled up for carrying out such operations, special crankpin grinding machines are used for production purposes. These are equipped with the appropriate truing device for shaping the periphery, corner radius, and the sides of the wheel, and with means to control the table traverse precisely to suit the distances between the pin positions. The grinding itself is a plunge process, sometimes supplemented by a controlled lateral movement.

(*Continued on next page.*)

Table 1-1.1 (*Cont.*) **Examples of Typical Cylindrical Grinding Processes**

DESIGNATION	DIAGRAM	DISCUSSION
Polygonal shape grinding		Cylindrical grinding generally produces surfaces whose cross-sectional contours are essentially circular, while the part's axial profile may vary according to process objectives. In its extended applications the basic system can be applied to produce shapes whose cross-sectional contours are polygonal, thus providing rotational locking when mated with a part whose bore has a complying contour. Implemented on special polygonal grinding machines.
Cam grinding		The noncircular contour of the surface produced by the principles of cylindrical grinding does not have to be symmetrical as in the preceding examples, but can have any particular shape as long as it is related to an axis around which the part is rotating during the grinding process. The radial displacement of the work head, in accordance with the changing orientation of the part, can be controlled to reproduce such non-regular, yet exactly specified surfaces like the periphery of cams, either self-contained or as elements of a camshaft.
Cambered cylindrical grinding		Surfaces produced in traverse grinding, whether cylindrical or tapered, have straight line side elements, ground along a path parallel with the table traverse. Exceptions are cambered cylinders, the form of rollstand rolls. These, generally long and heavy parts, often have a slightly convex profile which is produced on special cylindrical grinding machines which are equipped with mechanisms to approach or to retract the wheel head in relation to the table's traverse position.

Table 1-1.2. Performance Characteristics of Cylindrical Grinding

Dimensional Characteristics	Regular Cylindrical Grinding	Precision-type Grinding Machines	Super-precision, Instrument Type Grinding Machines
	Optimum Tolerance Limits, in microinches (micrometers)		
Size holding (diameter)	50 (1.25)	25 (0.65)	10 (0.25)
Geometric accuracy (straightness, parallelism, roundness)	75 (1.9)	25 (0.65)	6 to 10 (0.15 to 0.25)
Surface finish, AA (R_a)	8 (0.2)	2 to 4 (0.05 to 0.1)	1 to 2 (0.025 to 0.05)

Machine Operating and Control Elements

Cylindrical grinding, in the comprehensive sense of the term, embraces a large variety of processes that have the common characteristic of being carried out on cylindrical grinding machines. Consequently, a good starting point for understanding the potential applications for the process and for visualizing the manner in which cylindrical grinding is accomplished, might be a survey of the basic tools of the process: the cylindrical grinding machines.

Several major elements of cylindrical grinding machines have similar operational characteristics, even though the sizes, capacities, and capabilities of the machines may vary widely.

In the following discussion the emphasis will be on these common characteristics, including both the general principles and certain particular forms of execution. A general survey of the basic cylindrical grinder characteristics is presented in Table 1-2.1. The subsequent discussion will, in general, follow the sequence in which the major operating elements are considered in the first column of the table.

The Wheel Head

The wheel head comprises the housing for the grinding-wheel spindle and its bearings, as well as the drive motor. The power of the wheel-drive motor usually far exceeds the aggregate amount of all the other power sources used in cylindrical grinding machines.

In universal-type cylindrical grinding machines the wheel head may also contain the internal grinding spindle with its individual drive motor. This optional accessory can be swung out of the operating area when not in use, in order to avoid interference with external grinding operations.

The drive of the grinding-wheel spindle is usually through multiple V-belts, whose sheaves are either interchangeable or offer several steps, thus providing for changes in the rotational speed of the grinding-wheel spindle. These speed changes, commonly in two or three steps, serve the purpose of approximating the optimum peripheral speed of the grinding wheel when new, and also at various levels of used

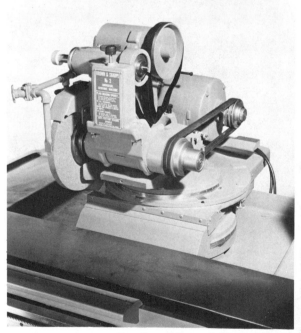

Courtesy of Brown & Sharpe Mfg. Co.

Fig. 1-2.1 Universal-type wheel head with swung-back, internal grinding attachment. Guards are removed for showing the belt arrangement.

Table 1-2.1 Principal Operating and Positioning Elements of Cylindrical Grinding Machines

<table>
<tr><td rowspan="5">OPERATIONAL MOVEMENTS</td><td>MOVEMENT</td><td>MACHINE ELEMENTS</td><td>CONTROLS, RATES, AND RANGES</td></tr>
<tr>
<td>Grinding-wheel rotation</td>
<td>Grinding-wheel spindle with bearings and drive motor.</td>
<td>Speed of rotation established for the safe operation of the max. permissible wheel diameter. To assure the use of worn wheels with efficient peripheral speed, provision is often made for higher rotational speeds, e.g., by change of belt sheaves. Bearings of special design and in various systems are used for assuring the true running of the wheel spindle.</td>
</tr>
<tr>
<td>Work rotation</td>
<td>Headstock with rotating driveplate or rotating spindle, drive-motor and speed-change device.</td>
<td>A wide range is usually provided, such as with ratios from 1:10 to 1:50 between lowest and highest speed, to accommodate different work diameters and to obtain optimum grinding conditions. Lowest speed should suit the max. swing diameter. Speed change by multiple step belt drive (incremental), or by variable speed drive (stepless). For grinding between dead centers either special headstocks or headstocks which can be adapted for either type of operation, are used.</td>
</tr>
<tr>
<td>Cross feed (infeed)</td>
<td>Cross slide moving normal or at an angle to the direction of the table traverse.</td>
<td>Hand wheel for positional adjustment and manual infeed. Rapid travel operating over set distance within the available range, e.g., from ¼ to 5 inches (6.35 to 127.0 mm), for approach and retraction. Automatic feed (a) incremental at table reversals in traverse grinding, e.g., from .0001 to .003 inch (0.0025 to 0.0762 mm), or (b) continuous for plunge grinding, e.g., from .004 to .070 inch (0.1016 to 1.778 mm) per minute. Machines with automatic cycle incorporate these and other movements in their controls, in sequential order.</td>
</tr>
<tr>
<td>Reciprocating table movement (for traverse grinding only)</td>
<td>Work table in its guideways. Exceptionally, for very large machines, the wheel head traverses. Not used in plunge grinding.</td>
<td>In most types of modern machines the table traverse is actuated by a hydraulic cylinder and has infinitely variable speeds, e.g., over a range of 15 to 240 inches (about 0.38 to 6.1 m) per minute. For grinding wheel truing a lower speed range, e.g., 3 to 25 ipm (76 to 635 mm/min), is generally available. Table dwell (tarry) at stroke ends is often also provided and may be set to the required duration.</td>
</tr>
</table>

(Continued on next page)

Table 1-2.1 (*Cont.*) **Principal Operating and Positioning Elements of Cylindrical Grinding Machines**

	MACHINE ELEMENT	CONTROLS	PURPOSE AND RANGE
POSITIONAL ADJUSTMENTS	Table positioning and limit setting for the traverse movement	Lateral positioning by hand wheel. Reversal points of power traverse movement can be set to any stroke starting from a minimum length (e.g., 0.1 inch).	Accurate table positioning and locking is required for plunge grinding. For traverse grinding, the reversal points are set with adjustable dogs acting on the reversing lever of the table traverse. Coarse setting by displacing the dogs along the rack, fine setting with micrometer screw. Manual lever flipping is used for bypassing the dog, such as in the case of wheel dressing.
	Cross-feed setting for the end of the infeed and also for intermediate points of rate change	Hand wheel with graduated dials and control elements on or around the hand wheel unit. Adjustable stop for end position and for changeover between different feed rates.	Hand wheel can be set either to coarse or to fine infeed at different rates. Cross feed can be disengaged for setting dials and stops. Incremental feeds at table reversals are often steplessly adjustable over a wide range, either for each or for alternate stroke ends. Continuous infeed needed for plunge grinding. Infeed usually stops prior to halting work rotation, to assure sparkout.
	Wheel head and cross slide swivel (for "universal" grinding machines)	Either the wheel head alone, or in combination with the wheel slide, are swiveled by hand after loosening of locking bolt. Adjustment guided by graduated ring scale.	For grinding steep tapers, compound tapers, or applying angular-approach plunge grinding on a universal-type cylindrical grinding machine. Used for both external and internal spindles. Wheel slide swivel controls the direction of the infeed. Turret base, where available, repositions the wheel head laterally.
	Table swivel setting and realignment	Through adjusting nut by an amount shown on the graduated segment.	To grind slender tapers in traverse grinding, usually holding the part between centers. Typical adjustment range: back 11 degrees, forward 3 degrees.
	Headstock swivel	By hand, guided by graduated ring scale showing angular units. Locked with clamping screw.	These adjustments are available on universal-type grinding machines and serve for grinding steep tapers or for carrying out face-grinding operations.

condition, down to about two-thirds of the original diameter. It is important that the correct rpm be restored when a new grinding wheel is installed.

The wheel head is mounted on the cross slide, either directly or through intermediate elements as used in univeral-type cylindrical grinding machines. Such intermediate elements serve for changing the orientation and location of the wheel head. A typical example of a universal-type wheel head is shown in Fig. 1-2.1, with the internal grinding attachment in a swung-back position. Two graduated rings also can be seen underneath the wheel head.

The lower graduated ring indicates the swivel setting of the cross slide which, in the case of the illustrated example, can be made over a range of 90 degrees. The direction of the infeed movement will be inclined to the table travel by the amount of that swivel adjustment.

The graduations of the upper ring indicate the swivel position of the wheel head with respect to the cross slide; in the basic position, the wheel spindle is perpendicular to the guideways of the cross slide. The wheel head can, in this model, be swiveled in either direction by 90 degrees, causing the wheel spindles, both external and internal, to be inclined by the set value with respect to the direction of the cross-slide movement.

The combination of these two swivel settings permits the grinding of steep tapers, compound tapers, and shoulders, thereby substantially extending the adaptability of the universal-type cylindrical grinding machines. Figure 1-2.2, Diagram B, demonstrates the manner in which the double swivel setting, that of the cross slide combined with the wheel head, extends the application potentials of universal grinding machines thus equipped.

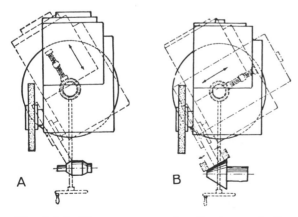

A   B

Fig. 1-2.2 Diagrams comparing the taper grinding potential of wheel heads with: (A) single swivel base (for plunge grinding); and (B) double swivel base (for traverse grinding).

In the illustrated model of the universal wheel head, a third level of support is inserted between the described swivel members. This consists of an eccentric platform, designated by the manufacturer as a *turret*. That turret can be swiveled on the cross slide and may be locked in any of four positions, located 90 degrees apart. Figure 1-2.3 shows the different locations of the grinding wheel, which can be accomplished with the aid of the turret, combined with the alternate end mounting of the grinding wheel. The special wheel spindle, providing for the alternate end mounting of the grinding wheel, represents an important design feature of wheel heads used in certain types of cylindrical grinding machines, because it permits the full utilization of the swivel adjustments, such as may be needed for particular workpiece configurations.

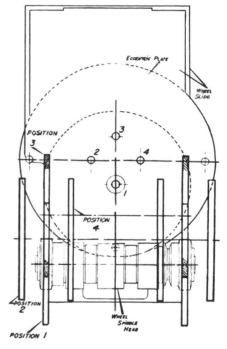

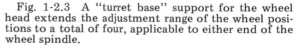

Courtesy of Brown & Sharpe Mfg. Co.

Fig. 1-2.3 A "turret base" support for the wheel head extends the adjustment range of the wheel positions to a total of four, applicable to either end of the wheel spindle.

Grinding-Wheel Spindle Bearings

For the support of grinding-wheel spindles in cylindrical grinding machines, plain bearings are generally used; the use of antifriction bearings may be considered exceptional. Because wheel diameters as large as the machine will accept are preferentially used in cylindrical grinding, the required rotational speed of the grinding-wheel spindles is not so high as to exceed the efficient operating limits of plain bearings. The essentially constant speed at which such spindles are run over extended periods of time, is also suitable for the effective operation of plain bearings.

Various types of bearings developed for grinding-wheel spindles offer excellent properties: they are smooth running, have resistance to heavy wheel pressures, are cool in operation, maintain low friction, and have an extended service life when reasonable care is applied.

Hydrodynamic Bearings

Most of the plain bearings used for wheel spindle support are designed to create (as the result of spindle rotation) an oil film with circumferentially spaced high-pressure areas. The favorable hydrodynamic conditions brought about by the development of high-pressure areas in the oil film which surrounds the spindle journal is, however, predicated on the rotation of the spindle within a specific speed range. Because of that interdependence between the dynamic component, the spindle speed, and the resulting hydraulic conditions, such bearings are designated as the *hydrodynamic* type.

From a rather extensive list of various types of hydrodynamic bearings used for grinding machine spindles, two types will be discussed here as they have proved their excellent performance over several decades of industrial use. The designations of these bearing types are the proprietary terms of the respective manufacturers.

The Microsphere bearing, Fig. 1-2.4, consists of a one-piece steel bushing which has a large number of

pad segments distributed circumferentially around the bore surface and interrupted by grooves which serve as oil reservoirs. The bushing has a spherical outer surface which is supported by spherical cups for self-alignment and has very precisely ground shoulders for taking up thrust loads. Both the bore and the thrust faces of the bushing are lined with babbitt metal composition.

The Filmatic bearing, Fig. 1-2.5, has five independent and uniformly spaced shoes which face the spindle journal. The inside surface of each shoe is designed to sustain the formation of a high-pressure, wedge-shaped oil film when the spindle is rotating at regular operating speed. Filtered lubricating oil is supplied by a separate system equipped with an automatic shut-off for the spindle drive. The wheel spindle rotation can only be started when the correct

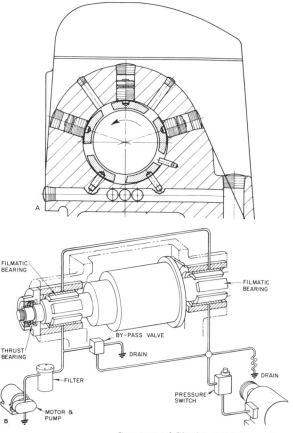

Courtesy of Cincinnati Milacron Inc.

Fig. 1-2.5 The "Filmatic" wheel spindle bearing. Diagram A shows the arrangement of the individual bearing segments which form multiple wedges drawing in oil and causing its pressure to rise. Diagram B shows the pressure interlock system which prevents the spindle from being started until the correct minimum pressure is built up in the bearing chamber by the action of a motor-driven oil pump.

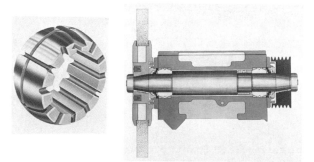

Courtesy of Landis Tool Co.

Fig. 1-2.4 The "Microsphere" wheel spindle bearing is a one-piece steel bearing with spherical outside surface for self-alignment, and a face exactly square with the bore for thrust support. A. View of the bearing sleeve. B. Diagram of the two bearings supporting the spindle close to the ends.

minimum oil pressure has been reached and will shut off promptly in case of a pressure drop.

Hydrostatic Bearings

Hydrostatic bearings were developed for machine tool applications as an improvement over the performance and operational properties of conventional hydrodynamic bearings. The effective spindle support by hydrostatic bearings is not dependent on the running, i.e., the dynamic, conditions of the supported member. Hydrostatic bearings do not require warm-up time, are equally effective over a wide range of spindle speeds, and possess a stiffness; that is, a resistance to deflection, superior to that attainable with a hydrodynamic system.

The diagram in Fig. 1-2.6 illustrates the principles of operation of hydrostatic bearings. Around the internal surface of the bearing, bushing pockets are machined, into which a continual flow of oil at high pressure is introduced. The pockets are designed to taper off at their edges which then extend to drain grooves, thus providing for the circulation of the oil, yet maintaining its high pressure in the pocket areas.

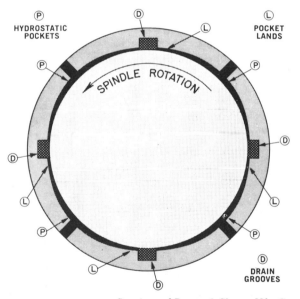

Courtesy of Brown & Sharpe Mfg. Co.

Fig. 1-2.6 The operating principle of the hydrostatic spindle bearing in which the oil delivered at high pressure fills the opposed pockets and holds the rotating spindle in a floating condition.

In alternative designs of hydrostatic bearings the fluid medium may be an inert gas or air, such as is applied for certain types of hydrostatic internal grinding spindles. While the operational principles are identical, the actual designs of hydrostatic bearings

using gas instead of liquid, differ from the illustrated type and are adapted to the characteristics of the applied medium.

When, as the result of external forces acting on a hydrostatic spindle, it is momentarily moved out of its central position inside the bushing, the oil pressure in the pockets of the approached side will suddenly increase as a consequence of reduced pocket space. At the same time, the volume of the pockets on the opposite side will grow, resulting in a pressure drop. The imbalance of forces thus created will cause the spindle to return to its central position, thereby equalizing the oil pressure within the bearing. Because of the substantial forces resulting from high oil-pressure in the system, even very small displacements of the spindle will trigger a prompt balancing action, resulting in an excellent bearing stiffness and assuring a high degree of consistent positional accuracy.

In their applications on grinding machines, particularly of the cylindrical grinder type, the use of hydrostatic bearings is gaining acceptance for operations on critical workpieces. Hydrostatic bearings may be used successfully on cylindrical grinding machines for wheel spindles, replacing the hydrodynamic type; and for work spindles in the headstocks, where antifriction bearings are also commonly used.

Cylindrical grinding machines equipped with hydrostatic spindles can produce ground workpieces with a geometric accuracy (roundness and straightness), surface finish and size (diameter) control, which may exceed by a factor of two, or even higher, the values obtained with the best alternate models of the same machines running on conventional bearings.

Hydrostatic bearings, however, due particularly to the required supporting equipment (high-pressure pumps, oil lines, reservoirs, filtering and cooling systems) are more expensive to make and to operate than conventional bearings. For that reason, at the present time, the selection of hydrostatic bearings for cylindrical grinding machines is generally limited to applications where the process requirements warrant the additional cost that such advanced equipment entails.

The Headstock

Since the rotation of the workpieces around a fixed axis is an essential motion in cylindrical grinding, the headstock, which provides that controlled motion, is a basic element of all types of cylindrical grinding machines. The following three major functions can be distinguished in analyzing the operation of the headstock:

1. *Locating the work* for rotation around a defined axis. This is accomplished either (a) in combination with the footstock, for parts held between centers, or (b) by the headstock alone, for parts held in a chuck, collet, or other holding device.

2. *Holding the work* either (a) by the restraining action of two opposing centers mounted in the headstock and in the footstock, respectively (a method known as "dead-center" grinding), or (b) by means of any of various holding devices installed on the rotating spindle of the headstock.

3. *Driving the work* for producing its rotational motion at a preselected speed either (a) by imparting that motion through a drive dog, an intermediate element which is attached to the workpiece supported between two dead centers, or (b) by rotating the headstock spindle on which the holding device for retaining the work is solidly mounted.

These functions must be carried out in conformance with the applicable process requirements, in reference to which the subsequently discussed design features of different types of headstocks may be appraised.

In universal-type grinding machines the headstock is usually resting on a swivel base which can be displaced, longitudinally, along the swivel table. The swivel base permits the headstock to be rotated, commonly up to 90 degrees in either direction, from the zero position where the work spindle is parallel with the direction of the table travel. By the swivel adjustment of the headstock, workpieces held in a chuck or any other similarly acting work-holding device, can be ground to the desired external or internal taper, or may be presented to the grinding wheel at a 90-degree setting for the purpose of face grinding. The swivel base of the adjustable type headstock has a graduated ring showing the angle of the swivel setting, in degrees. Figure 1-2.7 illustrates the swivelable headstock of a typical universal grinding machine.

Dead-centers and Rotating Spindle Types of Grinding

The dead-center type of grinding, in which the workpiece is supported by its center holes held between the nonrotating centers of the headstock and the footstock, respectively, offers the advantage of precise work-locating and holding accomplished by relatively simple means. In this case, the rotational movement is imparted to the workpiece through a drive dog attached to the work and kept in contact with the arm of the work-driving plate which is mounted in the headstock. That plate turns on its own bearings which are independent of the nonrotating work spindle. The diagram of a typical headstock with dead-center type work spindle is shown in Fig. 1-2.8. In this design, the drive is transmitted to the sheave keyed on the spindle of the drive plate, through an intermediate shaft. The selected rotational speed can be obtained by shifting the V-belt between the drive motor and the jackshaft. In the case of the illustrated design, the entire headstock must be exchanged when rotating spindle type grinding is to be applied.

Cylindrical grinding machines, particularly of the universal type, which are used alternately for dead-center type or rotating spindle type grinding, are

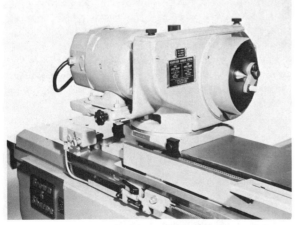

Courtesy of Brown & Sharpe Mfg. Co.

Fig. 1-2.7 Adjustable headstock of a universal type cylindrical grinding machine, with graduated base ring showing the angle of the swivel setting.

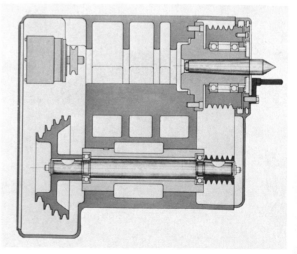

Courtesy of Landis Tool Co.

Fig. 1-2.8 Cross-sectional diagram of a headstock with dead-center work-spindle and four-step speed range for the independently supported drive plate.

usually equipped with headstocks which can be operated in either of the two types of cylindrical grinding. For changing over to rotating spindle type grinding commonly, a locking plunger, which prevents the spindle from turning during the dead-center grinding, has to be withdrawn and a spindle driving latch must be engaged. The drive plate is then exchanged for an appropriate work-holding device mounted on the nose of the now rotating type spindle. Changing over to dead-center grinding requires the same simple procedure to be carried out in reverse.

The Headstock Spindle

The headstock spindle has a major role in the rotating-spindle type cylindrical grinding, in that a chuck or similar work-holding device is mounted on the spindle for locating, holding, and driving the workpiece. In such applications the support of the work spindle in its bearings has a direct effect on the quality of the ground surface, including the obtainable geometric and dimensional accuracy. The true and smooth running of the headstock spindle is accomplished by properly designed bearings, either of the plain or of the antifriction type. Machine tool manufacturers usually select the bearing type in accordance with the offered speed range of the headstock and the expected weight of the workpieces which must be supported with a minimum of deflection.

For workpieces which have to be ground to very high dimensional and form accuracy, particularly with regard to roundness, hydrostatic headstocks are offered as optional features by a few manufacturers. Such spindles possess high stiffness in both the radial and axial directions; the latter minimizing the camming type runout which can affect the accuracy of face and shoulder grinding. The hydrostatic support virtually eliminates the wear of the spindle by avoiding metal-to-metal contact between the rotating spindle and its bearings. The principles of design of hydrostatic headstock spindle bearings are similar to those used for wheel spindles, which were discussed earlier.

Headstock Drive

One of the variables which control the effectiveness of the cylindrical grinding process is the peripheral speed of the rotating workpiece. In order to approach the best-suited work speed for a given set of operational conditions applied to any work diameter within the capacity limits of the grinding machine, the required rotational speed of the headstock spindle must be adjusted over a rather extended range. To permit close approximation of the optimum value, that speed should be adjustable in narrowly spaced steps or, preferably, in a stepless manner.

Headstocks supplied with various models of cylindrical grinding machines are designed to meet the requirements ranging from those of single-purpose equipment built for a specific type of workpiece, to those of the universal type of cylindrical grinding machines designed to accommodate widely different workpieces and grinding processes.

Headstock drives of the stepped-speed type permit changing the rotational speed of the spindle by multiple-step belt sheaves. The limits of the different speeds obtainable by shifting the drive belt are usually related in a ratio of 1 : 4 to 1 : 5, with the actual rpm values conforming to the mean range of the expected work diameters.

The infinitely variable headstock speeds can be obtained either (a) by means of mechanical drives, or (b) by using a variable speed dc motor regulated through a rheostat. The ratio of minimum to maximum speeds obtainable with the infinitely variable drives is usually about 1 : 10.

Jogging control is available on certain types of cylindrical grinding machines, for the fractional rotation of the work spindle. By jogging, the work can be brought to a position most convenient for loading and unloading or, in the case of certain types of workpieces with nonsymmetrical adjoining sections, for inspection while the work is still mounted in the grinding machine.

The power of the work drive motor is usually in the order of 1/10 to 1/15 of the wheel drive motor's power.

The Footstock

The footstock is an integral element in the setup of the dead-center type of grinding. Through the center which it holds, the footstock has the double function of, (a) participating with the headstock in locating the work, and (b) applying the force needed for the retention of the work mounted between centers.

The position of the footstock can be adjusted along the top of the machine table to suit the workpiece length. For obtaining that position the footstock is slid along its guide, frequently the rear edge of the swivel table, against which the lip of the footstock base rests, thereby maintaining the alignment of the footstock spindle with the headstock spindle.

The footstock spindle has an axial movement for permitting retraction when loading and unloading the work. The spindle can be guided in plain bushings or supported in preloaded ball sleeves, as in Fig. 1-2.9. The footstock spindle is held against the work by a controlled force, sufficient for keeping the stock centers securely in contact with the centerholes of the workpiece. At the same time, that force must not prevent the expansion of the work caused by heat generated in the grinding process; in the case of rigid constrainment, the workpiece might tend to buckle when its temperature rises.

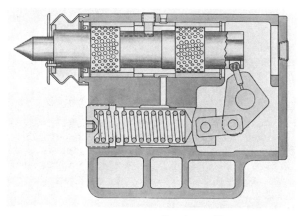

Courtesy of Landis Tool Co.

Fig. 1-2.9 Cross-sectional diagram of a footstock (tailstock), with pre-loaded ball sleeve and adjustable center force which is produced by a compressive helical spring. Quick acting retraction with level control is provided; the lever, which is mounted on the top of the footstock, is not shown in this view.

Commonly, the retaining action is provided by a spring, pressing the spindle toward the workpiece with a force that can be adjusted from the outside. For semiautomatic and automatic grinding, hydraulically operated footstocks are recommended, as their force can be adjusted with high sensitivity.

The retraction of the footstock spindle is usually actuated by a hand-operated lever, see Fig. 1-2.9, or,

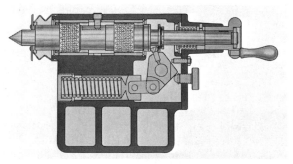

Courtesy of Landis Tool Co.

Fig. 1-2.10 Footstock for heavier workpieces with handwheel control. By locking the lever the center force can be adjusted with the handwheel. The lever is not shown.

for heavier workpieces this can be supplemented by handwheel adjustment, Fig. 1-2.10. The retraction of the hydraulic footstock can be actuated by the operator using a push button or foot pedal, also, it can be made a part of the automatic grinding cycle, thereby obviating manual controls. Power actuation of the footstock spindle with foot-pedal control will leave both hands of the operator free for handling the work.

The Cross Feed

The movement of the wheel head which is mounted on a cross slide causes the grinding wheel to approach the work or to be retracted from it.[1] These movements are needed (a) for the original setup of the machine in the case of work or wheel change, (b) for the approach of the wheel during the grinding process to establish contact with the work, (c) for the infeed, the rates and extent of which must be closely controlled, and (d) for the retraction of the grinding wheel after the termination of the operation. In addition to these basic operational elements there are various auxiliary functions as well, e.g., compensation for wheel wear and wheel dressing, temporary retraction for measurement, etc.

The setup adjustments, movements (a) of the wheel head are nearly always hand operated, except for machines that offer a jogging-type power travel of the cross slide. Both the approach and the retraction movements (b) and (d), are frequently carried out by manual control, either relying on direct visual observation of the wheel's position relative to the work, or guided by the graduated drum of the cross-feed handwheel, Fig. 1-2.11. Modern cylindrical grinding machines are generally equipped with power operated rapid advance and retraction which, after having been adjusted for the proper terminal positions, can be actuated by a single lever.

The infeed movement (c), for traverse grinding, is usually power actuated, in increments effected at the ends of the reciprocating strokes, and is generally adjustable to be operated at either or both end positions of the table traverse. The range of the incremental steps of the cross-feed varies according to the specific grinding machine model. A typical cross-feed range extends from 0.0001 to 0.0003 inch (0.0025 to 0.076 mm) at each single or alternate table reversal.

Movements (a), (b), and (d) will be continuous, while movement (c), the infeed, will either be contin-

---

[1]Different terms are used to designate that movement, such as "cross feed," "infeed," "wheel infeed," and "wheel feed."

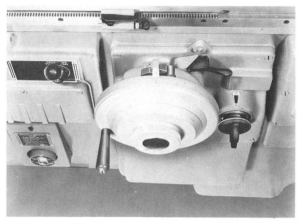

Courtesy of Brown & Sharpe Mfg. Co.

Fig. 1-2.11 Cross-feed handwheel with graduated drums for guiding the manual setting and also for adjusting the terminal point of the automatic cross feed.

uous, such as in plunge grinding, or incremental as in traverse grinding.

For plunge grinding, a continuous infeed movement is needed. This always should be power actuated for the dependably uniform advance of the grinding wheel. The rates of the continuous power infeed can, on modern machines, be steplessly regulated to suit the various workpiece diameters and other pertinent operational conditions. A typical range of cross-feed rates for plunge grinding is 0.0125 to 0.300 inch per minute.

When power controlled cross-feed is used, provision has to be made to terminate the feed at a specific point corresponding to the desired final size of the ground surface. On cylindrical grinding machines of advanced design such feed-terminating settings can be made with an accuracy of 0.0001 inch, and exceptionally, even to closer limits.

Modern cylindrical grinding machines are equipped with oscillating movement of the machine table,

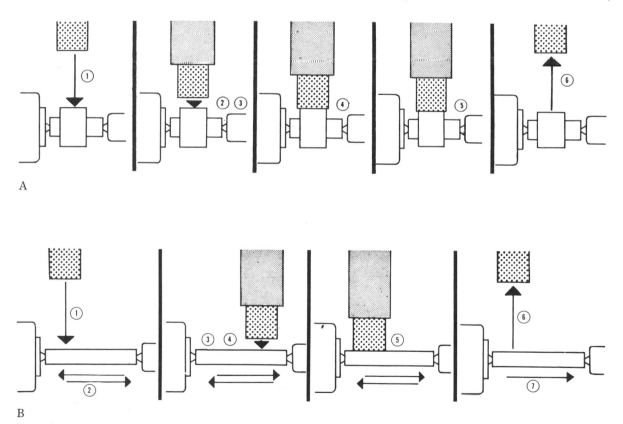

Courtesy of Brown & Sharpe Mfg. Co.

Fig. 1-2.12 Operation diagrams of the automatic grinding cycles on a cylindrical grinder. A. *Plunge Grinding*: (1) Wheel advances at rapid travel; (2) Feed reduced as wheel nears work; (3) Coarse feed starts before the wheel touches the work; (4) Fine feed; (5) Sparkout; and (6) Wheel retracts at rapid travel. B. *Traverse Grinding*: (1) Wheel advances at rapid travel; (2) Power table traverse starts; (3) Feed reduced as wheel nears work; (4) Wheel feed at each or alternate table reversal; (5) Timed sparkout; (6) Wheel retracts; and (7) Power table traverse stops.

which can be adjusted over a short range or entirely disengaged. Where the configuration of the plunge ground cylindrical work surface allows it, a short lateral movement of the work on the table, with a typical stroke length of 0.040 to 0.080 inch (1 to 2 mm), will improve the finish as well as the straightness of the ground surface by wiping off the wheel truing marks and reducing the harmful effect of the wheel corner breakdown.

Automatic cross feed cycle, when available, can provide a more comprehensive control of the grinding machine operation, making it to a great extent independent of the operator's attendance and skill. The operating principles of an advanced, yet typical system, are shown in the diagrams of Fig. 1-2.12 and explained in the pertinent captions.

Such systems can be used equally for plunge or traverse grinding, and once set, the routine steps of the operator's work are limited to loading the workpiece, starting the automatic cycle with a single control lever, and exchanging the work at the end of the operation. After the wheel slide has retracted, the rotation of the work and, in traverse grinding, the reciprocating movement of the table, comes to a stop. Often, the coolant supply control is tied in with the single-lever control and, in that case, the coolant flow also stops at the end of the cycle. Some models are equipped with adjustable timers to keep the cross slide in the end position of its transverse movement over a specific time span, for controlled spark-out.

The size control and the occasional wheel truing with wheel position compensation, still remain the responsibilities of the operator, unless automation is also provided for these functions.

The Machine Table and its Power Traverse

The machine table, mounted on the guideways of the bed, supports the work-holding members and, by its longitudinal displacement, provides for the lateral positioning of the workpiece in relation to the grinding wheel. In traverse-type grinding the table also carries out the reciprocating travel along a preset distance and at speeds which can be selected according to the conditions of the process.

Finally, in most types of cylindrical grinding machines, the table has an upper level which can be swiveled from its basic position parallel with the bed guideways. That upper level, often designated as the swivel table, carries the work-holding members, thus, the swivel movement will cause a conforming angular displacement of the workpiece held by the swivel table-supported members.

The length of the machine table is a major factor in the work-capacity of the grinder with regard to the maximum work-length which can be mounted between centers.

The traverse movement of the table along its guideways can be carried out manually, by turning a handwheel in front of the machine bed. In most cases the table traverse by handwheel rotation can be made either at the slow or at the fast rate, commonly established in a ratio of 1 : 10, such as 0.1 and 1.0 inch (2.5 and 25 mm) traverse for each turn of the handwheel.

For traverse-type grinding the power traverse movement of the table is usually accomplished by a hydraulic system (cylinder and piston) which provides smooth table travel and reversal. The hydraulic drive also permits infinite speed regulation over an extended range. Typical speed ranges for hydraulic table traverse are 3 to 150 inches (75 to 3800 mm) per minute for a universal-type machine and 2 to 240 inches (50 to 6000 mm) per minute for a modern production-type, plain grinding machine.

The location and the length of the power table traverse can be set with the aid of reversing dogs,

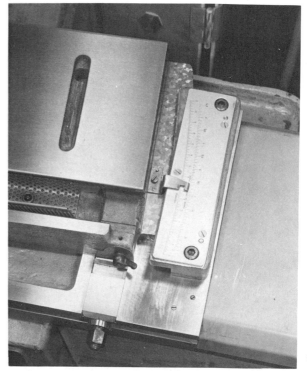

Courtesy of Brown & Sharpe Mfg. Co

Fig. 1-2.13 Swivel setting of the upper table guided by graduated scale. The locking pin and the adjusting nut are also visible.

having both coarse setting and fine adjustment. In traverse grinding it frequently proves necessary to delay the start of the return stroke for prolonging the contact of the wheel with the terminal sections of the ground surface, thereby assuring uniform stock removal over the entire length of the ground surface. To accomplish this controlled delay at reversals, most types of cylindrical grinding machines built for traverse grinding are equipped with a timer-regulated delaying device, producing a *dwell* or *tarry*, at the ends of the reciprocating table traverse. A typical dwell range is 0 to 2½ seconds, steplessly adjustable.

The swivel setting of the upper table is available on most types of cylindrical grinding machines for the purpose of grinding workpieces with shallow tapers between centers. On the basic machine table, often designated as the sliding table, an upper level is mounted, which can be swiveled around a fixed pivot point. The setting can be carried out sensitively with the aid of installed jackscrews, and guided by a graduated scale mounted on the sliding table, see Fig. 1-2.13. Such scales usually have two parallel graduations; one angular, in degrees, and the other showing linear units, representing the equivalent taper values expressed as diameter reduction per foot length. A typical range of settings is 11 degrees back and 3 degrees forward, the corresponding linear values being approximately 4 5/8 and 1¼ inches per foot (0.388 and 0.104 m per meter) taper.

When more accurate setting is needed than the graduated scale can provide, dial indicators may be used. Holding brackets and anvils for such swivel position indicators are sometimes available as optional accessories from the grinding machine manufacturer, together with information regarding the distance of the table pivot point from the gaging point. When the reliable value of that distance is not available, and also for error-free measurements over extended swivel distances, the use of indicators at both ends of the swivel table may be required; the indications of these gages will then have to be combined for obtaining the dependable value for the swivel angle.

The electronic gage is a more refined version of the swivel position indicator (Fig. 1-2.14). It permits

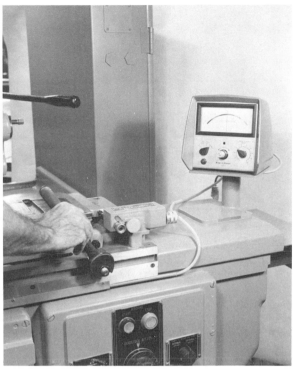

Courtesy of Brown & Sharpe Mfg. Co

Fig. 1-2.14 Swivel table position indicator with electronic gage which permits accurate position correction guided by direct readout.

the correction of the detected and accurately measured taper error by a single table adjustment. Such gages have two sensing positions, one at each end of the table, feeding into a common amplifier-meter, the indications of which will display the combined value of the displacements registered at the two sensing points. Another, similar gage, uses a double-pointer meter.

Swivel position indicator instruments are of great value either for the swivel setting of the table to a specific taper, or for aligning the swivel table with the direction of the sliding table travel in order to grind precisely cylindrical external or internal surfaces.

Basic Type Cylindrical Grinding Machines

Cylindrical grinding machines are manufactured in a wide variety of models which differ in the following two major respects: (1.) the general dimensions of the machine, which determine its work capacity, and (2.) the essential design characteristics, which define its operating capability.

While the general dimensions are relatively easy to define and to visualize, the design characteristics are more varied and their analysis is more complex. In establishing the design objectives of any particular model of grinding machine, the manufacturer is guided mainly by the application field, whose requirements the equipment is expected to meet most efficiently.

The systematic endeavor to design cylindrical grinding machines for particular operational purposes leads to the development of certain distinctive machine types. While the exact definition of boundaries among the major types of cylindrical grinding machines is not always possible, partly due to overlapping capabilities, the practice of designating cylindrical grinding machines by terms expressing the category to which the particular machine is considered to belong, is widely accepted. Such machine category designations, even when not perfectly descriptive, have proved to be useful in the preliminary selection of equipment for specific fields of application.

An attempt is made in Table 1-3.1 to present commonly used designations of cylindrical grinding machines, by pointing out the primary fields of application that machines belonging to the individual categories are intended to serve. The application purpose also defines the characteristic dimensional ranges of the grinding machines assigned to specific categories. Although machines belonging to a common category might actually have widely differing capacities and dimensions, the general range of such parameters is rather typical for the group. For that reason Table 1-3.1 also lists the ranges of characteristic dimensions for each of the presented categories.

Subsequently, a few characteristic features of typical machines for the major categories will be discussed in this, as well as in the following chapter which covers the limited purpose cylindrical grinding machines.

Machine models which have been selected to illustrate the different categories are standard products of prominent machine-tool manufacturers, which generally have supplied such types of grinders over a long period of years. However, the selection of specific models for the purpose outlined above, does not imply any technical superiority over similar types of machines manufactured by other, equally reputable, machine-tool builders.

Instrument-Type Cylindrical Grinding Machines

Very small parts, often in the miniature, or even micro-miniature, size category, when requiring grinding, should be machined on grinding machines specifically designed for that kind of work. It might be possible in exceptional cases to grind extremely small parts on precise universal tool and gage grinders, but probably less economically.

Figure 1-3.1 shows a typical representative of the very small universal cylindrical grinding machines designed for miniature parts, commonly known as instrument-type grinders. Besides the generally small dimensions and very precise execution, there are several typical design features which deserve mention.

The machine table is designed to accept different types of interchangeable elements, such as headstocks, a tailstock, steady rests, diamond dresser blocks for external grinding, and another for internal grinding,

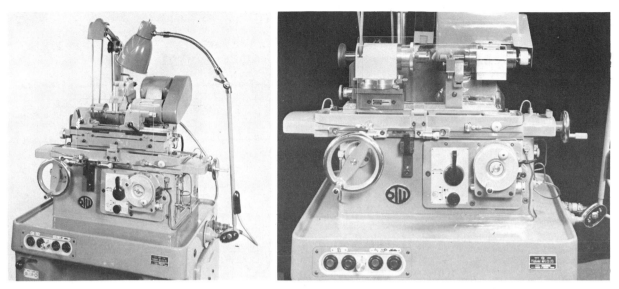

Courtesy of Louis Levin & Son, Inc.

Fig. 1-3.1 Instrument type cylindrical grinding machine for high-precision external and internal grinding of miniature and small-size parts. Interchangeable elements assure a great flexibility of adaptation to widely different work configurations. (*Left*) External model; (*Right*) Internal model.

etc. The flexibility of machine adaptation is comparable to that of precision instrument lathes, which are designed to accommodate a wide variety of workpieces of diverse configurations, without the need for custom-made fixturing.

The micro-feed table adjustment permits the precise longitudinal positioning of the table with respect to the wheel head. This adjustment is so sensitive that it can also be used for the "kiss-grinding" of shoulders with properly dressed wheels. The table also has a power drive for the reciprocating traverse, with steplessly adjustable stroke lengths from 0.4 inch to 3.2 inches (about 10 to 80 mm), with speeds infinitely variable from 0 to 106 inches (0 to 2.7 m) per minute. For grinding shallow tapers, the table can be swiveled four degrees in either direction with the aid of a sensitive adjusting screw, guided by a graduated segment.

The drive of the headstock spindle by means of an endless round belt through an idler pulley arm offers freedom from motor-associated vibrations and provides flexibility of adaptation by means of interchangeable headstocks. That drive system, combined with the general design of the machine table, makes it possible to mount a headstock in any position along the table and to swivel the upper portion of the headstock for operating the work spindle at an angle to the direction of the table traverse.

The interchangeable headstocks comprise a dead-center type for the most precise operations, the collet-holding type for repetitive work, and the universal type for providing a swivel movement — adjustable with gage blocks — in the manner used for setting up sine bars. The headstock speeds are infinitely variable from 0 to 1400 rpm.

The cross-feed can be carried out either manually or automatically; the latter applied in increments from 0.000050 to 0.0004 inch (0.00125 to 0.01 mm) at both ends, or at one end only, of the table reciprocating stroke.

Finally, some data regarding the operating capabilities of the instrument grinder, as stated by the manufacturer, follow:

Size holding of the diameter:	to 4 microinches (0.0001 mm) when between centers, and to 12 microinches (0.0003 mm) in collet work.
Straightness of the ground surface:	to 4 microinches/inch (0.00001 mm), or 6 microinches/3 inches (0.00015 mm /75 m).
Finish of the surface:	with 60-grit wheel, 4 microinches AA (0.1 micrometer R_a); in lapping type grinding, better than 1 microinch AA (0.025 micrometer R_a).
Internal grinding:	to 0.050 inch (1.25 mm), smallest diameter. Roundness, to 15 microinches (0.6 micrometer), and size holding, to 4 microinches (0.1 micrometer).

Table 1-3.1 Categories of Cylindrical Grinding Machines

DESIGNA-TION	MAJOR FIELDS OF APPLICATION	TYPICAL EXAMPLE, FIG. NO.	TYPICAL CAPACITY DATA AND CHARACTERISTIC DIMENSIONS				
			WORK SWING	DISTANCE BETWEEN CENTERS	GRINDING WHEEL DIA. $\times$ W.	WHEEL DRIVE	NET WEIGHT
			Inches (mm)			hp	Lbs (kg)
Instrument type Cylindrical Grinding Machine	Very small cylindrical or tapered parts for instruments, etc., requiring very close tolerances. Flexibility of adaptation for diverse configurations.	1-3.1	4 (100)	5 (125)	5 $\times$ 12 (125 $\times$ 13)	1/6	1200 (550)
Precision type Production Grinding Machine	Small and medium-size parts ground with high accuracy and to be manufactured in large or intermediate lots.	1-3.2	4 to 5 (100 to 125)	10 to 18 (250 to 450)	10 $\times$ 1 (250 $\times$ 25)	1	1600 (750)
Universal Tool Grinding Machine	Tools and fixture elements often utilizing adjustment flexibility and optional accessories (e.g., internal grinding spindles)	1-3.4	10 to 14 (250 to 350)	20 to 48 (500 to 1200)	10 to 14 (250 to 350) $\times$ ½ to 2 (13 to 50)	1½ to 5	5,000 to 10,000 (2,200 to 4,500)
Plain Cylindrical Grinding Machine	Mostly cylindrical or slenderly tapered regular features on parts with defined axes of rotation.	1-3.5	10 to 18 (250 to 450)	18 to 72 (450 to 1800)	24 to 36 (600 to 900) $\times$ 4 to 8 (100 to 200)	7½ to 25	7,000 to 18,000 (3,200 to 8,000)
Heavy Duty Plain Cylindrical Grinding Machines	Work of large dimensions and heavy weight, often ground with very high stock-removal rates.	1-3.6	18 to 24 (450 to 600)	48 to 240 (1,200 to 6,000)	30 (750) $\times$ 8 to 12 (200 to 300)	25 to 40	30,000 to 60,000 (13,000 to 28,000)
Plunge type Cylindrical Production Grinders	For grinding cylindrical or profiled round sections on generally short parts, also those having several coaxial features in different planes and diameters.	1-4.3 and 1-4.6	10 to 30 (250 to 750)	24 to 80 (600 to 2,000)	14 to 24 (350 to 600) $\times$ 2 to 6 (50 to 150)	10 to 30	15,000 to 60,000 (7,000 to 28,000)
Roll Grinding Machines	For large-size rolls such as are used in rolling mills; also for calender rolls in the paper, cloth, etc., industries.	1-4.7	36 to 60 (900 to 1500)	120 to 260 (3,000 to 6,500)	36 (900) $\times$ 4 (100)	40 to 80	150,000 to 200,000 (70,000 to 90,000)

(Continued on next page.)

Table 1-3.1 (*Cont.*) **Categories of Cylindrical Grinding Machines**

DESIGNA-TION	MAJOR FIELDS OF APPLICATION	TYPICAL EXAMPLE, FIG. NO.	TYPICAL CAPACITY DATA AND CHARACTERISTIC DIMENSIONS				
			WORK SWING	DISTANCE BETWEEN CENTERS	GRINDING WHEEL DIA. × W.	WHEEL DRIVE	NET WEIGHT
			Inches (mm)			hp	Lbs (kg)
Limited Purpose Cylindrical Grinding Machines	Typical examples are: Crankshaft (pin) grinders, Camshaft grinders, etc. Note: Characteristic dimensions are based on these types.	1-4.10 1-4.14 1-4.19 1-4.27 1-4.30	8 to 24 (200 to 600)	24 to 72 (600 to 1,800)	20 to 36 (250 to 900) × 2 to 4 (50 to 100)	25 to 40	25,000 to 40,000 (12,000 to 18,000)

High-Precision, Production-Type Cylindrical Grinding Machines

Small-size precision parts manufactured in medium or large batches and requiring cylindrical-type grinding are most economically and accurately produced on cylindrical grinding machines specifically designed for that kind of work. Such small workpieces often have several cylindrical or tapered sections of different diameters, and may require the grinding of shoulders as well; all surfaces to be produced precisely concentric, mutually, and possibly also in relation to the part axis, as represented by previously ground features, such as external or internal centers, etc.

A representative of this family of small, precision cylindrical grinders is shown in Fig. 1-3.2. In this illustration a few characteristic features can be seen; and these, as well as some others complementing them, will now be discussed.

The headstock is designed to accept different quills used for the rapid loading of parts which can be held (a) between centers, (b) in collets operated by lever or handwheel, or (c) mounted into a three-jaw chuck. The headstock bases accepting these quills are also interchangeable, comprising the fixed type, the 30-degree angled type, or one with a swivel adjustment, permitting angular setting from 0 to 45 degrees, or 45 to 90 degrees, respectively.

While the worktable may be swiveled for grinding shallow tapers, very small components with steeper tapers, which must be precisely concentric with cylindrical portions of the part, as are the tapered seats of injection needles, can also be ground with special attachments. Such a taper-grinding attachment is mounted on the machine bed, Fig. 1-3.3, has its own headstock with quill, and can be operated either with its footstock or with a special steady rest, one version of which is seen in the illustration.

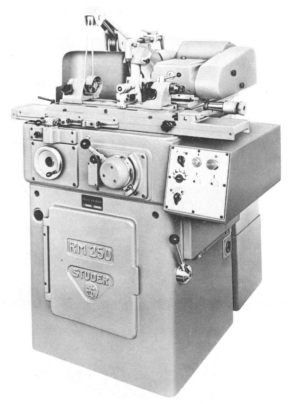

Courtesy of Fritz Studer Ltd./Cosa Corp.

Fig. 1-3.2. Small cylindrical grinder for high-precision type production grinding, with interchangeable headstock quills and a wide range of optional equipment.

One of the characteristic features of the discussed machine model, which makes it particularly well adapted for production-type work, is its single-lever operation acting on electromechanical controls to provide a semiautomatic grinding cycle. Depending

Courtesy of Fritz Studer Ltd./Cosa Corp.

Fig. 1-3.3   Bed-mounted taper grinding attachment with individual headstock and supplementary work-support elements.   This is a typical example of specialized accessories available for the small cylindrical grinder shown in Fig. 1-3.2.

on the position to which that control lever is set, one of six different programs can be instituted. These comprise both plunge and traverse grinding, including infeeds at different rates for roughing and finishing, position holding for spark-out, rapid approach and retraction of the wheel head and stopping the rotation of the headstock spindle. A special program is provided for automatic dressing with slow table traverse.

The accurate measuring of both the position of the adjustable machine elements and the size of the part to be ground, are essential prerequisites of the high-precision type work for which these machines are designed.   Accordingly, sensitive dial indicators can be mounted for setting the table swivel, for wheel truing, for adjusting the steady rest, and for table alignment.   Optional electronic gages with continuously contacting probes are available for measuring the workpiece diameter. The gage signals can be tied in with the infeed movement of the wheel head, thus assuring automatic size control. Similar types of electronic gages also can be used for match grinding, the general principles of which are explained briefly in item 10 of Table 1-5.3.

Universal Grinding Machines

The cylindrical grinding machine illustrated in Fig. 1-3.4 is one particular model in a family of universal grinding machines built in different sizes. The designation *universal* is a commonly used term to indicate great versatility, one of the major applicational purposes of this grinding machine category.

Another prominent design characteristic of universal grinding machines is a high level of accuracy, which is commensurate with the requirements of the primary field of application, that of grinding tools, gages, and fixture elements.

On the other hand, the attainable rate of stock removal is not considered a decisive factor in applications which these machines are intended to serve.

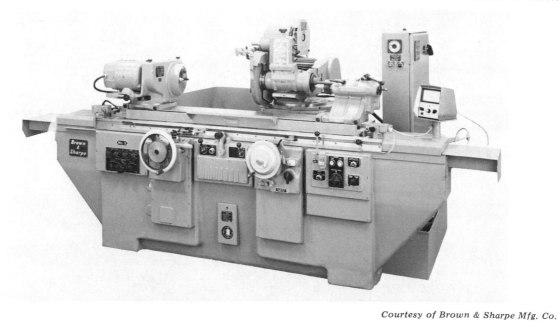

Courtesy of Brown & Sharpe Mfg. Co.

Fig. 1-3.4   Universal-type cylindrical grinding machine designed to accept internal grinding attachment.

The lesser emphasis on grinding performance is reflected by the relatively small wheel width (1 to 1½ inches) (25 to 38 mm), and by the output of the wheel drive motor which, at 5 hp for machines in the 4 tons net weight category, is lower than the power commonly selected for comparable size production-type grinding machines.

The *versatility* of the universal grinding machine is the result of design characteristics and an extensive variety of accessory equipment which enhances the machine's adaptability to different types of work-pieces and process requirements. A few examples which follow, will substantiate these design objectives:

(1) The headstock can be swiveled on its base, which can also be displaced longitudinally along the table. The infinitely variable speeds of the head-stock spindle have limit values which are related in a ratio of 1:20. The headstock can, by a few simple motions, be changed over from dead centers to rotating-spindle-type operations, or vice versa.

(2) The control elements can be operated entirely manually, or by applying various power movements ranging up to an entirely automatic grinding cycle.

(3) An internal grinding device is provided on the machine, and can be equipped with any of several different types of spindles.

(4) Wheel-truing devices, in addition to the basic straight type, are available for angle-wheel truing, for radius wheels, and also for generating continuous radius and tangent profiles.

(5) Work-holding and supporting devices belonging to the standard equipment of the machine or available as optional accessories, include scroll chucks, collet attachments, mechanical and magnetic face chucks, different types of steady rests, etc.

The attainable *work accuracy* is often of prime importance in applications for which universal grinding machines are selected. In order to achieve a consistently high degree of geometric and dimensional accuracy of the ground surfaces, particular aspects must be considered and conditions assured in the design and execution of the applied grinding machines. In addition to mentioning several of these aspects, the pertinent characteristics of the illustrated type of universal grinding machine are also indicated with the intent of pointing out important design characteristics which contribute to high work accuracy in cylindrical grinding, and highlighting typical features of advanced models of universal grinding machines:

(a) The design and execution must assure stability, freedom from vibration, smooth-running of the rotating elements and displacement movements without the sliding members being affected by stick-and-slip conditions. For satisfying these exacting requirements to an advanced degree, the illustrated model can be equipped optionally with hydrostatic bearings for one or all of the following spindles: the wheel head, the headstock, and the internal grinding device.

The best measure of a grinding machine's design and construction accuracy is its performance capability expressed in the geometric parameters of the

Roundness — (errors not exceeding the listed values):

Grinding between centers

Standard machine	0.000 020 inch (0.0005 mm)
Refined machine	0.000 010 inch (0.00025 mm)

Grinding in the headstock

Ball-bearing headstock	standard	0.0001 inch (0.0025 mm)
	refined	0.000 050 inch (0.00125 mm)
Hydrostatic headstock	standard	0.000 020 inch (0.0005 mm)
	refined	0.000 010 inch (0.00025 mm)

Straightness — (errors not exceeding the listed values):

	for the first 1 inch (25 mm)	for each further inch (25 mm), additional
standard machine	0.000 015 inch (0.00038 mm)	0.000 010 inch (0.00025 mm)
refined machine	0.000 010 inch (0.00025 mm)	0.000 005 inch (0.000125 mm)

Surface finish —

Grinding between centers	3 microinches AA (0.08 micrometer R_a)
Headstock grinding, ball bearing spindle	5 microinches AA (0.125 micrometer R_a)
hydrostatic spindle	3 microinches AA (0.08 micrometer R_a)
Refined machine with hydrostatic bearing spindles for both the wheel head and the headstock.	2 microinches AA (0.05 micrometer R_a), or better

ground surfaces. Typical values for such parameters, listed on page 27, from data supplied by the manufacturer of the universal grinder shown in Fig. 1-3.4.

(b) Sensitivity and precision of controlling the slide movements, both manually and automatically. Typical performance data, based on the illustrated model of a precise universal grinding machine, are quoted for *controlling the size* (diameter) of the ground surface within the following limits:

Standard machine   0.000 050 inch (0.00125 mm)
Refined machine   0.000 025 inch (0.0007 mm)
With electronic
   size control   0.000 010 inch (0.00025 mm)

(c) The accuracy of the actual size of the ground surface. While the repeat accuracy of the slide movements, as well as the sensitivity of limit setting for these movements are fundamental conditions of precise grinding, the accurate size of the ground surface can only be assured by measuring the actual size of the part. Such size measurements can be carried out on the part after it has been removed from the grinding machine. In the case of deviations from the specified value, adjustments are then made on the machine control elements.

There are, however, some conditions that will require the measurement of the part while it is held in the machine, or even while grinding is in progress. Examples of cylindrical grinding operations in which the measurement of the part should be carried out in the machine are:

(a) Single, or limited number of parts, which must have a dependably accurate size before being removed from the grinder, because rejects or the need for rework would be too costly.
(b) Parts produced with very high accuracy requirements, the grinding of which must be assured independently of variations from sources outside

the precisely controlled infeed movement. In such operations the direct measurement of the part, while in process, is mandatory. The continually sensed size changes of the part may then be used for controlling the last phase of the infeed movement, supplanting the fixed end-stop settings.

Plain Cylindrical Grinding Machines

The production grinding of workpieces with straight element surfaces which are concentric with respect to the axis of the part or with one of its features, is the primary design objective of plain cylindrical grinding machines, a typical representative of which is shown in Fig. 1-3.5. The illustrated example is a medium size member of a family of plain grinding machines which, in general design, embraces different models with swings from 6 inches to 24 inches (150 to 600 mm), and center distances from 18 to 120 inches (450 to 4000 mm). For each size group two or more different swing diameters and several standard center distances are available. The power of the wheel motor of the regular machines varies from 7½ hp to 25 hp, depending on the model size.

The photo shows the regular plain grinder with straight wheel mounted on the wheel head with cross slide normal to the table travel. The smaller sizes of this family of plain cylindrical grinding machines are also built with angular wheel heads, designed for angular approach cylindrical grinding. That method will be discussed in greater detail later in Chapter 1-4.

The design emphasis in this category of cylindrical grinding machines is more on productivity than on versatility, yet without compromising on work accuracy: this characteristic is comparable to best standard models of universal grinding machines.

Courtesy of Landis Tool Co.

Fig. 1-3.5   Plain cylindrical grinding machine for precision grinding in medium- or high-volume production.

The objective of high work accuracy poses various problems on large-size machines, in which great masses are moved and substantial wheel drive power is applied. The casting of the bed must be generously dimensioned in order to properly support the heavy moving members, dampen vibrations, and dissipate heat. To reduce the harmful effect of generated heat, a separate hydraulic power unit is used for storing the hydraulic oil and is mounted on the rear of the machine. A special system for lubrication of the table ways is designed to avoid the accumulation of excess lubricant that could produce a "floating" of the table, with harmful effect on the grinding accuracy.

These production grinding machines are equally adaptable for traverse and plunge grinding. In the latter version they may have profile-trued grinding wheels. In plain grinders the swivel-top table, although available, is usually not standard equipment. Considering the less varied applications of such grinders, many accessories, although supplied optionally, should be specified only when needed. Of particular importance are accessories which reduce operational time without affecting the consistent accuracy of the grinding operations. Sample devices, selected from a rather extensive list, are:

(a) those for controlling the infeed rate and travel, often to several specific points, for workpieces having various sections of different diameters

(b) those for controlling automatically, the rapid wheel approach

(c) those for fast and precise longitudinal positioning of the machine table, combined with sensing elements, when grinding parts with sections at different lateral locations, often involving shoulder grinding which calls for particularly accurate position setting.

Heavy Plain and Roll Grinding Machines

Large workpieces of essentially cylindrical form, such as shafts and spindles, when requiring precisely machined surfaces, can be ground with a high degree of form and dimensional accuracy on cylindrical grinding machines of the type shown in Fig. 1-3.6.

Courtesy of Cincinnati Milacron Inc.

Fig. 1-3.6 Heavy plain and roll grinding machine.

These grinding machines, while designed to accommodate comparatively large and heavy workpieces, are intended to perform very accurate work and to produce surfaces with an excellent finish, as well. Such capabilities make these grinders adaptable to very demanding tasks; such as the grinding of steel-mill rolls.

The range of sizes in which this type of heavy cylindrical grinders is manufactured is rather extensive, as indicated by the dimensional data in the table.

In the grinding of certain types of workpieces the truing of the grinding wheel may need to be carried out without the traversing of the table; for such applications a special truing device can be mounted on the wheel head. That device has a hydraulic traversing drive with rates infinitely variable over a wide range. The truing can produce either a straight or a profiled wheel face; the latter, accomplished with the aid of appropriate profile bars, is particularly useful in roll grinding.

The dependable control of the grinding speeds and feeds with regard to both the selection of the most suitable magnitudes and the maintenance of consistent rates, is an important prerequisite for obtaining

Characteristic Range of Dimensions of Heavy Plain and Roll-Grinding Machines

Swing Diameter		Max Weight of Work		Available Distances Between Centers		Net Machine Weight	
Inches	mm	Pounds	kg	Inches	mm	Pounds	kg
18 to 24	450 to 600	10,000 to 15,000	4500 to 6800	48 to 240	1200 to 6000	33,000 to 60,000	15,000 to 27,000

high surface quality.   In cylindrical grinding this means control of the traversing speed of the table and the rotational speed of the work.   In the type of grinders described both movements receive their drive from dc motors, regulated by rheostats, and are infinitely variable over extensive ranges.   Furthermore, the table traverse is electronically controlled for a high degree of uniformity, and has automatic deceleration before the reversals which are accomplished within 0.002 inch (0.05 mm) at any traverse speed. The dwell period of the table traverse at the end positions is adjustable from 0 to 18 seconds.

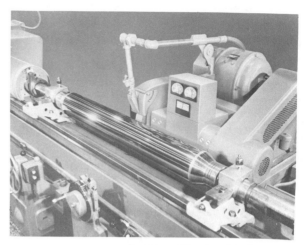

Courtesy of Cincinnati Milacron Inc.

Fig. 1-3.8   The grinding of a cambered roll with the aid of the cambering mechanism of a roll grinder type cylindrical grinding machine.

Courtesy of Cincinnati Milacron Inc.

Fig. 1-3.7   The gear box of a cambering mechanism which generates the tilt motion of the wheel head for producing a precisely controlled camber on the workpiece being ground.

For roll grinding, an important machine accessory is the *cambering mechanism* which permits the grinding of convex or concave roll forms with precisely controlled digression of the profile from the basic cylindrical form.   One type of cambering mechanism operates by means of a tiltable wheel head, whose sub-base pivots around supporting trunnions arranged in the front area.   When introducing a tilt motion, by either raising or lowering the rear end of the sub-base, the wheel is made to approach or retract from the work.   The tilt motion is accomplished by a lever system, actuated by an adjustable cam assembly, which, in turn, is synchronized to rotate with the table traverse.   Intermediate change gears, shown in Fig. 1-3.7, permit varying the ratio between the table traverse and the wheel-head tilt in order to apply the desired camber to any given roll length.   The finishing stage in the grinding of a cambered roll on a roll grinding machine equipped with cambering mechanism, is shown in Fig. 1-3.8.

Other types of cylindrical grinding machines built for roll grinding may have camber grinding mechanisms which operate by principles different from those just described.   Such mechanisms may be based on a swiveling table, with a vertical pivot axis in the center, using adjustable cam bars as actuating elements.

Limited Purpose Cylindrical Grinding Machines

In the high-volume production of parts requiring cylindrical grinding, a specific machine may be kept operating for the entire year on one type of part, or on a group of very similar parts. In those applications where the machine tool is kept "captive," emphasis is put on high productivity, simple setup, and operational controls to the extent of advanced automation, while the flexibility of adaptation is either of secondary importance, or not required at all.

With the expansion of industrial production, a continually growing demand developed for *limited-purpose cylindrical grinding machines* whose design was adapted to particular types of workpieces and operations. By limiting their purpose, the general dimensioning of these grinding machines, e.g., the center heights and distances, were determined by the sizes of the parts which would have to be accommodated. Also the range of movements and speeds could often be reduced, and many accessories not needed in a single or limited-purpose use were eliminated.

On the other hand, such limited-purpose machines, while compact in size, are often heavier in construction and equipped with higher-powered drive motors than their comparable, yet more versatile, counterparts. The accessories, although of reduced flexibility and of limited variety, are usually better adapted for the specific services needed in the planned operations.

Limited-purpose cylindrical grinding machines are predominantly, yet not exclusively, of the plunge grinder type. Plunge grinding, wherever the part's configuration and other specifications permit its application, is generally more productive, i.e., requires less operational time than grinding the same surface by traverse grinding. For that reason several of the grinding machine types reviewed in the following, as examples of limited-purpose cylindrical grinders, are of the plunge grinder type.

Cylindrical grinding, in the comprehensive sense of the term, is also applied to workpieces whose general configuration is definitely not of cylindrical shape. However, the surface to be ground has an axis that can be positively established in the grinding process and the surface itself, not necessarily a cylinder, can be generated in conjunction with the rotation of the part around the feature axis.

The grinding of such workpieces requires special grinding machines, of course, which are specifically adapted for the planned operation. Characteristic types of *special-purpose cylindrical grinding machines* will also be discussed subsequently: crankpin grinders, camshaft grinders, and the polygonal grinding machines being used as examples.

Angular-Approach Grinding

Many types of workpieces have design characteristics which require the grinding of both the periphery and the adjoining flange section, often to precisely controlled squareness and axial location. It is possible to carry out such operations with a plain cylindrical grinding machine in a setup as shown in Fig. 1-4.1 A, which involves the truing of the side of the grinding wheel to a relieved form, as also indicated in the tenth diagram of Table 1-1.1. In this case, the machine table must be traversed by the handwheel against a fixed stop or an indicator, thus removing exactly the amount of stock needed for assuring the specified axial location of the flange surface. When the shoulder is a bevel, therefore not normal to the axis of the work, a universal cylindrical grinding machine can be set up in the manner shown in Fig. 1-4.1B; this setup also provides for the traverse grinding of the shallow taper main section of the work. Grinding the bevel by advancing the swiveled

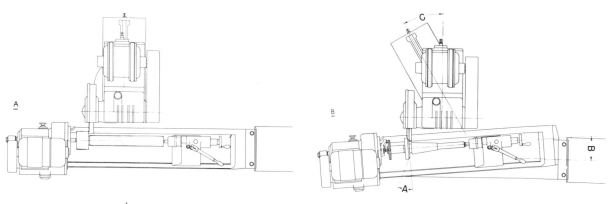

Courtesy of Cincinnati Milacron Inc.

Fig. 1-4.1   Grinding of shoulders on plain and universal cylindrical grinding machines. A. Setup for the grinding of a straight shoulder. The diagram indicates the process in which the cylindrical periphery of the work is first ground by power traversing the table; subsequently, the table is traversed by manual control to bring the face of the shoulder into contact with the grinding wheel, the side of which has been relieved. B. Setup diagram of a universal cylindrical grinder for grinding the tapered periphery and the beveled shoulder of a workpiece, by combining the effects of the table swivel and the wheel head swivel. The bevel angle A = angles C — B, or when swiveling the table in the other direction, bevel angle A = angles B + C.

wheel head produces a good finish and avoids wheel waste caused by excessive truing.

These methods, although applicable for single-piece, or low-volume, production are rather time-consuming and dependent on the operator's skill. More efficient and consistent grinding can be achieved with the grinding-wheel head swiveled to an angle with respect to the regularly used perpendicular to the work axis. In this case the infeed of the wheel slide is effected in an angled direction and the grinding wheel is trued on both its periphery and its side, providing two properly related, and in most applications, mutually normal operating faces, which are respectively, parallel with, and normal to, the work axis. The applicable operation will be plunge grinding, carried out without table traverse movement. The ratio between the wheel penetration into the contiguous sections, the cylindrical and the plane, will depend on the swivel angle of the grinding-wheel head, and must be appraised in planning the process. The diagram in Fig. 1-4.2 explains how these values are established in the case of a cylindrical surface adjoined by a perpendicular flange.

Angular-approach grinding can be carried out on universal type cylindrical grinding machines which permit swiveling the wheel head and operating the infeed in the direction of that swivel setting. In low-volume production on universal type grinding machines, the angular-approach grinding can also be combined with hand-actuated table traverse, thus increasing the length of the cylindrical work section ground in the operation, without being limited by the width of the grinding wheel.

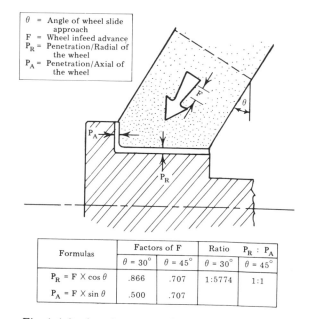

	θ =	Angle of wheel slide approach
	F =	Wheel infeed advance
	P_R =	Penetration/Radial of the wheel
	P_A =	Penetration/Axial of the wheel

Formulas	Factors of F		Ratio $P_R : P_A$	
	$\theta = 30°$	$\theta = 45°$	$\theta = 30°$	$\theta = 45°$
$P_R = F \times \cos \theta$	.866	.707	1:5774	1:1
$P_A = F \times \sin \theta$	.500	.707		

Fig. 1-4.2. Angular-approach plunge grinding; distribution of stock removal during the concurrent grinding of the work periphery and the shoulder. NOTE: The reduction of the work diameter is twice the amount of the radial wheel penetration.

For production-type grinding special, angular, plain grinding machines are available, commonly built with wheel-head angles at 30 degrees, as shown in Fig. 1-4.3. These machines are designed specifically for plunge grinding and the table movement by hand wheel serves only the purpose of machine setup. By offering the use of relatively wide wheels, the production-type angular grinding machines permit the machining of rather long cylindrical sections together with the adjoining flanges and, when needed, fillet areas also.

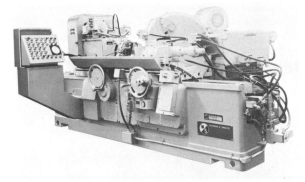

Courtesy of The Warner & Swasey Co., Grinding Machine Div.

Fig. 1-4.3 Angular-approach type cylindrical grinding machine designed for the semiautomatic plunge grinding of one or more diameters and a shoulder.

Aside from the process advantages resulting from the geometric conditions associated with an angular wheel approach, certain beneficial effects on the quality of the grinding should also be considered:

(a) The mutual perpendicularity of the ground surfaces is inherent in the shape of the trued wheel faces, and is not dependent on positional variables to which operations requiring more than a single loading of the work are exposed.

(b) The wheel contact on the flat shoulder surface is limited to a narrow area, a condition often referred to as "line contact." A narrow wheel contact on a flat work surface generally permits a more effective stock removal, without damaging the work surface, than does grinding with the face of the wheel.

(c) The contact of the wheel rotating around an axis inclined to both work surfaces produces an arcuate surface pattern, combined with some wiping action, producing a better surface finish value and a surface lay which can have beneficial effect on lubricant retention in the service of the ground part.

The form accuracy of the ground fillet, as well as its blending into the contiguous straight sections, must be assured by the correct truing of the grinding wheel. For that reason angular grinders are usually equipped with precisely operating and power-driven wheel-truing devices, which use profile bars to generate the required fillet configuration.

The theoretical length of the cylindrical section which can be ground in angular grinding with a grinding wheel of given width, results from multiplying the wheel width by the secant of the approach angle

θ ($L = W \sec \theta$). However, that portion of the nominal wheel width which is actually available for the grinding of the cylindrical section must be established by (a) making a small allowance for wheel breakdown at the corners, and (b) subtracting the wheel width needed for grinding the flange of height H. In the case of the commonly used 30-degree approach angle the nominal value of the additional wheel width is one-half of the flange height ($W = H \sin \theta$). These relationships between the wheel width and the dimensions of the contacted work surfaces are shown in Fig. 1-4.4.

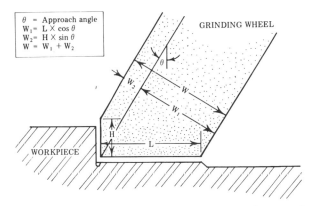

Fig. 1-4.4 The relationship between the total width of the wheel, its distribution by dressing into peripheral and face-grinding sections, and the corresponding lengths of the covered work-surface sections.

Angular-approach grinding can also be carried out with multiple grinding wheels, one or several of which are used for plain cylindrical sections of the part, thus combining various methods of cylindrical grinding. To illustrate the potentials of angular-approach grinding, Table 1-4.1 presents examples of typical workpieces which are particularly well adapted for being ground efficiently by this method.

Plunge Grinders with Crush-formed Wheels

The grinding of parts having essentially round shape but intricate profiles, when carried out on regular cylindrical grinding machines requires precisely reproducing form-truing devices that operate by using profile bars (copying templates) conforming to the designed contour of the workpiece. Form-truing with a single-point diamond traversed across the wheel face is, in the case of wide wheels, a rather lengthy process. Furthermore, when profiles with abrupt changes in the contour are needed, then the truing, which is guided by a profile bar, poses form limits due to the physical dimensions of the follower.

Table 1-4.1. Examples of Angular-Approach Cylindrical Grinding
(Showing automobile parts for illustrating a method of extensive applications.)

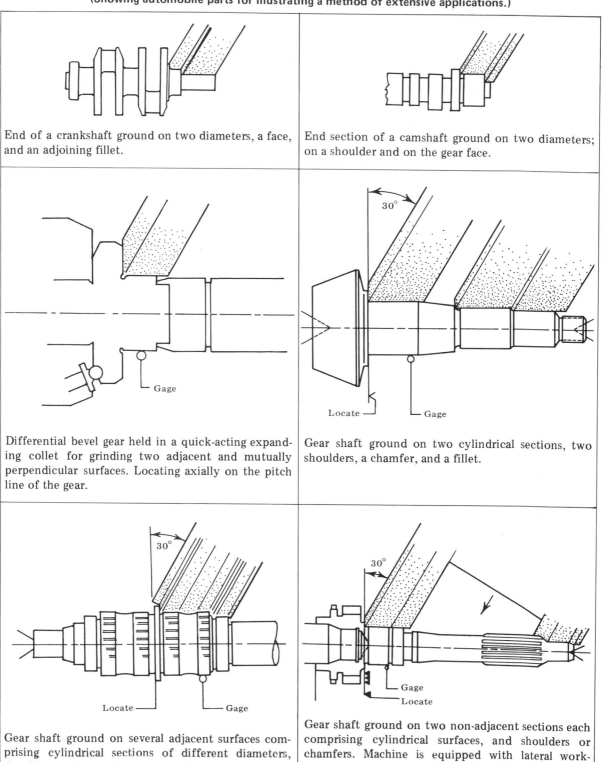

End of a crankshaft ground on two diameters, a face, and an adjoining fillet.

End section of a camshaft ground on two diameters; on a shoulder and on the gear face.

Differential bevel gear held in a quick-acting expanding collet for grinding two adjacent and mutually perpendicular surfaces. Locating axially on the pitch line of the gear.

Gear shaft ground on two cylindrical sections, two shoulders, a chamfer, and a fillet.

Gear shaft ground on several adjacent surfaces comprising cylindrical sections of different diameters, shoulders, and a profiled neck section.

Gear shaft ground on two non-adjacent sections each comprising cylindrical surfaces, and shoulders or chamfers. Machine is equipped with lateral work-locator.

Table 1-4.1. (*Cont.*) Examples of Angular-Approach Cylindrical Grinding
(Showing automobile parts for illustrating a method of extensive applications.)

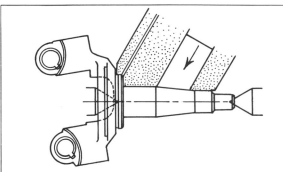

Automobile steering knuckle held between centers for grinding on three cylindrical sections, on a shoulder, and in the connecting fillet.

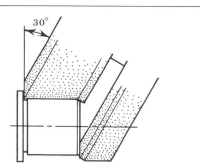

Bushing held on a quick-acting expanding collet for grinding a cylindrical section with adjacent shoulder, a chamfer, and the face.

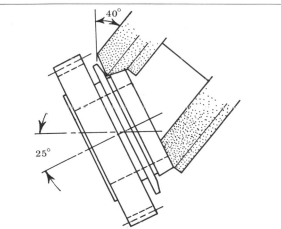

Gear ground on a taper section and the face by holding the work in a headstock swiveled at 25 degrees.

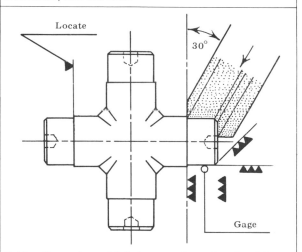

Spider for universal joint, ground on the shoulder, a cylindrical section, chamfer, and face, locating laterally from the shoulder of the previously ground opposite end.

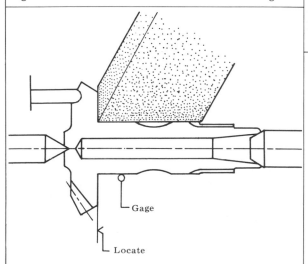

Gear with integral shank ground on two cylindrical sections and on the shoulder.

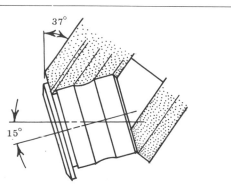

Bearing bushing held in a swiveled headstock for grinding several adjacent surface sections with a pro-file-trued grinding wheel and using an additional wheel on the same spindle for the concurrent grinding of the face and chamfer.

Crush-truing is a method by which even rather complex profiles can be produced on the wheel face; it is carried out very rapidly and with excellent repeatability. The method, the operational principles of which can be visualized from the close-up view in Fig. 1-4.5, consists in using a hardened steel or carbide roll made to the desired workpiece profile, and causing that roll to penetrate into the periphery of the grinding wheel. For that purpose the freely rotating roll is fed radially toward the wheel, by applying the necessary force to approach the roll at a controlled slow rate. The axis of the crusher roll is parallel with the grinding-wheel spindle and during the crushing process the grinding wheel is rotated at a very slow rate. When in contact, both the grinding wheel and the crusher roll rotate with the same peripheral speed, and under such conditions no grinding takes place. The continuously fed roll will gradually sink its profile into the operating face of the wheel, without harmful effect on the crusher roll itself, except for a small degree of attritious wear that can be corrected by occasional regrinding of the crusher roll.

The crush-trued wheel face will present a reasonably faithful inverse reproduction of the crusher-roll profile. An additional benefit of crush-forming the wheel face is that it is dressed to a sharp condition providing very good cutting properties, in distinction to the condition obtained by conventional truing methods. Furthermore, the excess grains removed by the crushing action leave minute open voids on the wheel face, which may act as lubricant reservoirs and provide the larger chip space needed when grinding with high stock-removal rates.

Crush-truing, in principle, can be carried out on many types of general-purpose grinding machines when designed for the eventual use of that method, and also equipped with a special truing device operating with a crusher roll. The actual application of crush-truing on general-purpose grinding machines is, however, rather rare. That limited usage becomes quite understandable when the basic machine design requirements, to be reviewed in their main aspects, are now considered.

(a) The grinding-wheel spindle and its bearings must be designed to withstand, without any harmful deflection, the substantial radial forces which act in both the crush-truing and in the wide-faced plunge-grinding processes.

(b) For the extended capacity and efficient operation of the method, wide grinding wheels with large diameters should be used, driven by adequately powered motors.

(c) A special, low, grinding-wheel speed must also be available for carrying out the crush-truing process.

(d) A crush-truing device, of particular design, with generously dimensioned bearings supporting a strong crusher-roll spindle is needed, mounted on a special slide whose power movement must have a variable feed range.

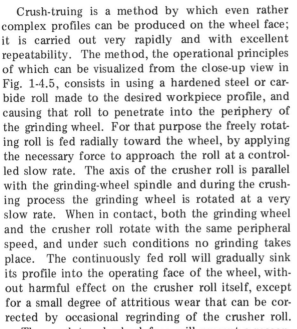

Courtesy of Bendix Corp.—Automation & Measurement Div.

Fig. 1-4.5 The crush-form process in cylindrical grinding: Close-up view of the crush-roll at top, the grinding wheel below, and the form-ground part in the foreground.

Courtesy of Bendix Corp.—Automation & Measurement Div.

Fig. 1-4.6 Special cylindrical grinding machine designed for crush-form grinding.

To satisfy the above mentioned, as well as various other pertinent design requirements, including automatic controls, of the efficient, dependable and accurate form plunge-grinding process, special cylindrical grinding machines, relying on crush-truing, have been developed such as that shown in Fig. 1-4.6, representing a small-size model of a series whose capacity range is indicated by the data in the table.

Capacity Range of Small Size Cylindrical Grinders Using Crush-Formed Wheels

	Workpiece capacity, Diameter × Length, Inches (mm)		Grinding-wheel dimensions, Diameter × Width, Inches (mm)
From:	4″ × 24″ (100 × 600 mm)	From:	16″ × 3″ (400 × 75 mm)
To:	24″ × 100″ (600 × 2500 mm)	To:	42″ × 12″ (1050 × 300 mm)

The productivity of crush-trued form grinding is uniquely high in the case of workpieces whose configurations are well adapted to this method. Parts are usually finish form-ground from solid bars or forgings and, depending on the amount of stock removal, 10 to 200 parts may be ground between crush truings. The crusher rolls may sustain crush-truing cycles in the order of 10 to 50, or even more, before the roll face needs regrinding.

Dimensional tolerances of ±0.0005 inch (0.013 mm) on the diameter, and ±0.0003 inch (0.008 mm) for lateral location may be achieved when needed, and the surface finish obtained may be in the order of 8 microinches AA (0.2 micrometer R_a). The actual performance and capability values for any given operation must, of course, be determined individually.

In addition to annular forms, crush-trued grinding may also be applied advantageously to such different shapes as screw threads, gear racks, etc.

Traveling Wheel Head Roll Grinding Machines

Very large rolls, such as are used in blooming mills and similar applications, require cylindrical grinding of the traverse grinding type, however, the heavy weight and the bulk of such workpieces defies the practical adaptation of the conventional concepts of the sliding table design.

For the cylindrical grinding of very heavy rolls, particularly in weight ranges above 7 to 10 tons, special grinding machines are built in which the relative traverse movement between the work and the grinding wheel is reversed. The movement of the work is limited to rotation, while the traverse movement is carried out by the carriage of the wheel head, travers-

ing in machine ways similar to those used for the sliding table on most other types of regular cylindrical grinding machines.

In the traveling wheel-head type cylindrical grinding machine, shown in Fig. 1-4.7, the drive of the carriage travel is provided by a hydraulic motor, driving, through speed reduction gears, the shaft of a helical pinion which engages the rack along the machine bed.

The hydraulic motor, controlled through a servo-handwheel, provides an infinitely variable speed range of from 1 to 150 inches (25 to 3800 mm) per minute of travel.

Courtesy of Cincinnati Milacron Inc.

Fig. 1-4.7 Traveling wheel-head type, heavy roll-grinding machine.

The cross-feed is actuated also by hydraulic motors, one of which moves the cross slide through a ball leadscrew for coarse adjustment in the setup. The second motor actuates the power infeed with very high sensitivity, by means of the longitudinal displacement of a wedge causing the tilt motion of the wheel head, which is pivoting around solid trunnions located near the front end of the head, close to the grinding-wheel spindle. The principles of an essentially backlash-free, very accurate infeed control which can be accomplished by this rather unique design, are shown in the diagram in Fig. 1-4.8.

A camber generating action can be superimposed on the infeed movement; it operates by a synchro motor-actuated constant load cam. The design of that device permits direct reading of the camber setting and easy changeover from convex to concave profile. Taper grinding may be accomplished by swiveling the wheel-head carriage.

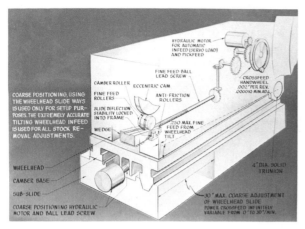

Courtesy of Cincinnati Milacron Inc.

Fig. 1-4.8 Diagram of the cross-feed mechanism of the traveling wheel-head type roll-grinding machine. Fine infeed by tilting the head provides very sensitive control.

Crankpin Grinding Machines

The crankshafts used to translate reciprocating motion into rotational motion, in general, and more particularly the crankshafts of internal combustion engines, are typical workpieces requiring the cylindrical grinding of pin sections whose axes, while parallel with the general-part axis, are not in alignment with it. Several of these pin sections may have to be ground in a single setup, locating each section in a specific longitudinal position, as well as displaced from the general part axis in different orientations.

Crankpin grinding is a rather extensive subject in view of the many different types and sizes, design varieties, and productivity requirements of crankshafts. While no attempt is made here for a thorough discussion, a few characteristic aspects of conditions and equipment are pointed out, essentially as examples of special cylindrical grinding applications on surfaces with offset axes.

Because of the differently oriented eccentric locations of the work surfaces, very large diameter grinding wheels must be used for the grinding of the individual pins, in order to avoid the interference of the wheel flanges with the throws of adjacent pins. Most crankpin grinding operations also involve the grinding of the throw walls, as well as of the fillets connecting the pin surface and the walls. To comply with that requirement modern crankpin grinding machines are equipped with special wheel-truing devices, guided by profile bars, and often using rotary diamond rolls for truing efficiency and excellent form retention. The operating principles and advan-

tages of such truing processes are indicated in item 9 of Table 1-5.2.

As the crankpins are ground consecutively, the workpiece has to be relocated for bringing the axis of the pin to be ground into the correct angular position, in alignment with the workhead axis. Special work-holding devices are used for radial indexing, which is done automatically on production-type machines.

Concurrently with the angular orientation, the proper axial positioning of the pin to be ground must also be assured in a movement often designated as *lateral indexing*. In order to assure that an equal amount of stock will be removed from both sidewalls bounding the pin being ground, the lateral indexing may be followed by an automatic location-adjusting function. In one system that adjustment is effected by symmetrically expanding contact jaws of a special device, the operating principles of which can be visualized from the cutaway view in Fig. 1-4.9.

Courtesy of Landis Tool Co.

Fig. 1-4.9 Automatic lateral locating in crankpin grinding. Cutaway view of a system in which expanding jaws assure equal stock removal from both side walls through axial positioning of the work.

In order to avoid nonsymmetrical torsional stresses acting on long crankshafts, some grinders are equipped with two driving workheads actuated from a common power source which imparts a balance rotating force to the workheads.

For assuring the maximum stability to the workpiece during the grinding process, the pin being ground is supported by a work rest arranged opposite the grinding wheel (see Fig. 1-4.10). Such work rests have double jaws and are often equipped with a hydraulic advancing mechanism acting in concert with the decrease of the work diameter. The work-rest jaws may also serve as datum points for automatic in-process gaging, the sensing head of the gage being mounted through the body of the work-rest.

Courtesy of Landis Tool Co.

Fig. 1-4.10 Crankpin grinding. Close-up view showing the arrangement of the self-adjusting work positioner opposite the grinding wheel.

Cam Grinding

Cams are mechanical elements with curved operating surfaces which, when in continuous contact with a follower passing along the operating surface of the cam, cause a controlled motion of the follower in a direction generally normal to the contact area. The operating surface is usually the periphery of a disc-shaped cam, but it may also be in the form of a curved groove located on the side of the cam, or on the face of a hollow, cylinder-shaped cam, to mention a few examples from a wide range of varieties. The relative motion between the cam and the follower is, in the majority of applications, produced by the cam rotating at uniform speed around its own axis.

The manufacture of plate or disc-shaped cams with the operating surface on the periphery, is carried out as a modified version of cylindrical grinding. The basic motions of cylindrical grinding, those of the workpiece rotating around its own axis while its periphery is maintained in contact with the rotating grinding wheel, are also present in this kind of cam grinding. One additional kind of motion, the varying approach and withdrawal of the workpiece, resulting in controlled changes in the distance between the axes of the wheel and of the work, is produced by special equipment which characterizes the cam-grinder type cylindrical grinding machines.

In high-volume production, such as in the manufacture of automobile engine camshafts, an example of which is shown in Fig. 1-4.11, the controlled variations of the axis distance between the workpiece and the grinding wheel are most conveniently produced by the rocking motion of a pivoting work support. Such a system reduces significantly the harmful effects of inertial forces at direction reversals, which would result from a reciprocating linear motion.

Courtesy of the Warner & Swasey Co., Grinding Machine Div.

Fig. 1-4.11 Automobile engine camshaft — a typical example of a multiple-cam workpiece for the high-volume production of which automatic cam-grinding machine are used.

It is with this goal of a vibration-free workpiece support in mind that the work-holding devices, called "work cradles," of modern camshaft grinding machines are designed, as exemplified by the work cradle shown in Fig. 1-4.12. This work holder member is of particularly sturdy design, and has dual cradle pressure function which provides, under automatic control, high pressure during the heavy stock-removal cycle, and low pressure during the finish grinding cycle.

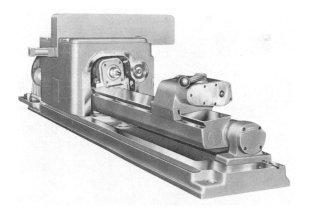

Courtesy of Landis Tool Co.

Fig. 1-4.12 The cradle, which is a work-holding device with rocking motion, used in modern cam-grinding machines.

For increasing the stability of the workpiece support, work rests of recent design, such as that shown in Fig. 1-4.13, have a pre-loaded upper jaw to accommodate variations in the diameter of the camshaft main bearings, thereby maintaining the workpiece in a straight axial position.

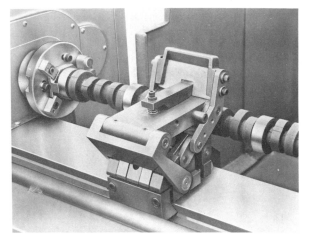

Courtesy of Landis Tool Co.

Fig. 1-4.13 Workpiece support with pre-loaded upper jaw, for accommodating and holding the main bearing section of the camshaft in a well-aligned position during the grinding of the cam sections.

An example of an automatic cam contour grinder is shown in Fig. 1-4.14, which represents a recent version of a well-proven basic model of automobile engine camshaft grinder. This machine carries out roughing and finishing in one continuous cycle, including the wheel truing. The wheel surface speed used for roughing is 50 per cent higher than that used for finishing. Wheel feeds for roughing and finishing are programmed and produced by a precisely functioning stepping electric motor which rotates the feed screw, thus avoiding feed rate variations which could result in a hydraulic system due to unstable oil viscosity.

The rocking motion of the work cradle is derived from a master cam which, in modern cam grinders, is rotated at a uniform speed by means of a pivoting belt drive, replacing the former universal joint. One possible drawback of contour duplication through varying the distance between the axes of the workpiece and the wheel by means of a rocking motion, is the effect of wheel diameter changes on the position of the point of contact between the work and

Courtesy of The Warner & Swasey Co., Grinding Machine Div.

Fig. 1-4.14 Automatic cam-grinding machine of advance design for the high volume production of automobile camshafts.

the grinding wheel. While the magnitude of the duplication error due to the shifting level of contact can be calculated and the acceptable limits definitely established, staying within these limits would require either exchanging the master cam as the wheel diameter is reduced by wear (a procedure not suitable for high-volume production), or the replacement of the wheel as soon as it wore down to a specific diameter (which is expensive due to wheel cost and time lost due to frequent wheel changes). Modern camshaft grinding machines are designed to operate with double type master cams which have two sets of lobes for each cam position, as shown in Fig. 1-4.15. The roller follower of the grinder shifts from one lobe to the other when the wheel wear has progressed to a predetermined diameter. This type of master cam is mounted in superprecision antifriction bearings, in which it is also ground to its final shape. The master cam, with its bearings and housings, constitutes a unit assembly which can be readily interchanged from machine to machine.

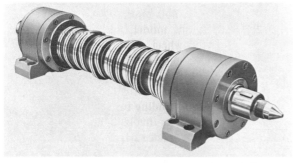

Courtesy of Landis Tool Co

Fig. 1-4.15 Double type master cam with two cam elements for each individual cam section on the workpiece, offering alternate positions for the cam follower as a means of reducing the effect of grinding-wheel wear on the accuracy of the ground cam shape.

Cams of other types than those contained in a camshaft are also produced in high volume; an example of these are the distributor cams of internal combustion engines. In the modern manufacture of such single-element cams the emphasis is on the total automation of the process, which comprises rough and finish grinding, as well as the unattended loading and unloading of the workpieces. In a special type of cam grinder, the work area of which is shown in Fig. 1-4.16, two grinding wheels of different grades are mounted side-by-side on the wheel spindle. After the roughing operation which has ground the work close to its final size, the wheel head shifts laterally and the finishing wheel completes the process. An automatic loader with U-shaped notches accepts the

finished part and moves the new one to be ground in line with the grinding position, where an expanding arbor picks up and holds the part during the grinding, which is followed by automatic ejection.

Courtesy of Landis Tool Co.

Fig. 1-4.16 Work area of an automatic cam-grinding machine with rotary indexing type loader and discharge chute. The machine uses two grinding wheels with different grit, mounted side-by-side, for the successive rough- and finish-grinding of distributor cams.

Multiple-Wheel Cylindrical Grinding Machines

Several individual, but in-line diameters of a workpiece can be ground simultaneously on cylindrical grinders equipped with multiple wheels mounted on the wheel spindle. In this statement the shop term "in-line" refers to surfaces with a common axis of rotation and "diameters" expresses round outside surfaces which are generally of cylindrical shape, but may be also, e.g., conical or barrel shaped.

The regular cylindrical grinding machines are built with wheel spindles accommodating a single wheel of a specific maximum width which, in the case of multiple-wheel mounting, limits the span containing the combined widths of the installed wheels and, when not adjacent to each other, also the widths of the separating spaces.

There are many types of workpieces with widely spread-out coaxial round surfaces which are manufactured in quantities warranting the simultaneous grinding of these sections of the work surface. Typical examples of such machinery components are shown in Fig. 1-4.17. For the concurrent grinding of such surfaces a special type of cylindrical grinding machine, known as multi- or multiple-wheel grinders,

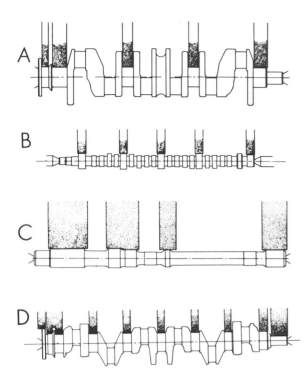

Courtesy of The Warner & Swasey Co., Grinding Machine Div.

Fig. 1-4.18   Long wheel spindle with seven different grinding wheels mounted in controlled locations, to operate simultaneously on a multiple-spindle cylindrical-grinding machine.

Courtesy of The Warner & Swasey Co., Grinding Machine Div.

Fig. 1-4.17   Examples of typical workpieces manufactured in high-volume production and ground economically on multiple-wheel cylindrical-grinding machines: A. Automobile engine crankshaft — five in-line diameters are plunge-ground simultaneously; B. Camshaft, the main bearing surfaces are plunge-ground with the wheel spindle reciprocating for finer finish; C. Five diameters of a transmission shaft are finished in a single plunge-grind operation, using four wheels, one of which is trued for step-grinding; D. Ten in-line diameters of a long crankshaft are ground in a single plunge-grinding operation.

is needed.   These grinders are designed to accommodate several large grinding wheels spaced at distances corresponding to those at which the functional surfaces of the workpiece are located.   An example of a particularly long wheel spindle with several large-size wheels mounted at controlled distances apart, is shown in Fig. 1-4.18.

It is evident that a very special category of cylindrical grinding machines is needed to accept and operate such a long and heavy wheel spindle supporting several large grinding wheels.   In order to reap the potential advantages of grinding simultaneously a substantial number of round surfaces, the multiple-wheel grinding process involves, of course, many other technical characteristics, in addition to accommodating and operating several grinding wheels installed on a common spindle.

A representative example of a modern multiple-wheel cylindrical grinding machine is illustrated in

Fig. 1-4.19.   This particular grinder can accept workpieces up to a maximum length of 36 inches (about 900 mm), with a maximum ground work length of 30 inches (about 750 mm), and up to a work swing of 14 inches (about 350 mm).   Some of the design characteristics of this model, which are typical of up-to-date multiple-wheel cylindrical grinding machines, are discussed in the following.

The headstock and the footstock are of particularly rugged design, adapted to the heavy workpieces and the considerable grinding forces which have to be supported in processes involving substantial stock removal from an extended surface area.   These work-

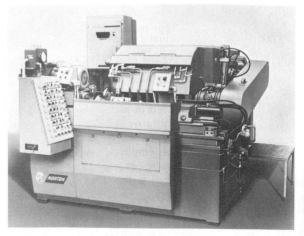

Courtesy of The Warner & Swasey Co., Machine Tool Div.

Fig. 1-4.19   Large-size multiple-spindle cylindrical-grinding machine with electric feed.   This grinder can use wheels up to 42 inches (about 1065 mm) in diameter.

support members are mounted on a swivel table whose only purpose is to eliminate taper. For accommodating workpieces of different lengths both the headstock and footstock can be moved along the table with rack and pinion.

Truing of the grinding wheels is most efficiently accomplished with the heavy-duty, rotary diamond truing device which is mounted on the wheel slide in a position easily accessible to the operator. The rotary diamond roll in engagement with the grinding wheels is shown in a close-up view in Fig. 1-4.20. Rotary diamond truing, in multiple-wheel grinding operations, eliminates the need for frequent diamond changes and diamond wear compensation; it acts fast in a cool operation and produces a wheelface with good cutting properties. A form bar is also available for each wheel, to be used in cases where the workpiece design requires the contour truing of the wheel. Single diamond side-truing devices, with individually adjustable diamond positions, are used in operations for which the side of the wheel must also be trued.

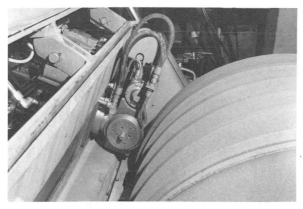

Courtesy of The Warner & Swasey Co., Machine Tool Div.

Fig. 1-4.20 Close-up view of the rotary diamond truing roll in engagement with one of the large-diameter grinding wheels in a multiple-wheel cylindrical-grinding machine.

The lateral or endwise location of the work in relation to the grinding wheels is an essential function in multiple-wheel grinding, particularly for workpieces which offer only very limited side clearance for the wheel and/or the wheel width just covers the surface section to be ground. Adjustment to assure the proper endwise location of such workpieces can only be avoided by drilling the centerholes to a very accurately controlled depth in relation to a reference surface on the workpiece. Meeting such requirements may involve an expensive auxiliary operation.

To avoid the need for the time-consuming manual adjustment to provide the proper lateral positioning

of successive workpieces ground in continuous production, automatic endwise positioning is provided using locator gages with probes sensing the actual lateral location of the workpiece.

The locator gage feeds the information on the workpiece position to the automatic endwise locating mechanism, which can be provided in either of two systems:

(a) A sliding spindle headstock (Fig. 1-4.21), moving the spindle on which the work center is mounted, until the workpiece held on that center reaches its proper endwise location, whereupon the spindle is securely locked to prevent lateral movement of the workpiece during grinding.

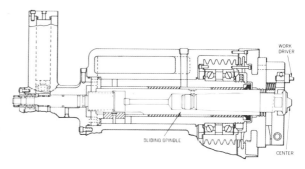

Courtesy of The Warner & Swasey Co., Grinding Machine Div.

Fig. 1-4.21 Diagram of the sliding spindle headstock used for the lateral alignment of the workpiece with respect to the grinding wheel, in automatic operation controlled by the side locating gage.

(b) Sliding wheel spindle (Fig. 1-4.22), in which a hydraulic piston and rack turns a screw causing the wheel spindle to move laterally until the proper wheel-to-work position is accomplished. This position is detected by the locator gage, whose probe acts on the limit switch for stopping the sliding movement of the wheel spindle.

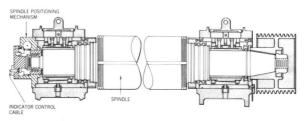

Courtesy of The Warner & Swasey Co., Grinding Machine Div.

Fig. 1-4.22 Diagram of the sliding wheel spindle of a multiple-wheel cylindrical-grinding machine, providing an alternate means for the lateral alignment of the wheels with the work surfaces being ground.

The wheel feed system of the multiple-wheel grinder is electronically controlled utilizing an electric pulse motor to rotate the feed screw at the pre-set rate. This system also provides for wheel wear compensation after the truing of the wheel. Furthermore, it actuates, at a pre-selected point, the steady rest and the in-process gage. The latter operates a feedback system for the final phase of the wheel feed, thereby assuring accurate size control of the ground parts.

Other control functions which are available optionally include:

(a) Automatic truing at pre-selected intervals

(b) Automatic speed control of the wheel spindle rotation for maintaining a constant peripheral speed of the wheel, thus assuring uniform grinding performance unaffected by changing wheel diameter

(c) Wheel spindle reciprocation can also be provided for workpieces requiring this supplemental motion which produces a smoother finish on the ground surface. An interlock mechanism assures the suspension of the wheel spindle reciprocation during the truing process.

Polygon Grinding Machines

Surfaces ground on cylindrical grinding machines have, in general, round forms; that is, any cross section in a plane normal to the axis is bounded by a circle. That condition is equally valid for figures of cylindrical or tapered shape, as well as for those having formed contours in the axial direction of the part.

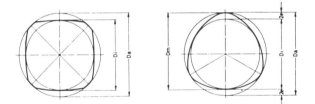

Courtesy of Fortuna-Werke/Stoffel Grinding Systems

Fig. 1-4.23 Standardized four- and three-lobed polygonal profiles, designated as PC 4 and P 3 types. D_a = circumscribed circle; D_i = inscribed circle; D_m = mean diameter of the profile, which is constant.

Polygonal shafts have surface elements parallel with the axis, yet the radial contour is not circular, but of some other regular shape, consisting of blending sections which are either all curved or a combination of straight and curved. Examples of polygonal shapes are shown in Fig. 1-4.23. Such shapes are used for mechanical connections, replacing keyed shafts or splines, for example, in certain applications

where the greater torque capacity of polygonal shafts is considered of distinct advantage.

The grinding of polygonal shafts is carried out on a special kind of cylindrical grinding machine, mounting the workpiece essentially in the same manner as for plain cylindrical grinding, as shown in Fig. 1-4.24, which illustrates a three-lobed polygonal shaft in the final stage of its grinding. The machine used for this operation is a polygon grinder, in which the conventional regular rotary and reciprocating motions are supplemented by a cyclic movement of the wheel axis along an elliptical path contained in the plane of the infeed movement.

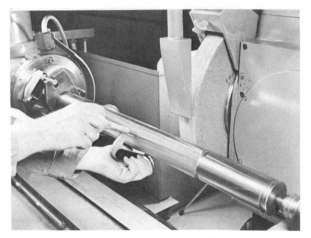

Courtesy of Fortuna-Werke/Stoffel Grinding Systems

Fig. 1-4.24 A three-lobed polygonal shaft ground between centers on a polygon grinding machine.

The diagram in Fig. 1-4.25 illustrates the sequence of typical points during a continuous change in the position of the grinding-wheel center, resulting in the generation of a polygonal shape on the surface of the workpiece which is rotating around a fixed axis. The path described by the grinding-wheel center is an ellipse, produced by combining a vertical rise and fall, with the horizontal approach and retraction of the grinding-wheel spindle. The parameters of these movements can be modified by appropriate settings of the drive elements. The workpiece is rotated in synchronism with these motions of the wheel spindle. The ratio between the cycle of wheel-spindle movements and the work rotation determines the general shape of the generated surface.

Figure 1-4.26 is a diagram showing the operating principles of the drive used to produce the polygon-generating motions of the grinding-wheel spindle. The grinding wheel (1) is mounted on a spindle (2), which is supported in parallel guides receiving horizontal movement through a connecting rod (3). The

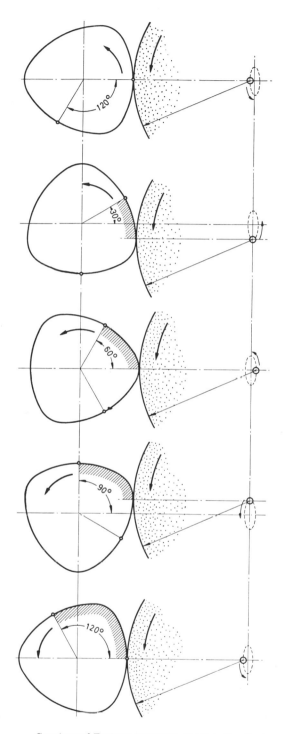

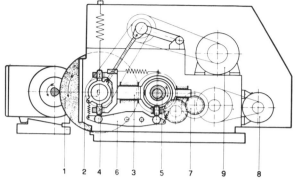

Courtesy of Fortuna-Werke/Stoffel Grinding Systems

Fig. 1-4.26. Diagram of the basic polygon generating mechanism, employing a single eccentric to control both the vertical and the horizontal motions of the wheel spindle. (See text for key to numbers 1 through 9.)

concurrent imparting of a vertical motion to the spindle carrier (4) from an infinitely variable eccentric (5) through a rocker arm (6), results in an elliptical path of the spindle axis. The gear drive (7) permits adjustment of the transmission ratio and the motor (8), which is driving the whole wheel-spindle-positioning device, is electrically interconnected with the workhead drive for assuring the precise synchronization of the work rotation with the wheel-spindle motions. The grinding wheel motor (9), transmits its drive through a countershaft which can follow the elliptical movements of the wheel spindle.

Figure 1-4.27 shows a polygon grinding machine which, in its external appearance, resembles a plain cylindrical grinder. By disconnecting the wheel spindle displacing mechanism, the machine can be operated as a regular cylindrical grinder.

The mating internal surfaces forming the "holes" into which the polygon shafts are assembled, are often produced by broaching. Internal polygonal surfaces can, however, also be ground on a polygon grinder equipped with internal grinding attachment which is mounted on the front of the wheel head. High frequency, electric motor driven spindles are used to produce the appropriate peripheral speed needed for the small diameter internal grinding wheels. Figure 1-4.28 is the close-up view of a sleeve with polygonal hole shown in the process of being ground with the internal attachment of a polygon grinder.

Numerically Controlled Cylindrical Grinding

Workpieces which are particularly well adapted to the economical use of numerically controlled cylin-

Courtesy of Fortuna-Werke/Stoffel Grinding Systems

Fig. 1-4.25 Sequential phases in generating a three-lobed polygon by imparting kinematically superimposed vertical and horizontal motions to the grinding wheel, which results in an elliptical path. This system assures the controlled accuracy of the profile shape, independently of the actual diameter of the grinding wheel.

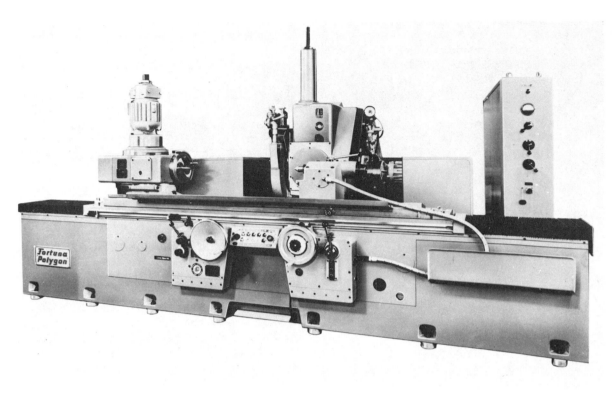

Fig. 1-4.27 Large-size polygon grinding machine with electrically controlled synchronization of the work rotation and of the wheel spindle motion along an elliptical path.

drical grinding are those with the following basic characteristics:

1. They have several coaxial cylindrical sections of different diameters and lengths, which must be ground to close tolerances, such as for multiple-diameter shafts of electric motors, gear boxes, machine tools, etc.

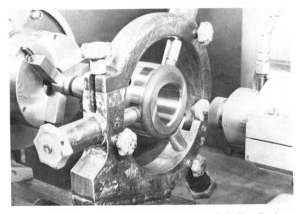

Fig. 1-4.28 Work area view showing the internal grinding process applied for producing a polygonal bore.

2. They are produced in small lots, either at a single occasion or in recurrent production runs. Numerically controlled cylindrical grinding provides even greater benefits for this latter alternative.

The elements whose functions are generally controlled numerically on cylindrical grinding machines equipped for such applications are:

1. Lengthwise movement of the machine table for the purpose of:
   (a) Lateral positioning of the workpiece to present the section to be ground to the action of the grinding wheel, according to the programmed sequence of operations. Such positioning may be carried out, depending on the equipment, with an accuracy of from 0.004 inch (0.1 mm) to 0.0004 inch (0.01 mm); these values having been selected from a range of available systems.
   (b) Reciprocating traverse movement of the machine table on N/C cylindrical grinding machines which are also adapted to traverse grinding, in addition to the more generally used numerically controlled plunge grinding.
2. Transverse movements of the wheel head slide which usually comprise the rapid advance for approaching the workpiece section to be ground, the

feed motion during the actual grinding process, the dwell for sparkout, and finally, the rapid retraction. Complementary functions which are frequently tied-in with these basic wheel head movements comprise:

(a) The automatic gaging of the ground work diameter, for gathering information which is utilized by the wheel head advance control system to initiate a microfeed sequence based on the measured size of the work, thus substituting an adaptive control for the finishing phase of the programmed advance movement

(b) Compensation for the wheel diameter changes caused by wheel dressing.

3. The control of the sizing gage whose function includes the setting of the attainable work diameter to the programmed value, contacting the work surface, and relaying to the feed system of the grinding wheel slide the size information obtained by continual gaging. Advanced types of automatic gaging systems can control the work diameter in increments of 0.0001 inch (2.5 micrometers) or even finer, and have sizing ranges in the order of 4 inches (100 mm), which may be taken as a mean value of different systems. The adaptive control of the final phase of the grinding operation by means of in-process gaging and feedback, eliminates the potentially harmful effects of such non-programmable variables as workpiece deflection caused by wheel pressure, temperature changes affecting the machine members, nonuniform grinding wheel conditions, etc. Figure 1-4.29 shows an example of such a sizing gage installed on an N/C cylindrical grinding machine which is set up to grind a multiple-diameter shaft.

Courtesy of Landis Tool Co.

Fig. 1-4.29 Multiple-step shaft ground in a numerically controlled cylindrical-grinding machine, employing a sizing gage; seen in engagement with the shaft section whose grinding is in progress. The dimensions of the ground diameters are controlled through the size information provided by this gage.

4. The rotational speed of the work, adjusted to the actual diameter of the work section being ground.

The programming of the machine operation. The programmed instruction for carrying out these, and possibly other functions, such as the automatic truing of the grinding wheel in conjunction with a cycle counting arrangement, are introduced into the machine system in the form of punched data-processing cards, punched tape, or magnetic tape, or manually by dial setting, depending on the applied data-processing system. These control elements are usually incorporated into a separate console, installed alongside the N/C cylindrical grinding machine, as shown in Fig. 1-4.30.

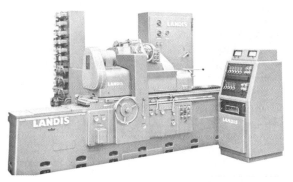

Courtesy of Landis Tool Co.

Fig. 1-4.30 Numerically controlled cylindrical-grinding machine designed for fully automatic operation by following the programmed instructions contained in a punched card or set manually on the control console.

The control console of this particular type of N/C cylindrical grinding machine is also equipped with hand-set dials, to permit the operation of the grinder without the use of punched cards.

The constructional characteristics of numerically controlled cylindrical grinders differ according to the manufacturer and the particular machine model, which may be a basic type grinder equipped with N/C devices, or a cylindrical grinding machine designed expressly for numerically controlled operations. A few constructional features found in specifically developed N/C grinders are mentioned as examples of mechanical design characteristics which may find applications to varying degrees for complementing and supporting the operation of the numerical controls.

1. Infeed movement with direct drive by electric stepping motor acting on a precision ball screw.
2. Wheel head of exceptionally massive design, supported on large bearings located close to the wheel flange and the drive sheave; such wheel heads are

designed to accept very wide wheels, when required, for certain plunge-grinding operations.

3. Table lubrication system with uniform pressure oil flow, to assure stick-free movement, accurate tracking, vibration dampening and to eliminate the possibility of table flotation.

4. Wheel-head slide ways with pressurized oil pockets for reducing the effect of the wheel-head weight, thereby assuring sensitive movement control, and extended machine life.

The application of a digital feed system to basic cylindrical grinders can offer significant advantages over purely manual control, without incurring the substantial investment in a complete N/C cylindrical-grinding machine.

Figure 1-4.31 shows a machine equipped with supplementary electronic feed controls which are housed in a console seen at the side of the grinder. Such machine and control systems serve to simplify the step-grinding operation required for multidiameter shafts which are to be ground in a single loading and are manufactured in small size lots. Control systems of this type are available in three basic arrangments, namely (a) for plain or traverse grinding; (b) for plunge grinding, and (c) in a combination arrangement for the selective use of traverse and plunge grinding.

This control system employs oscillators to generate the pulse frequency for the motor and the position counters. A translator amplifier transfers the generated pulse train to the servomotor which drives

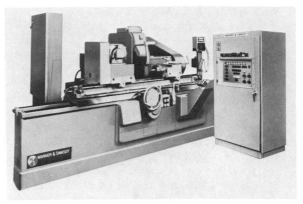

Courtesy of The Warner & Swasey Co., Grinding Machine Div.

Fig. 1-4.31 Electronically controlled feed system applied to a basic cylindrical-grinding machine. After having set the required work diameters by dialing the thumbwheel switches, the machine automatically repositions itself for the selected diameter, while the operator only initiates the cycle.

the feed screw for imparting the feed motion to the wheel slide. Different variable feed rates are available, in addition to the variable feed rate for power positioning and return. These rates are established by dial settings on the feed control station.

In plunge feed the retraction of the wheel slide at the termination of the controlled cycle can be accomplished manually by push button, by a timer, or by gage signal. In traverse grinding the set feed increments applied on table reversals can operate at either one or both ends of the table reciprocation.

Supplementary Functions and Equipment

The principal elements and movements of cylindrical grinders, discussed previously, comprise the basic functions which, however, must be supplemented to permit the actual implementations of the cylindrical grinding process. Certain additional functions may be needed for every operation, others will have to serve particular process conditions only. The generally or frequently needed supplementary functions and their equipment will now be reviewed.

Work-holding in Cylindrical Grinding

A general characteristic of cylindrical grinding is the rotation of the workpiece around an axis which will also become the common reference element of all the surfaces ground in the process. That fundamental concept is the easiest to visualize by considering an essentially round part, with centerholes at the two ends of its axis, which is mounted between the centers of a grinding machine. Because of these basic conditions of part configuration and mounting, the cylindrical grinding process is occasionally referred to by the term *center-type grinding*.

Although the simple conditions of a round part with distinct axis, which is physically established by means of centerholes, are not always present on workpieces which are processed by methods designated as cylindrical grinding, some of the outlined principles remain common. The axis of rotation of the part will always be identical with the reference axis of the ground surfaces, even when the location of such an axis is not made evident by directly related physical elements.

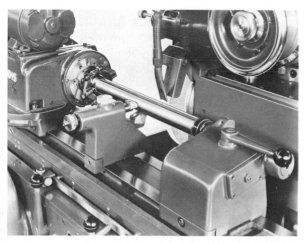

Courtesy of Landis Tool Co.

Fig. 1-5.1 Basic cylindrical-grinding setup holding the work between centers, rotating it with a driver, and providing additional support by a steady rest.

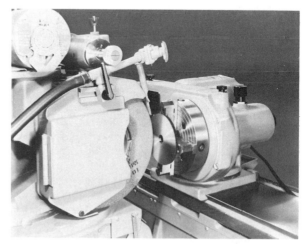

Courtesy of Brown & Sharpe Mfg. Co.

Fig. 1-5.2 Four-jaw chuck holding a disc-shaped part in the headstock at 90-degree swivel position, for grinding the face of the work with the periphery of the wheel.

49

Courtesy of Landis Tool Co.

Fig. 1-5.3 Collet attachment with handwheel operation. The collets are interchangeable and can accommodate work from 1/8 to 1-1/16 inches (about 3 to 27 mm) in diameter.

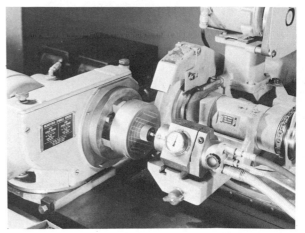

Courtesy of Brown & Sharpe Mfg. Co.

Fig. 1-5.4 Permanent magnet chuck holding a ring gage whose bore is ground with the internal attachment of a universal type cylindrical-grinding machine.

It is therefore an important requirement of work holding in cylindrical grinding to conform with that fundamental relationship between the rotational axis of the mounted workpiece and the produced work surfaces.

For accommodating parts having widely varying locating and mounting surfaces, the methods and devices of work holding for cylindrical grinding differ in many respects. The major systems are reviewed in Table 1-5.1. In this review the emphasis is on the principles, while deliberately disregarding details of design and dimensions.

To supplement the information and diagrams of the table, a few of the listed work-holding systems are illustrated by photos showing typical applications. The pertinent items of Table 1-5.1 refer to the applicable illustrations of Figs. 1-5.1 to 1-5.4

Many of the work-holding devices adapted for cylindrical grinding are similar, or even identical with those used on lathes in turning operations. Differences in the operational conditions and requirements are present, however, and often affect the choice and design of the applicable equipment. While in turning, the main machining force is imparted to the rotational movement of the workpiece; in cylindrical grinding, that force is much smaller — perhaps 1/10 or 1/20 part of the total machine-drive power. It follows that in cylindrical grinding, less powerful clamping is needed and holding methods which are generally not adaptable for turning can be used effectively for grinding, i.e., magnetic plates, diaphragm-actuated chuck and collets, etc.

Some holding devices, which, in principle are more or less common to turning operations, are used in cylindrical grinding under substantially different conditions. Typical, yet not unique, examples are the steady rests. In turning, the steady rests support the part on a surface which has been previously machined, either in a preceding or in in the same operation, and which will not be altered while being supported by the jaws of the steady rest. In cylindrical grinding, steady rests are sometimes used under similar conditions, yet the most frequent type of application consists in supporting the surface area actually being ground, and consequently, subjected to progressive size reduction. Because of these particular application conditions a more detailed discussion of steady rests used in cylindrical grinding seems indicated.

Steady Rests

Steady rests are supplementary work-holding devices used for providing additional support to workpieces held between centers, or for operating in combination with single-end type work holding by means of chucks, collets, or fixtures. Such additional supports are needed particularly on long and slender parts which, by their own weight, are prone to sag between the two primary support points at the ends, unless supported by supplementary means. Those parts which may be deflected by the force of the grinding wheel, or by centrifugal force due to inherent imbalance, and those which do not offer sufficient bearing surfaces for center-type work holding,

Table 1-5.1 Examples of Work Holding Methods and Devices for Cylindrical Grinding

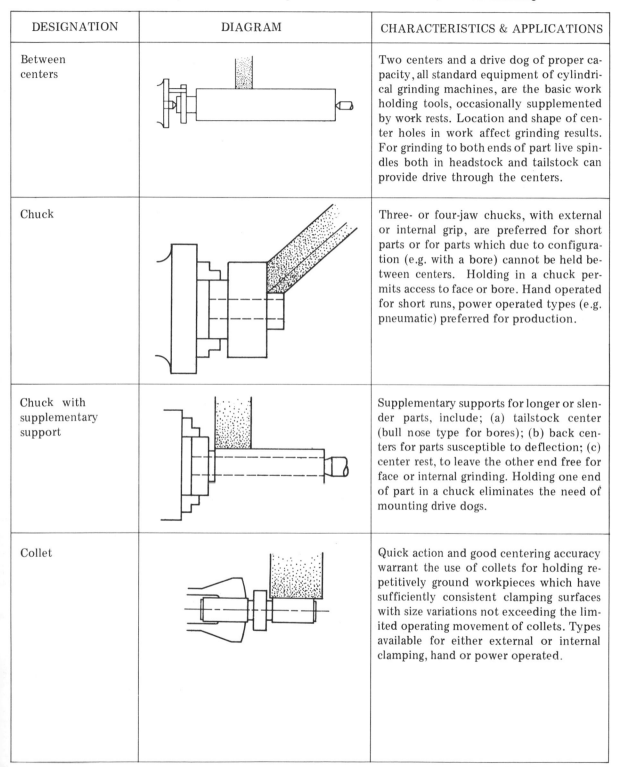

DESIGNATION	DIAGRAM	CHARACTERISTICS & APPLICATIONS
Between centers		Two centers and a drive dog of proper capacity, all standard equipment of cylindrical grinding machines, are the basic work holding tools, occasionally supplemented by work rests. Location and shape of center holes in work affect grinding results. For grinding to both ends of part live spindles both in headstock and tailstock can provide drive through the centers.
Chuck		Three- or four-jaw chucks, with external or internal grip, are preferred for short parts or for parts which due to configuration (e.g. with a bore) cannot be held between centers. Holding in a chuck permits access to face or bore. Hand operated for short runs, power operated types (e.g. pneumatic) preferred for production.
Chuck with supplementary support		Supplementary supports for longer or slender parts, include; (a) tailstock center (bull nose type for bores); (b) back centers for parts susceptible to deflection; (c) center rest, to leave the other end free for face or internal grinding. Holding one end of part in a chuck eliminates the need of mounting drive dogs.
Collet		Quick action and good centering accuracy warrant the use of collets for holding repetitively ground workpieces which have sufficiently consistent clamping surfaces with size variations not exceeding the limited operating movement of collets. Types available for either external or internal clamping, hand or power operated.

(Continued on next page.)

Table 1-5.1 (*Cont.*) **Examples of Work Holding Methods and Devices for Cylindrical Grinding**

DESIGNATION	DIAGRAM	CHARACTERISTICS & APPLICATIONS
Face plate		Parts which tend to be distorted by clamping in radial direction or with configurations not adapted to standard types of holding devices, can often be properly mounted on face plates, equipped with appropriate rest buttons and clamps. Mainly used in tool work or other jobs involving a limited number of parts only.
Fixtures		For parts of unusual configuration, or for those with non-identical axes for the part and for the section being ground, special fixturing is needed to assure proper part location and holding. Typical examples are the crankpin sections of crankshafts.
Mandrels (Arbors)		Parts with through bores can be ground along the entire cylindrical outer surface by mounting the part on a mandrel which then may be held between centers. Method also assures good concentricity between the bore and the ground outer surface. Mandrels may be cylindrical (with clamping nuts), slender tapers (e.g. 1:100) or expanding type.
Magnetic plate (Magnetic chuck)		The configuration of certain parts, usually of short length, makes them adaptable for mounting on a magnetic face plate, which is equipped with locating stops. Method offers rapid mounting and centering, assures good squareness between mounting and ground surface and leaves end open for face or bore grinding.

Courtesy of Brown & Sharpe Mfg. Co.

Fig. 1-5.5   Universal back rest (steady rest) with adjustable shoes. (*Left*) Close-up view for distinguishing the principal elements; (*Right*) Application example showing the use of two back rests supporting a long and slender workpiece.

are also dependent on the additional support provided by the use of steady rests.

Two basic types of steady rests are most frequently used in cylindrical grinding operations:

The two-shoe type, providing supporting surfaces essentially at the bottom of the workpiece and at a point opposite the grinding wheel contact. Steady rests of this type are also referred to as *work rests* or *back rests*, as seen in Fig. 1-5.5.

The three-shoe type, known as the *center rest*, with three individually adjustable shoes arranged as corners of an erect isosceles triangle, Fig. 1-5.6. The upper shoe is mounted into a hinged bracket which can be swung out when inserting or removing the work. Center rests are commonly used for long workpieces held at one end only, as in a chuck, but requiring additional support which cannot be provided by a tailstock center. Such is the case, for example, when the part does not have a center hole at its opposite end, or when that end must be kept accessible for face grinding or internal grinding.

Back rests are made in several basic types, selected in accordance with the purpose to be served. The *single bracket type* is made with either a solid or an adjustable shoe and provides a V-support for the workpiece, Fig. 1-5.5. The position of the single-bracket shoe can be adjusted both vertically and horizontally, to conform with the diameter of the finished work. To accommodate the grinding allowance, the shoe can deflect against spring action, the force of which is also adjustable to suit the operational conditions.

Courtesy of Brown & Sharpe Mfg. Co.

Fig. 1-5.6   Center-rest type steady rest with three jaws for concentric work support on a section which has previously been ground.

The double-shoe type in Fig. 1-5.7 has individually adjustable supports for both the bottom and the back-up shoe. In most cases the adjustment is first done on the lower shoe for establishing the proper vertical position of the supported workpiece, and then the horizontal shoe is moved to a position corresponding to the finished diameter of the work.

Courtesy of Brown & Sharpe Mfg. Co.

Fig. 1-5.7 Double-shoe back rest with a single adjusting screw. Preferred for grinding with coarse infeed.

These positions can be fixed by means of adjustable stop pins, and then a shoe can be retracted for inserting the unground work.

During the grinding process the shoes can gradually be brought to their final position by actuating the knob of an adjusting screw. This kind of steady-rest operation is often referred to as the follow-up type. The shoe advance movement can also be actuated by hydraulic means, such equipment being used on large cylindrical grinding machines built for heavy workpieces. The operation of the hydraulic steady rest may even be integrated into the automatic wheel infeed cycle, the steady rest shoes reaching their final position shortly before the wheel advance movement is completed.

Figure 1-5.8 shows diagrammatically, the operating principles of one type of hydraulically adjusted work rest. A spring-loaded wedge moves the work-rest jaws forward as the work diameter is reduced. The force by which the jaws are actually moved against the workpiece is adjusted by varying the hydraulic pressure exerted through a cylinder which opposes the action of the spring on the wedge.

For long workpieces more than a single steady rest may be needed. As a rule of thumb, for work-pieces of about one-inch diameter, one steady rest is needed for every ten inches of work length. Accordingly, for a workpiece of one-inch diameter and 30 inches length, three steady rests may be installed;

that number increasing with thinner, less rigid parts, and decreasing for more rigid workpieces.

The operation of cylindrical grinders with steady rests requires a certain degree of experience, particularly for long parts involving the use of several supports, which may need individual adjustments according to their locations along the part. The selection of the shoe material, e.g., hardwood, bronze, carbide, etc., must be based on a judicious balance between high wear-resistance and the avoidance of shoe marks on the ground work surface. Shoe contact surfaces of slightly convex shape prevent the scoring of the work surface by the shoe edges.

Steady rests, usually belong to the standard equipment of a cylindrical grinding machine and are of the type selected by its manufacturer. The number of the rests furnished depends on the maximum distance between centers of the particular machine.

Courtesy Landis Tool Co.

Fig. 1-5.8 Cross-sectional diagram of a follow-up type work rest with spring pressure counterbalanced by the regulating action of a hydraulic cylinder.

Dressing and Truing

The two terms designate processes with different, yet complementary, objectives. In precision grinding, which includes the majority of cylindrical grinding processes, the objectives of dressing and truing are usually accomplished in a single operation.

The purpose of dressing is to restore the original cutting capacity, the "sharpness," of the grinding wheel, should that become reduced by embedded particles or dulled grains which were not released from the bond. However, when the general grinding conditions permit, the composition of the grinding wheel and the applied machining data should be such as to utilize the self-dressing capacity of the grinding wheel, thus reducing the frequency of wheel dressing.

Truing, as the term implies, serves to assure the true running of the grinding wheel. Truing has as a further objective the establishment of the correct geometric form, designated also as the profile of the operating wheel face. In the case of plain cylindrical grinding these requirements are limited to a profile which is straight and parallel with the grinding-wheel axis. The same wheel profile is used for taper grinding when carried out with the aid of the swivel table. For grinding steep tapers by the plunge method, as well as in angular approach grinding, the wheel profile should be straight but not parallel with the wheel axis, or it may be composed of two straight elements, bounding adjoining wheel surfaces which are commonly at right angles to each other.

The required wheel profile also may include corners with specific radii, usually blended tangentially with the adjoining straight element. In other applications the grinding wheel truing produces composite profiles, comprising either a single curved element, or several elements in succession, having straight or curved forms. A typical application for this latter type of profile is *cylindrical form grinding*.

In truing, a layer is removed from the operating face of the wheel, for the purpose of producing or restoring the required profile. The removed surface layer may have contained dull grains (glazing), or embedded particles (loading), thus the purpose of dressing is also met, provided that the requirements of a properly dressed wheel are observed in the truing process.

Considering the variety of profiles which may be needed for different types of cylindrical grinding operations, as well as the range of wheel sizes, accuracy requirements, wheel-face characteristics, and operational conditions, no single system of truing could be universally satisfactory. For that reason many methods and means of grinding-wheel truing and dressing are used for different types of cylindrical grinding operations. Table 1-5.2 presents a survey of the more commonly used methods, indicating the pertinent equipment and discussing the major applications.

Complementing the diagrams of the table a few photographs are shown in Figs. 1-5.9 to 1-5.12, illustrating typical truing and dressing devices of cylindrical grinding machines. References are made

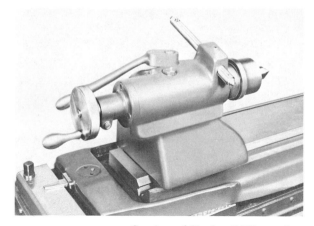

Courtesy of Cincinnati Milacron Inc.

Fig. 1-5.9 Tailstock type truing device for cylindrical grinder.

in the pertinent items of the table to the applicable photographic illustrations.

The speed and feed in truing must be selected and maintained at rates which will produce the required wheel surface with regard to dimensional accuracy and finish, as well as free cutting properties. Assuming the use of single-point diamonds as truing tools, the following conditions should be considered.

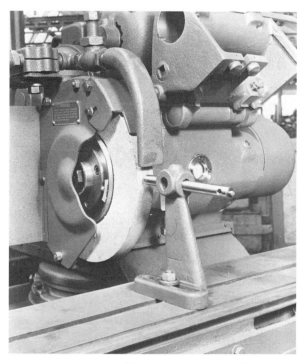

Courtesy of Cincinnati Milacron Inc.

Fig. 1-5.10 Table type truing device mounted on a cylindrical-grinding machine.

Table 1-5.2 Methods and Equipment for Grinding Wheel Truing and Dressing in Cylindrical Grinding

DESIGNATION	DIAGRAM	CHARACTERISTICS & APPLICATIONS
Table-mounted wheel dresser		Using the reciprocating movement of the machine table for traversing the diamond across the face of the wheel is the simplest and most widely used method in general-purpose cylindrical grinding. The diamond holder may be installed into the tailstock or into a stand mounted on the table. Application requires traversing the table from the work to the dresser position and reducing the reciprocating speed of the table movement to that needed for wheel dressing. (See also Figs. 1-5.9 and 1-5.10.)
Wheel head mounted grinding wheel dresser		Preferred for production type plain cylindrical grinding because it permits retaining the lateral position of the work in relation to the wheel. Device has its own guideways and traversing drive, usually hydraulic and steplessly variable. Diamond advance either by hand or mechanically, by preset increments; the latter may, in modern machines, be tied in with the corresponding readjustment of the wheel head position, thus accomplishing automatic wheel-wear compensation.
Multiple-track truing device		Various methods of wheel truing may be used for the simultaneous grinding of cylindrical sections having different diameters, located either adjacently or at positions separated by unground sections. Examples are stepped templates guiding a single diamond point, or devices using separate diamond nibs for each of the wheels mounted on a common spindle; the simultaneous truing of several wheel diameters reduces the total time of truing.

Table 1-5.2 (*Cont.*) **Methods and Equipment for Grinding Wheel Truing and Dressing in Cylindrical Grinding**

DESIGNATION	DIAGRAM	CHARACTERISTICS & APPLICATIONS
Form-duplicating truing devices		These wheel head mounted devices operate by following the contour of a template which corresponds to the desired wheel profile. The follower is held against the template by spring pressure or hydraulic force and the traverse of the truing slide is actuated hydraulically at steplessly variable speed. The direction of the truing slide traverse may be parallel with, or at an angle to, the work axis, in order to provide the optimum climb angles for the follower along the contour of the template.
Angle-truing device (Table mounted)		Intended for low-volume operations and built as a table-mounted fixture with slide movement operated by hand wheel. The slide can be rotated on its base for setting to the desired angle from 0 to 90 degrees in either direction. Adjustable end stops for the swivel movement of the base permit the consecutive truing of two different angles. For truing alternate sides of the wheel the diamond may be turned end for end. (See also Fig. 1-5.11.)
Radius truing device (Table mounted)		A pivoting bracket held in a base mounted on the machine table carries the diamond nib holding arm, whose extension can be adjusted to suit the required radius length. By adjusting the position of the diamond point in relation to the pivot axis the resulting radius will be either convex or concave. The length of the arc swept by the bracket swivel can be controlled by adjustable end stops.

(Continued on next page.)

Table 1-5.2 *(Cont.)* Methods and Equipment for Grinding Wheel Truing and Dressing in Cylindrical Grinding

DESIGNATION	DIAGRAM	CHARACTERISTICS & APPLICATIONS
Continuous radius and tangent truing device		Designed for generating grinding wheel contours in which a radius must be blended, without discernible point of transition, into a straight line. Such devices can be adapted for either convex or concave radii and for tangent lines at either end of the radius; the central angle and the radius length of the curved section can be set to the desired value within the range of the instrument. Such devices are usually mounted into a stand which is resting on the machine table. (See also Fig. 1-5.12.)
Internal grinding-wheel dressing device		For grinding inside surfaces with an internal grinding attachment mounted on a cylindrical grinding machine, a special wheel-truing device is needed, which can be swung out of the way after wheel truing. Such devices are designed to have the diamond contacting along a line which is also the center of contact between the wheel and the work. The precise setting of the diamond point, e.g., with an indicator, permits the bore size to be controlled by means of the proper wheel truing.
Rotary diamond-truing device		Rotary diamond dressers using a motor-driven diamond-impregnated roller can be operated on cylindrical grinding machines either as replacements for single-point diamonds in a wheel-head mounted traversing type truing device (the type of arrangement shown in diagram), or as form truing tools, equipped with profile shaped diamond rolls corresponding in width to that of the grinding wheel, to which the diamond roll will advance in a radial direction (as in crush-truing). Efficient wheel-truing method for high volume production.

(Continued on next page.)

Table 1-5.2 (*Cont.*) Methods and Equipment for Grinding Wheel Truing and Dressing in Cylindrical Grinding

DESIGNATION	DIAGRAM	CHARACTERISTICS & APPLICATIONS
Crush-truing device		It is the standard method of wheel truing for special crush form grinding machines, but may also be applied on general-purpose cylindrical grinding machines when adapted to this type of wheel truing. Such adaptation involves a low crushing speed for the grinding wheel spindle and spindle bearings capable of sustaining the relatively high radial loads caused by crush truing. Efficient truing method for complex shapes to be ground in quantities warranting the cost of special crushing rolls.

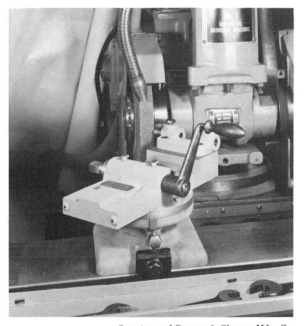

Courtesy of Brown & Sharpe Mfg. Co.

Fig. 1-5.11 Angle wheel-truing device for a universal type cylindrical-grinding machine.

Traverse speed across the face of the wheel will primarily determine the finish of the trued surface, and can also affect the form-trueness in profile grinding. Modern automatic truing devices with hydraulic drive can be adjusted steplessly to any traverse speed within the available range, e.g., 2 to 20 inches (50 to 500 mm) per minute. The diamond tip in a continuously traversing truing device will produce a screw-thread-like helical path on the wheel face, the pitch (p) of which, in inches, will be a function of the traverse speed (t), expressed in inches per minute, and the grinding-wheel speed (s), in revolutions per minute ($p = \frac{t}{s}$). Consequently, for roughing, a higher traverse speed of the truing device is used, e.g., that producing a 0.005-inch (0.13 mm) pitch, whereas for finishing operations the pitch is kept at 0.002 inch (0.05 mm), or less.

The feed rate, expressed as the depth of diamond penetration per pass, will again depend on the desired structure of the trued wheel face. With few exceptions the depth per pass should not exceed 0.001 inch (0.025 mm), but for very fine surface finishes, or when duplicating intricate profiles, that rate may be reduced to 0.0002 inch (0.005 mm), or even 0.0001

Courtesy of Brown & Sharpe Mfg. Co.

Fig. 1-5.12 Continuous radius and tangent truing device.

inch (0.0025 mm). The number of passes, resulting in the total thickness of the removed wheel surface layer, will depend on the conditions of the wheel with regard to both geometry and cutting capacity. Four passes per truing operation may be considered an average value, sometimes reduced to two, but more often exceeded. For producing very smooth surfaces, in rare cases, a final diamond traverse pass with no infeed rate is added to the truing operation, although the danger of creating a glazed wheel surface must be considered.

Measurement and Gaging in Cylindrical Grinding

The primary purpose of cylindrical grinding, as a precision machining process, is to impart a specific size, commonly a diameter, to a ground part or feature. In order to accomplish that objective, the size of the ground surface must be determined during the process, as well as at its termination. This can be done by measurement, supplying the actual size of the inspected dimension; or by gaging, comparing that dimension to a master which incorporates the desired nominal size. These basic concepts point out the major systems of size determination and control in the precision grinding process.

Modern grinding machines are equipped with very precisely operating control elements for the slide movements which produce the size of the ground surface. These elements can control both the amount of infeed for reducing the size of the work and the precise duplication of a defined slide position. These actions, however, must be preceded by the measurement of the work for establishing the reference base of the subsequent machine slide movements.

Courtesy of Federal Products Corp.

Fig. 1-5.14 Operating electronic in-process gages in engagement with the workpiece on cylindrical-grinding machine. The gages feed the size information to the remotely located amplifier which is equipped with: (*Top*) meter showing the deviations of the work size from the nominal value or (*Bottom*) digital readout, which covers a much wider size-range than a meter.

The machine slide movement, even when controlled with very high precision, will not guarantee that parts ground over an extended period of time will be produced to the same exact size. First, there is the wear of the grinding wheel, a continuous process which, although generally quite slow, particularly in finish grinding with light stock removal rates, is still a factor to be considered. Second, the periodic truings of the grinding wheel will result in more substantial size variations, unless corrected by compensating adjustments of the machine slide travel. Finally, size variations, commonly in a rather small order of magnitude, may result from such factors as temperature changes, nonuniform grinding stock allowance, declining sharpness of the wheel, etc.

Due to the potential effect of these factors on accurate size control, the grinding to very light dimensional tolerances often requires the actual mea-

Courtesy of Federal Products Corp.

Fig. 1-5.13 Indicator type grinding gage with hydraulic mounting for easy engagement and retraction. (*Left*) In contact with the workpiece; (*Right*) In retracted position.

surement of the workpiece while in process, either throughout the entire operation, or during its last phase. The indications resulting from in-process gaging can either be used for guiding the operator when grinding with manual control, or they can generate command signals for power-actuated slide movements in automated processes.

In view of the many application varieties, different objectives, and the wide range of tolerance requirements which occur in cylindrical grinding operations, the methods and means of measurement are many and varied. While the analysis of the different instruments is beyond the scope of this book, a survey of various systems of measurement used in cylindrical grinding is warranted. Such a survey is presented in Table 1-5.3 which, although not all-comprehensive, can prove a helpful guide in selecting the general method best suited to the given operational conditions.

To provide a better visual conception of instruments which are particularly well adapted to measurements in cylindrical grinding, a few illustrations, Figs. 1-5.13 to 1-5.15, complement the diagrams of the table. The listed gaging systems, with a single exception, serve the measurement of part diameters, that being the most frequently controlled dimension in cylindrical grinding. Of course, other dimensional conditions, e.g., length of the ground surface, axial location, contour, perpendicularity of adjoining sections, etc., as well as different parameters such as roundness, straightness, surface finish, etc., will also have to be inspected in various cylindrical grinding operations. These measurements, however, are usually carried out on parts removed from the grinding machine, and are considered post-process type inspection operations.

Center Hole Grinding

In one of its basic applications, known as dead-center grinding, the cylindrical grinding process produces a round surface around an imaginary part axis, which is represented by two opposed center holes; these are of conical shape, commonly with 60 degrees included angle. In order to be truly representative of the part axis the sides of the center holes must be precisely concentric to it.

During the cylindrical grinding process the center holes are in direct contact with the machine centers, consequently the geometric form characteristics, such as roundness and angle of these mating elements must have excellent compliance, and should be free of blemishes, nicks, burrs, and other surface irregularities, waviness or chatter marks, for example; either

of these conditions could affect the quality of the ground surface. In conclusion, the center holes of the part in dead-center type cylindrical grinding must satisfy the following requirements:

(a) Concentricity to the part axis
(b) Alignment with the opposite-end center hole
(c) Roundness, with regard to both the general shape and circumferential waviness
(d) Correct angle (generally 60 degrees, included)
(e) Clean and smooth surface.

Center holes are usually produced by center-hole drilling in the soft state of the part, with the use of special center drills. However, in the course of heat treatment which usually precedes grinding, scale may develop on the hole surface and distortions (warpage)

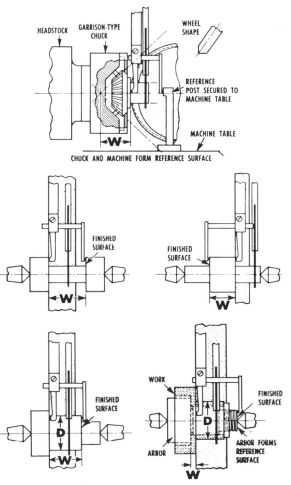

Courtesy of Federal Products Corp.

Fig. 1-5.15 Lateral locating gage for setting and controlling the correct axial location of a workpiece surface in relation to a finished shoulder. Diagrams show several possible applications.

Table 1-5.3 Methods and Instruments of Gaging in Cylindrical Grinding

DESIGNATION	DIAGRAM	DESCRIPTION & DISCUSSION
Outside Micrometer Caliper (Screw Micrometer)		The traditional micrometer calipers are still widely used for measuring the work diameter while the part is mounted on the cylindrical grinder, particularly for toolroom operations and short-run jobs. The wide range of sizes in which micrometers are made and the ready availability of these basic measuring instruments supports their acceptance as grinding gages, although the handling of micrometers for larger diameters requires skill and the reading is to be made carefully to avoid errors.
Snap Gages— Fixed or Indicating Types		Snap gages of the fixed type, whether rigid or adjustable to the required size, are used in pairs or with consecutively contacting jaws to supply GO and NOT GO information. Snap gages of the indicating type, while requiring regular checking with the setting masters, have the advantage of indicating actual size in relation to a fixed nominal dimension, thus can guide the machine operator with regard to the amount of stock removal still needed. Snap gages are preferred for larger lots to be ground to the same size.
Swing-In Type Grinding Gages		These are essentially indicator snap gages mounted on special counterbalanced brackets, permitting the gage to be swung into measuring position and then to be retracted to clear the work surface. For protection from grit and coolant and also to facilitate reading, the indicators of these gages are mounted at a distance from the work contact area, the movements of the sensing element being transmitted through a reed-supported linkage, or operate as electronic gages with remotely located meters for their setting these gages require an appropriate master.

Table 1-5.3 *(Cont.)* **Methods and Instruments of Gaging in Cylindrical Grinding**

DESIGNATION	DIAGRAM	DESCRIPTION & DISCUSSION
Continuous Contact Grinding Gages		Such gages can have fork type or single-point type contact members, the latter often operating in conjunction with the steady rest and gaging the changes in the height of a chord subtended by the contacting jaws of the work rest. The sensing elements in contact with the work are usually electronic gage probes connected with a remotely located amplifier and meter. Such gages are frequently used as the sensing members of automatic sizing devices.
Continuously Indicating Grinding Gages for Interrupted Surfaces		Interrupted surfaces, such as those of spline shafts, reamers, milling cutters, etc., are difficult to measure on the diameter, particularly when an odd number of lands constitute the elements of the surface. Special grinding gages for noncontinuous surfaces are designed to lock the probes while passing over gaps and have very fast measuring response for sensing the size of the momentarily engaged lands, thus providing uninterrupted indications.
Automated Compensation for In-Process Variations by Slide Position Correction		Size control systems guided by sensitive in-process electronic or air-electronic gages and operating by extra-fine incremental infeed for the final sizing. In order to keep the number of fine-sizing increments on an essentially even level, a device counts the number of the actually needed fine-feed increments, and corrects accordingly the starting position of the wheel-head cross slide, causing it to recede in case of work expansion due to increased temperature, or to advance for balancing wheel wear.

(Continued on next page.)

Table 1-5.3 (*Cont.*) **Methods and Instruments of Gaging in Cylindrical Grinding**

DESIGNATION	DIAGRAM	DESCRIPTION & DISCUSSION
Chordal Height Measuring Indicator Gages		Method using a single sensing point and referencing from two points on the work surface, arranged to symmetrically straddle the gage point. The reference points are considered the end of an imaginary chord of known length, and the size of the chordal height, indicated as deviation from a nominal value, is translated into diametrical dimension. Method applicable to both in-process gaging and also for hand gages, preferred for large diameters, which exceed the range of regular micrometer calipers.
Lateral Position Gages		In cylindrical grinding, particularly for operations involving shoulders, faces, fillets, and chamfers, the axial location of the part in relation to the profile-trued grinding wheel must be precisely controlled. Unless the holding device locates the part axially, the traverse of the table must be guided by an instrument, also known as "flagging gage" which indicates the position of the section being ground in relation to a reference plane which contains the face of the grinding wheel.
Axial Length (Width) Indicator Gages		For grinding operations which must produce precisely controlled axial distances, either between two ground face surfaces, or of the face surface being ground with respect to a reference surface on the part, special indicator gages which carry their own referencing probe, must be used. The distance between the reference contact and the indicator probe must be set by appropriate masters or measuring instruments. Mechanical indicators of such gages are connected with the probe by a linkage, but electronic gages with remotely located meters may also be used.

(Continued on next page.)

Table 1-5.3 *(Cont.)* **Methods and Instruments of Gaging in Cylindrical Grinding**

DESIGNATION	DIAGRAM	DESCRIPTION & DISCUSSION
Diameter Gaging with Varying Reference Dimension (Match Gaging)		For components which must be assembled with specific and tightly toleranced fit conditions, the frequently applied segregation can be avoided by grinding individual parts precisely to the size needed for mating with a particular component. The mating finished part is mounted in a remotely located gage, presenting the actual size of the fit surface as a target dimension which, with the specified allowance, should be matched by the part being ground and gaged while in process.

in the part can occur, which will affect the originally correct shape, position, and orientation of the center holes.

Considering, on the one hand the fundamental operational role of the center holes, and on the other, the potential damages which these functionally critical elements of the workpiece can suffer, it is a rule that high-quality, dead-center grinding should be preceded by retouching or, when needed, regrinding of the center holes.

Such preparatory operations can be carried out to different degrees of thoroughness, the choice depending on both the actual condition of the center holes prior to the cylindrical grinding and the accuracy requirements of the ground surface. The most frequently used methods of center-hole reconditioning in preparation of dead-center type cylindrical grinding are, in the order of increasing thoroughness:

(a) Grinding or lapping with an abrasive pin shaped to a 60 degree angle and rotated while in full contact with the conical surface area of the center hole. Such a method can clean the hole surface and improve both its roundness and finish, but will have practically no effect on the position and orientation of the hole axis, consequently, will not correct it, should that action be needed.

(b) Grinding with a conical shaped mounted wheel making line contact with the hole surface while the part is kept in rotation around its theoretical axis. A special center-hole grinding machine is generally needed to carry out this operation, in which the part must first be positioned with respect to its axis and then rotated around that axis by an appropriate drive. For centering the part in the machine several methods are available, two examples of which are:

(1.) Dead center opposite the end being ground used in combination with a center-rest type device contacting a precisely machined outside area of the part;

(2.) A chuck in the headstock as a single means for

Courtesy of O. Klein/Stoffel Grinding Systems

Fig. 1-5.16 Center-hole grinding machine with rotating wheel dresser.

short parts, or used in combination with a center rest for long workpieces.

The mounted wheel needs regular truing to assure its dependable geometric form. The truing can be carried out with a device reciprocating along a path inclined at 30 degrees in relation to the grinding wheel spindle and holding a diamond nib. For increased efficiency of the truing, the single-point diamond can be replaced by a rotating abrasive dressing wheel driven by an individual motor as in Fig. 1-5.16.

(c) The best results with regard to the listed critical parameters can be expected by applying a center-hole generating method which, however, requires special equipment for its implementation. Figure 1-5.17 illustrates the operating principles of the generating-type center-hole grinding, and Fig. 1-5.18 is the general view of a special machine built for carrying out this method. In the generating type of center-hole grinding the part does not have to rotate, because an equivalent effect is accomplished by the planetary movement of the grinding-wheel spindle, circling the part axis at a constant speed of 400 rpm. Substituting the planetary movement for the part rotation is particularly useful for heavy-

Courtesy of Bryant Grinder Corp.

Fig. 1-5.18   General view of a special center-hole grinding machine with generating movements.

weight parts or parts whose volume is not symmetrical with respect to the axis selected for cylindrical grinding. The grinding-wheel spindle also carries out a reciprocating motion along a path at 30 degrees to the part axis, and the stroke length of that movement can be adjusted from 0.005 to 0.100 inch (0.127 to 2.540 mm). The mounting of the workpiece on the generating type center-hole grinder is similar to the methods described previously in paragraph (b). As optional equipment a drive can be provided to impart a rotation to the part for attaining extremely accurate concentricity, claimed to be in the order of 0.0004 inch (0.01016 mm) under optimum conditions, such as dead-center support and outside locating surface of comparable accuracy.

It should be remembered that the beneficial effects of accurate center holes on the quality of the

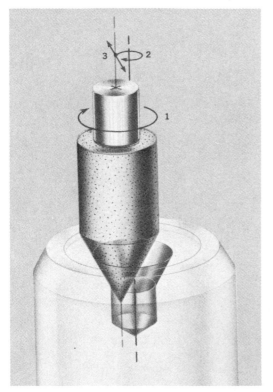

Courtesy of Bryant Grinder Corp.

Fig. 1-5.17   Operating principles of the generating type center-hole grinder, combining (1) rotating, (2) planetary, and (3) reciprocating movements.

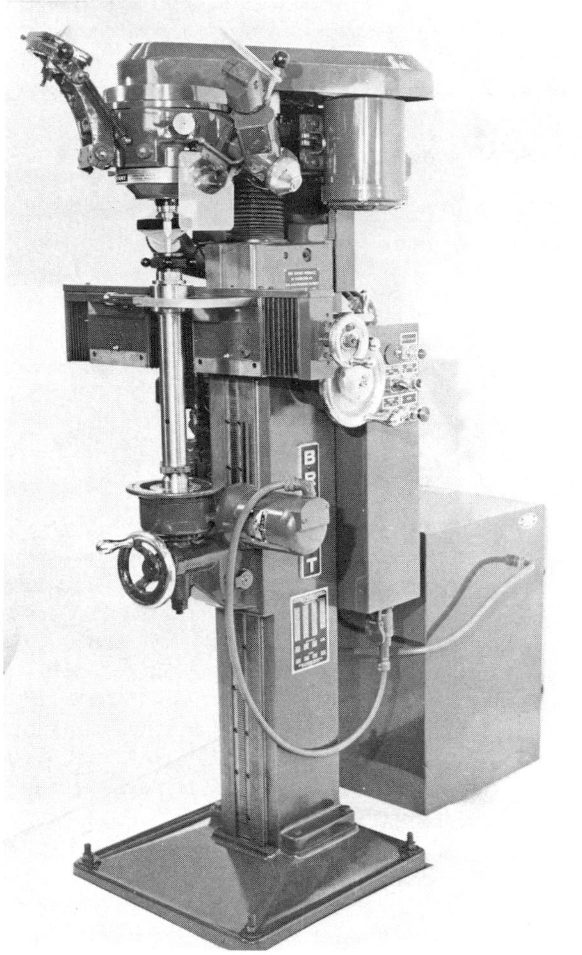

ground surface is predicated on the use of appropriate *machine centers*. Machine centers used in cylindrical grinding machines are generally made of tool steel, preferably high-speed steel or with carbide tips — the conical point accurately ground and concentric with the tapered mounting surface of the center. Proper lubrication of the center surface in contact with the center hole must be assured to reduce friction and wear. Worn centers must be reground, an operation which can be carried out in the headstock of universal grinding machines, while for plain cylindrical grinding machines, center grinding devices can be supplied as optional equipment.

Grinding Internal Surfaces on Cylindrical Grinders

The grinding of internal cylindrical and tapered surfaces is generally carried out by applying principles most of which are essentially identical with those of external grinding. The rotation of the workpiece around its axis, the reciprocating table movement of traverse grinding, and the infeed control by cross slide adjustment are equally present in both the external and internal grinding operations. Differences in the two systems exist, however, due to: (1) the limited access of internal surfaces, requiring an additional axial movement of the grinding wheel for introduction into and for retraction from the hole area, and (2) the limitations of the wheel size imposed by the diameter of the internal area.

It is possible to carry out internal grinding by utilizing the existing motions and adjustments of cylindrical grinders equipped with the necessary supplementary accessories, although for regular production purposes machines expressly designed for internal work as discussed in Section 3 on Internal Grinding, are preferred and used predominantly.

The use of cylindrical grinding machines for internal grinding is limited to single parts or to very small volume production, such as required, e.g., in toolroom operations. Apart from substituting for an internal grinding machine, for which not enough work might be available to justify its procurement, technological reasons, too, could justify the use of cylindrical grinders for internal work. Such is the case, for example, when the bore of a part must be ground exactly concentric with the outside surface, without having to rely on expensive special manufacturing equipment. On a cylindrical grinding machine equipped with an internal grinding attachment, workpieces of a certain size range can be ground consecutively, yet in a single set-up, both on the external and internal surfaces.

Internal grinding devices used on cylindrical grinding machines have individual drives and are designed to accept different types of internal spindles, which are interchangeable to accommodate a range of hole diameters and depths. The individual drive is needed to assure the high rotational speed of the small wheels used for internal grinding. On the other hand, such wheels require much less power than that of the main machine drive used for external work, which is carried out with a large, and often wide, wheel.

The universal type of cylindrical grinding machines which is primarily intended for toolroom work, is usually equipped with an internal grinding device as a regular attachment. It is mounted on a hinged base on the wheelhead, thus permitting the device to be brought easily into operating position, or swung out of the work area while external grinding is done. Figure 1-5.19 illustrates a typical example of an internal grinding device mounted on a universal type cylindrical grinder. The device is shown swung into operating position in front of the external grinding wheel; the large wheel has been brought out of engagement with the work by retracting the cross slide.

The operation of internal grinding devices on universal cylindrical grinding machines also requires the use of a special wheel-truing fixture which can be swung out of the way during grinding (see item 8 of Table 1-5.2).

Courtesy of Brown & Sharpe Mfg. Co.

Fig. 1-5.19 Internal grinding attachment mounted on the wheel head of a universal type cylindrical-grinding machine.

The Operation of Cylindrical Grinding Machines

The purpose of this section is to present a concise review of the many functions which are comprised under that general title. A detailed discussion would, by far, exceed the objectives of the review and also enter into aspects which might apply to certain types of cylindrical grinding machines only. For concrete instructions regarding the handling and care of any particular grinding machine the operator's manuals of the respective machine tool manufacturers should be consulted.

The following topics will be discussed:

(a) Initial installation
(b) Maintenance
(c) Setup and operation
(d) Corrective actions for improving the machine performance and/or the resulting product.

Initial Installation

The competence and care applied in installing a cylindrical, or any other precision type of grinding machine can affect, to a significant degree, the subsequent performance of the equipment. A few of the major aspects to be considered are the following:

The location of the machine should be selected to be free from transmitted vibrations, and not exposed to sudden or excessive temperature variations which may result, e.g., from drafts, direct sun rays, vicinity to furnaces, etc. Proper access to all control elements and areas essential for regular maintenance should be provided. The location of the cylindrical grinding machine should not be too close to other equipment, thus avoiding a crowded work area which could affect the operator's attitude, a factor of particular importance in manually operated equipment.

Foundation. In order to consistently preserve the original alignment of a well-built grinding machine, it should rest on a foundation capable of providing stable support for the fully equipped and operating machine. For general-purpose cylindrical grinding machines in the light- or medium-weight range, a properly constructed floor made of reinforced concrete of not less than 8-inch thickness will prove adequate. When the shop floor does not satisfy these conditions and for extremely heavy machines, the construction of an individual foundation will be necessary.

Generally, the simple placing of the grinder on a floor of adequate stability will prove sufficient. However, when particularly high performance accuracy is required, the firm bolting of the grinding machine to the foundation is usually recommended. Anchor bolts, combined with leveling jacks, are considered useful means for bolting the machine without affecting the execution of subsequent leveling adjustments.

Leveling, when properly carried out, can serve the double purpose of contributing to the accurate operation of the machine and also of establishing a reference plane for the occasional inspection of alignment accuracy with regard to machine elements and setup. Examples of conditions to be inspected are the freedom from deflections, particularly of the bed and guideways of larger machines; the planes of the table and cross slide movements; and the alignment of the workhead and tailstock axes. The precisely established horizontal machine position can also provide reference planes for checking the accuracy of certain movements which are only related to the horizontal.

The most frequently used instrument for guiding and checking the leveling of machine tools is the spirit level which, for this purpose, should be not less than 12 inches (300 mm) and possibly 18 inches (450 mm) long, of the precision type with graduations equivalent to 0.0005 inch per foot, (0.04 mm per meter) or 10 seconds of arc, as a minimum degree of resolution. For more accurate or advanced measurement of the level conditions, electronic levels with about one second of arc sensitivity, are used.

The use of optical tooling, by applying telescopes, autocollimators and/or laser alignment instruments, in combination with special mirrors and prisms, offers very dependable and sensitive methods for the inspection of positional or translational conditions, which are either related to, or are entirely independent of a horizontal reference plane.

Maintenance of the Cylindrical Grinding Machine

In this group, again, only examples of areas requiring regular care in the course of cylindrical grinding machine operations will be mentioned.

Lubrication. Correct and consistently performed lubrication is probably the most important maintenance function in the operation of precision machine tools, such as cylindrical grinders. The manufacturers of machine tools supply detailed lubrication instructions which generally contain the following information:

(a) List of lubrication stations, stating the parts to be lubricated, and usually complemented with diagrams showing the locations on the machine of these stations and the drains

(b) Frequency of the required lubrications (e.g., at daily, weekly, etc., intervals)

(c) Specifications of the lubricant to be used

(d) Instructions regarding access, quantities required, and the method of lubrication.

The bearings of the grinding-wheel spindles are particularly sensitive to damage in the case of inadequate lubrication. The critical role of correct lubrication in that area results from the great driving power and high rotational speed imparted to the spindle supported by these bearings and the tight clearances used for reasons of excellent running accuracy.

In the smaller types of cylindrical grinding machines, up to about 12-inch wheel diameter, the static pressure of an oil column is used to introduce oil into the spindle bearings. Larger machines have forced spindle-bearing lubrication using a motor-driven pump to produce the needed oil pressure. Safety devices, e.g., gravity floats for the static-pressure type lubrication and pressure sensors for the pumped oil flow are generally used to securely avoid bearing damage resulting from inadequate lubrication: a drop in the oil level or pressure will actuate a protective switch turning off the drive motor.

In traverse grinding, the table ways support the table, headstock, tailstock, various fixtures, and the work which, in addition to its own weight, also transmits the grinding pressure. Notwithstanding that often quite substantial weight, the table must slide smoothly at the selected uniform speed, react promptly to the reversing control and be guided with an accuracy unaffected by excessive lubrication which might cause "floating." In modern cylindrical grinding machines the table ways are generally pressure lubricated with continuously circulated filtered oil, supplied by a special pump, which is sometimes tandem mounted with the pump of the hydraulic oil system. The volume of the table way lubrication oil flow, as well as its pressure, can be regulated. As shown on the diagram in Fig. 1-6.1, the pressurized oil enters the table ways in their center portion and flows away toward the ends, thus it contributes effectively to preventing the entrance of foreign matter into the machine ways area.

The preceding examples should indicate the care applied in the design of modern cylindrical grinding

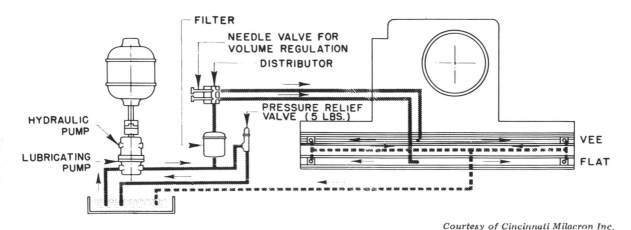

Courtesy of Cincinnati Milacron Inc.

Fig. 1-6.1 Diagram of the table way lubricating system of a universal type cylindrical-grinding machine.

machines for assuring the correct lubrication of all moving elements. The conscientious execution of the pertinent lubrication instructions is, however, a necessary complement of the inherent design features in the interest of long life and continued performance accuracy of cylindrical grinding machines.

The hydraulic system. Hydraulic actuating devices are used in most types of cylindrical grinding machines for two major, and various auxiliary, movements. Major movements powered by the hydraulic system are the reciprocating table traverse and the cross slide movements for positioning and feeding the wheelhead. Examples of auxiliary functions for the hydraulic system are the grinding wheel truing device and the power-tailstock.

The hydraulic system of grinding machines has the same principal elements which are used in all other fluid power transmission devices, namely (a) a pump to generate flow with pressure, (b) a piston in a cylinder which reacts to the pressure converting it into power movement, and (c) control elements to direct and meter the flow of the fluid.

The hydraulic system also comprises a reservoir from which the pump sucks the oil and to which the drained oil is returned. The operation of the hydraulic system causes the warming up of the oil, with potentially harmful effect on the accurate control of the hydraulically actuated movements. For that reason the oil reservoir is usually separated from the machine bed which should be protected from the transmission of extraneous heat. Hydraulic oil reservoirs are often designed for permitting air circulation to assist in cooling the oil. In grinding machines designed for very accurate work the hydraulic oil reservoir may be equipped with a cooling system.

The satisfactory operation of the hydraulic systems which are designed to assure very sensitive control in cylindrical grinding machines, is dependent on the proper selection and the cleanliness of the oil. The hydraulic systems of grinding machines are equipped with oil filters, often of the cartridge type, to permit easy periodic replacement.

Adjustments. Distinction should be made between adjustments needed in the setup for regular operations and those serving to correct conditions which affect the proper operation of the machine. These latter are regarded only as maintenance operations and are outlined in the following by a few examples:

The wheel-spindle bearings of cylindrical grinding machines are designed to be adjusted to a very small clearance, in order to assure the proper bearing stiffness, a precondition for producing chatter-free surfaces with excellent finish. A cool bearing during regular operation may indicate excessive clearance,

while a hot running bearing, unless that condition is caused by inadequate lubrication, results from a too-tight clearance. In each case the appropriate adjustment of the spindle bearing will be required.

The wheel spindle-end thrust support may also need occasional adjustment, particularly when precise shoulder grinding is to be carried out. Any axial clearance of the grinding wheel spindle in its support is usually avoided by applying spring pressure to force the spindle against a thrust plate. When axial looseness is noticed, judging by the quality of the ground surface, or should the actually measured end play of the running spindle be in excess of 0.0001 inch (0.0025 mm), then the tightening of the nut against which the spring is resting will be needed.

The cross-slide movement must be controlled particularly in two respects: (1) the tracking accuracy of its travel, and (2) the backlash-free translation of the control motions into the proportional displacement movements of the slide. When the travel of the cross slide begins to lose its accuracy due to wear, it can be corrected by adjusting the cross-slide gib. Backlash in the motion transmission commonly originates from looseness between the infeed worm and the worm wheel, due to wear which exceeds the limit covered by the backlash-eliminating device. These elements are usually designed to permit tightening in the user's plant as an occasionally required maintenance operation.

The drive belts, both in the main drive and in the headstock drive, require adjustment at intervals to compensate for wear and stretch. When tightening the belts in accordance with manufacturer's instructions, the originally set tightness should not be exceeded; however, the belts must be tight enough to prevent slippage when the driven machine elements are running at full load.

Setup and Operation

Establishing the process for cylindrical grinding will comprise essentially similar aspects considered by process engineering for other types of machining by operations, such as, e.g., the sequence of the operational steps, the locating and holding of the work, the selection of tooling, and determining the machining data. For process engineering purposes, in connection with the preparation of cylindrical grinding operations, the contents of the preceding chapters, as well as the subsequent one on "Grinding Wheel and Machining Data Selection," may provide useful guidance. The engineered process, usually spelled out on form sheets, together with the toleranced product drawing, will also serve as the basis for the machine setup.

The setup of the cylindrical grinding machine will include the installation of the selected grinding wheel, the work-holding devices, and auxiliary fixtures when needed. Also included in the machine setup functions are the adjustments of the required speed and feed rates, of the limit positions for the power controlled movements, as well as of the wheel-truing device and gages.

More intricate setup procedures may be involved when work other than plain cylindrical grinding is planned. Typical examples of such setup conditions on plain or universal cylindrical grinders are the following:

(1.) The grinding of tapers, which may be shallow and carried out by the swivel setting of the upper table, or of the steeper type, requiring either the swivel adjustment of the workhead or of the wheel head. In such cases the angle setting may either be guided by graduated scales producing an approximation which may require subsequent correction, or carried out with the use of gage blocks, when the machine is designed for this latter type of highly accurate angle adjustment.

(2.) The grinding of shoulders with precise axial position control with the aid of end stops, often combined with special "flagging" gages, for the actual measurement of the work position. When shoulder grinding is combined with traverse grinding, then the relieving of the grinding wheel side will be needed, unless angular approach grinding is being applied.

(3.) Internal grinding, equipping the grinding attachment with the proper type of quill and wheel, as well as installing the special truing device for internal grinding wheels.

(4.) Long and/or slender parts will require the installation of steady rests or center rest, including the necessary adjustment of the jaws.

(5.) Face grinding, when carried out, exceptionally, on universal type cylindrical grinding machines by swiveling the workhead at 90 degrees, mounting the part in a chuck, and applying the reciprocating table motion over a distance to cover half of the work diameter.

In addition to the setup adjustments carried out on general-purpose grinding machines, substantially more complex setup functions will be needed on automatic cylindrical grinders, such tasks usually requiring the assignment of a specialized setup man.

For setups involving a large number of adjustments it is often found useful to prepare detailed setup procedures, listing the essential actions, pointing out the critical requirements and stating the applicable dimensions or other magnitudes when necessary. The advantage of written procedures is the ready duplication of the setup independently of the person assigned to its implementation.

The execution of the cylindrical grinding operation even when guided by detailed process instructions and supported by precisely built, reliably functioning and sensitively adjustable modern grinding machines, is dependent on the skill and conscientious care of the experienced operator. The dependence on operator's skill and performance is even more pronounced when carrying out the process on machines of older vintage, and for complex or very tightly toleranced workpieces. The subsequent topic, dealing with grinding deficiencies and their potential causes, is also indicative of the role assigned to the operator's judgment.

The reliance on the human factor, although always present to a certain extent, can be substantially reduced by providing properly prepared and sufficiently detailed process sheets, preferably of the illustrated type, using a diagram to convey the following information:

(1.) Location on the workpiece of the surface to be ground

(2.) The principles of work holding, indicating the orientation of the work, as well as the locating and clamping surfaces

(3.) Dimensions and form, both prior and after grinding

(4.) Tolerances for dimensions, geometric parameters (roundness, straightness, perpendicularity, etc.), and surface finish.

The process sheets should also contain the grinding-wheel specifications, the list of tooling — particularly for work holding and gaging — together with the applicable machining data on speeds and feeds. For grinding operations in general, the proper grinding-fluid selection and the truing operation are also significant factors, therefore the spelling out of pertinent instructions on the grinding-process sheets is advisable. Some operational aspects of these functions will be discussed in the following paragraphs.

The use of grinding fluids may be considered a rather general practice in cylindrical grinding, in view of the advantages of wet grinding over dry in the majority of applications. Grinding fluids, whose composition and properties must be selected in accordance with the conditions of application, assist the grinding process in several respects, such as: absorbing and dissipating heat; lubricating the area of contact between wheel and work; as a rust preventive to retard oxidation, thus protecting the product and equipment; transporting the removed metal swarf and

Table 1-6.1 Common Faults in Cylindrical Grinding

TYPE OF COMPLAINT	FAULT DESIGNATION AND DESCRIPTION	PROBABLE CAUSES	POTENTIAL REMEDIES
Surface Deficiencies	Chatter Marks — axially directed parallel areas bounded by sharp lines.	Intermittent contact between the work and the grinding wheel, caused by vibrations, runout, or improper work support.	— Check, and when needed, repair wheel spindle bearings. — Check and correct center holes and machine centers. — Adjust steady rest, when needed. — Use wheels of softer grade, well-balanced and properly dressed. — Check balance and speed or workpiece (speed may have to be reduced). — Investigate environmental conditions for transmitted vibrations.
	Closely Spaced Wave Marks — produce different reflections of light, yet boundaries between adjacent areas are not distinct.	Not consistent, yet continuous contact between the work and the wheel—distinguishable from chatter by the gradual transitions between the high and low sections.	Generally resulting from vibrations within the system. Origins, which will need correction, may be located in any of the following elements: — Drive motors — Hydraulic system — Belts, chains, sprockets — Gear trains — Headstock; seal may be too tight. Also check for transmitted vibrations.
	Widely Spaced Wave Marks — observable as low number lobing when inspected by means of roundness tracing.	Grinding wheel out-of-balance, resulting in periodic approach of the wheel periphery to the work. The wave pattern usually reflects the ratio between the rpm of the wheel and of the work.	— Improperly balanced wheel; to be checked on a balancing stand and corrected by the proper adjustment of the balance weights. — Dress new wheels after mounting on grinding machine. — Prevent coolant collecting on one side of wheel; let wheel run after having turned off the coolant. — Assure free rotation of the workpiece; excessive center pressure can cause work binding.
	Mottle Marks — discontinuous, widely spaced wave marks.	Caused by conditions similar to those resulting in widely spaced wave marks, but when affecting a traversing workpiece the marks along the surface will not match up, but appear in straddling positions.	— Careful balancing of the wheel, preceded by a rough dressing and followed by a final truing. — Run wheel without grinding to throw out coolant which could have accumulated on one side. — Remove grease spots from wheel surface when present, by re-truing — Check wheel surface for embedded foreign material or hard spots; discard by heavy truing.

Table 1-6.1 (*Cont.*) **Common Faults in Cylindrical Grinding**

TYPE OF COMPLAINT	FAULT DESIGNATION AND DESCRIPTION	PROBABLE CAUSES	POTENTIAL REMEDIES
Surface Deficiencies (*cont.*)	Rough Finish — inadequate smoothness or the presence of random surface irregularities.	Principal causes may be of two sorts: — Grinding conditions not properly chosen to assure the specified finish (expressed in microinch average —AA— values). — Interfering conditions which deteriorate the general finish.	— Use wheel with finer grit or of different grade. — Adjust truing traverse to the sharpness of the diamond. — Lower work speed or traverse speed. — Check cleanliness and consistency of coolant.
	Burlap Finish — closely spaced and randomly distributed pattern resembling burlap cloth in appearance.	Reflection on the work of an irregular wheel surface which resulted from improperly controlled truing.	— Check diamond and replace when cracked or chipped. — Secure diamond nib should it have gotten loose. Be sure that diamond nib is rigidly supported and held in a shaft which is not vibrating during wheel truing. — Check whether the movement of the traversing element, the machine table, or truing device slide, is even during truing.
Work Surface Injuries	Scratches on the Work Surface	Randomly distributed deep marks on the work surface can result from incorrectly selected wheel type. Continuous marks on the work are often due to improper wheel dressing or, in traverse grinding, sharp edges on the wheel face.	— Grain type of the wheel may have to be changed, or finer grit must be used. — Check conditions and rigidity of truing tool, also the feed and traverse speed, as well as the frequency of the truing operation. — Round off or chamfer wheel edges — Reduce or vary traverse rate in relation to work rotation.
	Feed Lines — helical marks on the work surface.	These surface defects are reflecting the table traverse in relation to the work rotation and may be caused either by nonaligned work or improper wheel truing.	— Assure proper alignment of the work supporting members: the headstock, the tailstock, the centers, and when used, the work rest. — Produce even wheel surface in truing, to be carried out with uniform coolant distribution.

(*Continued on next page.*)

Table 1-6.1 (*Cont.*) **Common Faults in Cylindrical Grinding**

TYPE OF COMPLAINT	FAULT DESIGNATION AND DESCRIPTION	PROBABLE CAUSES	POTENTIAL REMEDIES
Work Surface Injuries (*Cont.*)	Diamond Truing Lines — on the work surface	Appear on the work surface as very fine thread and are usually the reflection of a similar pattern of the wheel face as produced in truing.	— Don't use too sharp diamond points and direct nib off from the radial to the wheel center. — Reduce truing feed rate and/or depth of diamond penetration, perhaps finishing truing without feed for the last pass.
	"Barber Pole" Finish — pattern of alternating spiral stripes	Intermittent approach and retraction of the work with respect to the wheel during the grinding process.	— Check condition and matching forms of machine centers and work center holes. — Assure solid work support — Check lubrication of table ways.
Metal-lurgical Damages	Burning of Work — discoloration of the work surface	Results from overheating the work surface by applying improper grinding data using inadequate grinding wheel, or insufficient volume of coolant.	— Decrease rate of stock removal and keep it continuous, avoiding work stoppage while in contact with the wheel. — Use softer grade or make the wheel act softer. — Increase volume of coolant and assure its uniform distribution over the work surface.
	Check marks on Work Surface — appear as a fine line network of very shallow cracks.	May originate from inadequate heat treating, a condition aggravated by grinding, or improper grinding may also cause it on originally sound work surface.	— Check heat-treating procedure — Determine whether excessive heat is generated in grinding and correct conditions by adjusting wheel composition and/or grinding data.
	Cracked Work Surface	Stresses originating in heat treatment, when not properly relieved by tempering, can lead to cracks during grinding. Excessive heat and improperly applied coolant during grinding can also cause cracks on adequately heat-treated parts.	— Investigate heat-treat process and also design of the part (to avoid stress raiser points). — Reduce heat generated during grinding by such measures as increased work speed, avoidance of too high wheel speed, copious coolant well distributed, free cutting wheels, producing sharp wheel surface in truing.

Table 1-6.1 (*Cont.*) **Common Faults in Cylindrical Grinding**

TYPE OF COMPLAINT	FAULT DESIGNATION AND DESCRIPTION	PROBABLE CAUSES	POTENTIAL REMEDIES
Size and Form Inaccuracies	Out-of-Roundness	Usually the result of the part not rotating around a constant axis, causing runout or vibrations during the grinding process.	— Check shape, angle, alignment, and surface finish of work center holes and machine centers. — Assure rigid retainment of the machine centers in their sleeves. — Control condition of center holes and machine centers, avoiding burrs, dirt, scored surfaces, etc.
	Non-concentric Adjacent Surfaces	Lack of concentricity between subsequently ground surfaces of a part is generally due to inadequate retainment of the work's axis of rotation.	— Check alignment of centers. — Assure proper fit between centers and spindles. — Lock firmly the elements involved in holding the work, like headstock, tailstock, swivel table, etc.
	Size-Holding Difficulties — Variations in the size of subsequently ground parts.	Improperly prepared parts with excessive size variations or bent, may introduce defects which are difficult to correct in a single grinding operation. Inadequate work support, improper infeed or traverse rates, or even unclean coolant can cause size variations in the work.	— Check semifinished work before mounting on cylindrical grinder. — Define and verify tolerances for material allowance and form errors. — Firmly secure positions of machine elements holding the work. — Verify evenness of infeed and traverse movements; have faulty members repaired.
	Lack of Parallelism — Taper, concave, or convex shapes instead of the sought cylindrical.	Generally due to alignment errors in the work-holding elements, or insecure work support.	— Check and correct as needed, the alignment of headstock and tailstock with respect to the direction of table travel. — Check conditions of work centers and whether center holes in work are deep enough to clear the points of the centers. — Adjust, when needed, the force of the center pressure in the tailstock.
Unreliable Machine Operation	Intermittent Cutting Action in Traverse Grinding	Non-consistent penetration of the grinding wheel into the material of the rotating and traversing work can result from out-of-roundness, too wide rough-size variations, excessive load on centers, distortions in the machine or foundation.	— Check conditions of semifinished workpiece and conditions listed as causes of out-of-round. — Inspect pressure of tailstock center; should be adjusted to the weight of work. — Verify traverse movement of machine table and condition of foundation.

(Continued on next page.)

Table 1-6.1 (*Cont.*) **Common Faults in Cylindrical Grinding**

TYPE OF COMPLAINT	FAULT DESIGNATION AND DESCRIPTION	PROBABLE CAUSES	POTENTIAL REMEDIES
Unreliable Machine Operation (*Cont.*)	Uneven Infeed or Traverse Movement	Inadequate mechanical condition of involved machine elements due to wear or distortions (binding), also insufficient lubrication of guideways, defective operation of the hydraulic system, imbalance in the workpiece or in the rotating machine elements, are examples of possible causes.	— Inspect general condition of guideways and verify movements while the machine is running idle. — Check the hydraulic system thoroughly — Re-balance grinding wheel, when needed, and eliminate all possible sources of imbalance in the machine, like loose pulleys, uneven belts, etc.
Inefficient Grinding Wheel Performance	Irregular Form or Unevenness of Trued Wheel Surface	Difficulties in producing a consistently regular wheel shape with uniform surface.	— Diamond may not hold up; change to larger size. — Assure copious flow and even distribution of coolant during truing. — Ascertain whether diamond holder is rigidly attached. — Avoid table float caused by too heavy oil or excessive supply.
	Wheel Loading or Glazing	Caused by metal from the workpiece getting lodged in the pores, or even on the grains of the wheel; recognizable by the shiny appearance and slick feel of the grinding wheel.	— Alter wheel specifications to more open bond, softer grade, or coarser grain. — Produce sharper and more open wheel face by using faster dresser traverse, deeper cut, and sharper diamond. — Increase volume of coolant, make it thinner, add soda and soften water when needed.
	Improper Effective Hardness of the Grinding Wheel	Hard acting wheel develops loading, causes chatter or burning of work.	— Increase work and traverse speed. — Decrease peripheral speed of the wheel. — Dress more often and produce a sharper wheel surface. — Select softer wheel, coarser grain for soft work and finer grain for hardened material.
		Soft-acting wheel, causes rapid wheel wear and impedes the holding of form and size of the work.	— Decrease speed of work rotation and table traverse, also rate of infeed. — Increase wheel speed and diameter within safety limits, use wider wheel. — Apply heavier coolant. — Change wheel to harder bond and/or less friable grain.

wheel grit; and keeping the wheel surface cleaner, thereby improving its cutting ability.

Grinding fluids are available in different basic types and compositions, to be used straight or diluted with water. For the latter, the concentration of the fluid will be selected according to the requirements of the work, the ratio of the soluble oil to the water added usually being between 1:25 and 1:100. The type of fluid to be used depends on the relative priorities assigned to the expected properties, such as cooling effect, lubricity, corrosion prevention, etc. Sometimes a combination of two different fluid types is used to provide benefits from several properties which are offered at widely different degrees by the selected constituants.

Modern cylindrical grinding machines are equipped with pumps dimensioned to supply sufficient amounts of fluid for operations within the capacity range of the machine. It is, however, the operator's responsibility to direct the discharge nozzle to the spot where the fluid is most effective, generally at the work-wheel interface. Kept in continuous circulation, the returning fluid is collected in a reservoir which, on smaller machines, is an integral part of the machine base casting, while larger machines have a fluid tank installed adjoining the grinder.

The fluid containers are usually equipped with deflectors to prevent the recirculation of the sludge which will accumulate at the bottom of the reservoir. At intervals, depending on machine usage and stock removal, sometimes as frequently as weekly or even twice a week, the sludge has to be removed after first draining the reservoir. It is a good practice to combine the sludge removal with the cleaning of the reservoir, as well as of the entire grinding fluid system. Swarf and grit in the grinding fluid which is directed at the work, whether due to recirculation from a sludge-filled tank or originating in the uncleaned system, including, e.g., the inner surface of the wheel guard, can seriously interfere with producing a smooth finish on the work surface.

Truing the grinding wheel in cylindrical grinding is always a part of the new setup when it involves wheel change, and is also needed at intervals during the grinding process. In plain cylindrical grinding the primary function of routine truing may be to dress the operating wheel face for restoring its free-cutting properties. In precision grinding the regular wheel truing may be a part of the size control, by compensating in the cross-slide setting for the exact amount by which the truing operation reduced the wheel diameter. Truing, sometimes at rather short intervals, is necessary in form grinding in order to consistently assure the true profile of the grinding wheel face.

In view of the need of regular grinding wheel truing, the standard equipment of cylindrical grinding machines includes truing devices, at least those needed for the basic operations. Special truing devices, designed for less common applications, are usually made available by grinding machine manufacturers as optional equipment.

A few basic aspects of proper truing in cylindrical grinding were already indicated in the preceding Chapter 1-5 under the heading, "Dressing and Truing."

Common Faults and Their Correction in Cylindrical Grinding

Harmful effects on the work quality will result when the equipment, tools, and operational elements are not properly selected or controlled. The great variety of causes and effects precludes a complete survey. Therefore, only a few of the more common defects resulting from improper cylindrical grinding conditions are reviewed in Table 1-6.1, listing the most frequent sources of deficient operations and pointing out both the possible sources and the probable cures of improper grinding performances.

Getting acquainted with typical interrelations between inadequate performance, the elements prone to be the sources of such shortcomings, and the characteristics which can be their cause, should be helpful in carrying out similar analyses when grinding defects other than those listed arise. Furthermore, such understanding of interacting factors in cylindrical grinding should lead to establishing process conditions that are more apt to produce satisfactory results from the beginning.

Grinding Wheel and Machine Data Selection

Most cylindrical grinding machines offer a rather wide range of work rotation speeds, table traverse speeds, feed rates, and, sometimes, different wheel speeds as well. The proper selection of these variables, in combination with the choice of the grinding wheel, has a great bearing on the results and the economy of the grinding operation.

The conditions which may have to be considered in determining suitable machining data are multiple and may be assigned to one of the following groups:
(a) Workpiece characteristics — material, hardness, general configuration and dimensions, rigidity, adaptability for holding, and, of course, the stock allowance to be removed in the grinding operation.
(b) Requirements for the ground surface — geometric and dimensional accuracy, surface texture, surface integrity, etc.
(c) Economic objectives — grinding performance, optimum cost, desired quality levels, as well as the priorities for any conflicting objectives.
(d) Operational conditions — construction and capabilities of the grinding machine, the composition of the grinding wheel, grinding fluid, frequency and methods of truing and dressing, operator's skill, degree of automation, etc.

The preceding listing of factors to which machining data will have to be adjusted, although incomplete, indicates why experience, often even experimentation, is so widely relied upon for the selection of machining data in cylindrical grinding. It is obvious that recommendations for machining data can, in most cases, represent an approximation only of the optimum, which may not be attainable without subsequent testing and adjustment.

Nevertheless, such recommendations can prove useful in two major respects:

1. To indicate the initial values for setting up a specific grinding operation, even when subsequent refinements of the starting data appear warranted and will be carried out; and
2. To determine approximate values of operational performance, for the purpose of time and cost estimates, and also for manufacturing capacity assessment.

To provide starting guidance in the selection of process data for cylindrical grinding the following information will be supplied:

(a) A brief list of grinding wheel recommendations for various commonly used work materials
(b) Average machining data for basic cylindrical grinding operations under common conditions, established for the grinding of widely used types of work materials
(c) A concise review of frequently occurring grinding process variables, and their effects.

Grinding Wheel Selection for Cylindrical Grinding

The listing of grinding wheel specifications for commonly used materials (on page 80) serves the purpose of general information only. It may be used as a first approach in grinding wheel selection for cylindrical grinding operations, should more pertinent data, as from comparable past operations, not be available.

Machining Data Recommendations

The data listed in Table 1-7.1 are intended to be used as starting information when establishing process

Table 1-7.1 Basic Machining Data for Cylindrical Traverse Grinding

WORK MATERIAL	MATERIAL CONDITION	WORK SURFACE SPEED (v), feet per min (m/min.)	INFEED, Inch/pass (mm/pass)		TRAVERSE FOR EACH WORK REVOLUTION IN FRACTIONS OF WHEEL WIDTH	
			ROUGH	FINISH	ROUGH	FINISH
Plain Carbon Steel	Annealed	100 (30)	0.002 (0.05)	0.0005 (0.013)	1/2	1/6
	Hardened	70 (21)	0.002 (0.05)	0.0003 to 0.0005 (0.0075 to 0.013)	1/4	1/8
Alloy Steel	Annealed	100 (30)	0.002 (0.05)	0.0005 (0.013)	1/2	1/6
	Hardened	70 (21)	0.002 (0.05)	0.0002 to 0.0005 (0.050 to 0.013)	1/4	1/8
Tool Steel	Annealed	60 (18)	0.002 (0.05)	0.0005 or less (0.013 or less)	1/2	1/6
	Hardened	50 (15)	0.002 (0.05)	0.0001 to 0.0005 (0.0025 to 0.013)	1/4	1/8
Copper Alloys	Annealed or Cold Drawn	100 (30)	0.002 (0.05)	0.0005 or less (0.013 or less)	1/3	1/6
Aluminum Alloys	Cold Drawn or Solution Treated	150 (46)	0.002 (0.05)	0.0005 or less (0.013 or less)	1/3	1/6

NOTE: These are recommended starting values which may prove to be too conservative in many applications and could then be modified for improving the productivity of the process.

Grinding Wheel Specifications For Commonly Used Work Materials

FERROUS METALS		NONFERROUS METALS	
Work Material	Grinding Wheel Specifications	Work Material	Grinding Wheel Specifications
Steel, hardened		Aluminum	A 46 I 8 V
small wheel	A 60 L 5 V		
large wheel	A 54 L 5 V	Brass	C 36 K 5 V
Examples of a few			
uncommon work-		Copper	C 46 K E
pieces:			
spline shafts	A 60 O 6 V	Bronze	A 46 L 5 V
plug gages	A 80 K 8 V		
gages, high finish	C 500 J 9 E	Cemented carbide	
		rough	C 60 K 8 V
Stainless steel		finish	C 100 J 8 V
300 series	A 46 J 8 V		
400 series, hard-			
ened	A 54 K 5 V		
High-speed steel			
small wheel	A 60 L 5 V		
large wheel	A 54 K 8 V		
Cast-iron bushings	C 36 J 5 V		
	C 46 K 5 V		

Note: In many cases refinement of, or deviations from the above indicated general recommendations will be warranted and could contribute to improving both the efficiency and quality of the cylindrical grinding operation.

specifications for *cylindrical traverse grinding* operations. They are predicated on the use of grinding wheels of proper specifications, the operation being carried out on grinding machines of adequate rating and condition, with the required supply of grinding fluid provided. Furthermore, the listed values apply to workpieses with diameters not exceeding about 2 inches (50.8 mm), and are sufficiently rigid to withstand the grinding force without harmful deflection, well-balanced and properly mounted. In cases of deviations from the assumed conditions, modification of the basic data may be required, according to the expected effect of pertinent factors, a few examples of which will be discussed later.

Grinding wheel speeds should be maintained within a range of 5000 to 6500 feet per minute (1525 to 1980 meters per min), unless special conditions prevail or equipment and wheels designed for substantially higher wheel speeds are used. Higher wheel speeds, such as employed in high speed grinding, are predicated on the use of special grinding wheels and on machine design, including reinforced wheel guards, adapted to such operations.

The infeed rates listed in Table 1-7.1 refer to the penetration of the grinding wheel into the work material. In the case of cylindrical grinding the diameter of the workpiece will be reduced by twice the amount of wheel advance. It should be noted, however, that many makes of cylindrical grinding machines have cross-slide handwheels with graduations indicating double the amount of actual wheel advance, thereby expressing the amount of the expected work diameter reduction.

To establish the rotational speed of the workpiece in rpm to agree with the work peripheral speed values (*v*) of Table 1-7.1, expressed in feet per minute (meters per minute), considering the largest work diameter (*D*) to be ground in inches (millimeters):

$$\text{rpm} = \frac{v \times 12}{D \times 3.14} = \frac{v \times 3.82}{D}$$

Equivalent equation in metric form, where *v* is in m/min and *D* is in mm:

$$\text{rpm} = \frac{v \times 1000}{D \times 3.14} = \frac{v \times 318}{D}$$

EXAMPLE: Hardened tool steel part with D = 1.75 inch. Work peripheral speed (v) as per table: 50 feet per minute

$$rpm = \frac{50 \times 3.82}{1.75} = 109$$

To determine the table traverse speed (T) in inches per minute values, consider the wheel width (W) and proceed as shown by the following example:

Assuming a 2-inch-wide wheel is used for the rough grinding of the hardened tool steel part rotated at 109 rpm, and applying a traverse factor (t) of $\frac{1}{4}$, as per table:

$$T = W \times t \times rpm = 2 \times \frac{1}{4} \times 109 = 54.5 \text{ in. per min.}$$

In *plunge grinding* the applicable feed rates per revolution of the part are less than in traverse grinding. The moderating effect on the specific chip removal resulting from the gradual penetration of the wheel into the work material along a helical path, which is taking place during the traverse grinding, is not present in plunge grinding; here the operating face of the wheel is in simultaneous contact with the entire length of the ground section. On the other hand, the infeed during the plunge grinding operation is continuous, as compared with the intermittent feeds at the end of the table strokes in traverse grinding. Consequently, the total machining time might be less in plunge than in traverse grinding, notwithstanding the reduced feed rates per part revolution.

Typical infeed rates for plunge grinding are listed in the chart.

quently occurring variations in process conditions, as well as the wheel performance changes which can be expected to result, are reviewed in the following.

Changes caused in the action of the wheel will be indicated as trends only, pointing out the sense, but not the magnitudes. Several of the discussed variables have opposing effects and, when concurrently present, may tend to balance out. Other variables can have similarly directed effects which then should be expected to work additively.

To avoid duplications, variations in the process and of the participating elements were considered to occur in a single sense, that is, increasing only. Of course, such changes from an assumed base level can occur in the opposite, decreasing sense also, in which case the expected results are the inverse of those indicated above.

Being familiar with the directions in which process variables interact with grinding wheel performance, permits the deliberate modification of certain easily controllable variables in the interest of improving the overall operational conditions.

Grinding wheel diameter. Increasing the diameter causes the grinding wheel to act harder, because the expanded contact area distributes the stock removal over a larger number of grains, thus reducing the specific force expended on each individual grain.

Peripheral speed of the grinding wheel. Increasing the peripheral speed will also cause the grinding wheel to act harder, for reasons similar to those mentioned above; in this case the greater number of grains is brought into action consecutively, in distinc-

Typical Infeed Rates For Plunge Grinding

Work Material	Roughing	Finishing
	Infeed per Revolution of the Work, in. (mm)	
Steel, soft	0.0005 (0.013)	0.0002 (0.005)
Plain carbon steel, hardened	0.0002 (0.005)	0.000050 (0.00125)
Alloy and tool steel, hardened	0.0001 (0.0025)	0.000025 (0.0006)

Note: In this case too, the resultant reduction of the work diameter will be twice the amount of cross slide movement causing the wheel infeed.

Process Variables and their Effect on Grinding Wheel Performance

Grinding wheel recommendations, unless prepared for specific cases, are predicated on conditions which may be considered as average. Deviations from such basic conditions will generally affect the manner in which grinding wheels actually act. Therefore, fre-

tion to the simultaneous contact resulting from increased wheel diameter.

NOTE: In cylindrical grinding, variations of the process conditions occur regularly when the wheel diameter decreases due to usage. To balance the resultant effect, the reduced-diameter wheel should be operated at a higher rpm, when available on the grinder. Two important rules must, however, be observed:

(a) In no case must the wheel surface speed in fpm be raised above the peripheral speed of a new wheel when operated at the prescribed rpm.

(b) When a new wheel is mounted on the spindle the lower rotational speed of the wheel head must be restored before starting the drive motor.

Workpiece diameter. When greater, it:

1. will cause the wheel to act harder — resulting from larger area of contact
2. will reduce the development of heat in the part — assisted by improved heat distribution.

Workpiece peripheral speed. An increase causes the wheel to act softer because the specific load on the individual grains is raised.

NOTE: Varying the work speed is, perhaps, the most convenient and effective way to modify the operational conditions in cylindrical grinding. The implementation of work speed variations is facilitated by the infinitely variable workhead speeds, extending over a wide range, which are available on modern cylindrical grinding machines. Varying the peripheral speed of the workpiece has to be applied, however, within limits, which depend on (a) the potentially adverse effects of higher work speed on the finish, and (b) the development of chatter caused by excessive work speed.

Table traverse speed. When increased it will speed up the operation, but can affect the obtained finish adversely and interfere with the required size control. Its effect on the apparent wheel hardness is to make the wheel act softer, for reasons similar to those indicated in the preceding paragraph.

NOTE: (a) Applies to traverse grinding, and to some extent, to plunge grinding with a small oscillating movement which can cause the wheel to act softer.

(b) Table speed is also a movement whose rate can be varied steplessly on most types of cylindrical grinding machines.

Infeed. When its rate is increased, whether by applying larger increments per stroke in traverse grinding, or by increasing the speed of the continual cross-slide advance in plunge grinding, higher stock removal will result, thereby improving the effectiveness of the process. On the other hand, high infeed rates cause greater heat development in the grinding area and increase the pressure on the work, which may result in deflections and impede size and finish control. The effect of increased feed rate on the wheel is to make it act softer.

NOTE: In cylindrical grinding, for reasons of productivity, the infeed rate is usually set as high as possible and still avoid interference with the required quality of the process. The specifications of the grinding wheel should be adapted to the selected feed rate.

Estimating Operational Times in Cylindrical Grinding

Major Elements of the Total Operating Time

The total operational time is usually composed of four major elements:

(1.) *Machining time*, comprising the time of actual cutting, as well as that needed for wheel approach and retraction, table overtravel, dwells, and sparkout.

(2.) *Handling time*, which comprises the routine functions of the operator, carried out while the machine is not actually operating. Elements of handling time can be reduced to different degrees, even to the level of complete elimination, depending on the completeness of the applied automation.

(3.) *Setup time*. This has to be distributed over the number of parts corresponding to the intervals within which the functions comprised in this group need to be carried out.

(4.) *Lost time*, due to operator's personal reasons, instructions, repairs, waiting for tools, work or setup adjustment, etc. Usually, for estimating purposes these delay sources are consolidated into a single item, considered to be proportionate in value to the sum of the preceding three groups, and expressed as the expected performance rate occasionally also referred to as the productivity of the process. As an example, a 75 per cent performance rate means that three-quarters of the total clocked time will be spent on machining, handling, and setup all closely predictable; while the rest is expended on other activities considered lost time. Sometimes the performance rate is established to include the time spent on making parts which later prove to be rejects.

For estimating purposes the machining times can be calculated on the basis of average values shown in Table 1-7.1, unless more concrete machining data, actually developed for the work on hand, are available.

To arrive at the estimated machining time the following elements of the actual work conditions, which may also define the applicable process system, must be considered:

(a) Work material and its condition (e.g., hardened, annealed, etc.)

(b) Dimensions of the work surface to be ground — particularly diameter and length

(c) The amount of material (stock allowance) to be removed in the grinding operation. Depending on the part's dimensions, and other pertinent conditions, the selected amount of stock allowance will differ. The following average values may be considered generally as the order of magnitude for stock allowance on workpieces up to about 2 inches (50 mm) in diameter:
  1. for roughing: 0.015 to 0.020 inch (0.38 to 0.50 mm) on the diameter
  2. for finishing: 0.0015 to 0.0025 inch (0.038 to 0.063 mm) on the diameter.

(d) Whether the applied grinding data should be those of roughing or finishing operations, or possibly on an intermediate level, will depend — in addition to the primary factor, the stock allowance — on several other conditions also. Examples of such additional factors are: the desired surface finish, the rigidity of the workpiece, the sensitivity of the work material to heat, the required form and size accuracy, etc. Any of these and various other factors might call for reduced speed and feed rates, even when the amount of stock to be removed would indicate the use of values recommended for roughing operations. When the part is ground from the rough state to the finished condition in a single operation, a *two-rate feed*, also known as *dual-feed*, is often used.

(e) *Dwell* or *tarry*, in traverse grinding, refers to the controlled delay of the table traverse reversal at the end of the strokes. The purpose of this is to extend the engagement time of the wheel with the work surface at the end position of the traverse movement. Dwell is employed in areas which do not have the benefit of being contacted twice during the to-and-fro movement of the table and is regularly used when grinding close to a shoulder. The dwell time is set, according to the process requirements, to about 2 to 6 seconds, sometimes even in excess of this range.

(f) *Spark-out*, also called *lap-out*, is applied to reduce the effect of part or machine deflections during grinding and also to improve the finish. Spark-out is carried out, after the part has attained final size, by discontinuing the infeed yet operating all the other movements such as work rotation, coolant supply and, in traverse grinding, the reciprocating movements of the table, making "dead passes." The time required for spark-out must be considered in estimating the operational time, and the following approximate values may be used:

Traverse grinding — roughing: 2 dead passes
                finishing: 6 dead passes.
Plunge grinding — Roughing: 6 work revolutions
            Finishing: 14 work revolutions

(g) Approach and retraction of the wheel head can be operated manually. However, in repetitive work for which detailed time estimates are really needed, these movements are commonly carried out by the rapid travel of the wheel slide and require only a few, e.g., 2 to 4 seconds. An additional time element, which will have to be considered besides the actual approach movement, results from the gap between the end of the rapid approach and the actual work surface. Although in that phase of the operation, the wheel is "cutting air," the ending of the rapid approach before the wheel actually reaches the work, is needed at a point based on the largest expected part diameter, to which about 0.003 to 0.004 inch (0.075 to 0.1 mm) is added for safety reasons.

In high-volume grinding *gap eliminator* devices are sometimes used for reducing the time spent on bridging that gap. Such devices cause the cross slide to continue its advance at an intermediate speed until the wheel actually contacts the work.

Approximate values for typical handling times and setup times will be given in the following section.

Regarding *lost time*, a reasonable starting point for estimating average kinds of work could be one-third

Handling Times for Cylindrical Grinding

Work loading, Clamping and Unclamping			
Between centers Weight range of the part	to 10 lbs (to 5 kg)	over 10 to 20 lbs (over 5 to 10 kg)	over 20 to 30 lbs (over 10 to 15 kg)
Handling time per part, in minutes	0.5	0.8	1.0
NOTE: Use of steady rest is not included.			
Single-end clamping Weight range of the part	to 7 lbs (to 3 kg)	over 7 to 12 lbs (over 3 to 5.5 kg)	over 12 to 20 lbs (over 5.5 to 10 kg)
Handling time for three- jaw chuck, in minutes	0.3 to 0.6	0.5 to 1.0	0.8 to 3.0
Handling time for power clamping, in minutes	0.2 to 0.5	0.4 to 0.8	

(*NOTE:* Handling time increases for parts of irregular configuration, for clamping on face plates, or in special fixtures with locating pads and individually operated holding elements.)

of the total operational time representing the sum of all the specifically calculated time elements; that value is based on an assumed performance rate of 75 per cent.

Handling Times for General-Purpose Cylindrical Grinding Operations

Time study, often detailed to the degree of motion studies, is sometimes used as a reliable method for establishing handling times in general manufacturing practice, and also in cylindrical grinding. Records of such studies, applied directly or by means of interpolation, represent information on which dependable time estimates for new jobs can be based.

For cases where such recorded information is not available, or is not applicable to the planned job, general values, tabulated according to workpiece characteristics, can prove useful for estimating purposes. The values in the chart, "Time Studies for Cylindrical Grinding," represent approximations only and are, of course, not valid for all circumstances.

Gaging

In most hand-controlled operations the gaging of the part while it is on the machine is a necessary element of the process. While the micrometer caliper is the basic measuring instrument for grinding operations in the toolroom, diameter measurements for production work are usually carried out with snap gages of the fixed, or preferably, of the indicator type. The time values in the chart. "Handling Times for Checking Work Diameters," refer to checking the work diameter with snap gages and distinction is made between two frequently used product accuracy grades.

Grinding-wheel Truing and Dressing

In cylindrical grinding the truing of the wheel is needed at intervals whose length depends on factors related to the performance and wear of the wheel, and also on the accuracy requirements of the operation. Disregarding the rather exceptional operations which require wheel truing for each single piece, truing will usually be carried out after a certain number of pieces have been ground. Consequently, the time spent on truing will have to be distributed over the applicable number of pieces, each carrying its share of the periodically executed truing.

For calculating the truing time, the pertinent process factors have first to be established, namely the feed rate and the depth-of-cut (infeed) of the diamond. The wheel surface layer to be removed in the truing process divided by the depth-of-cut will then determine the number of passes needed.

Unless the wheel face is damaged or has a profile form causing uneven wear, the average thickness of the layer to be removed in each regular truing will be in the order of 0.002 to 0.004 inch (0.05 to 0.10 mm). A thicker layer will have to be removed in the first truing operation of a newly installed wheel, but that is usually considered to be a part of the machine setup process.

The depth-of-cut in truing, that is, the penetration of the diamond into the grinding wheel, is usually not greater than 0.001 inch (0.025 mm), frequently substantially less, down to about 0.0002 inch (0.005 mm), for producing a very fine finish on the wheel surface or duplicating an intricate profile. For improving the finish or form compliance of the wheel face, a "dead pass" is frequently added; that is, a pass without diamond infeed.

The advance rate of the diamond along the face of the wheel can best be specified in inches per wheel revolution. Its actual value will depend on the desired finish, improved by lower feed rate, or on the desired free-cutting properties of the wheel, which may be assisted by increasing the feed rate of the diamond. In general practice the feed rate of the truing diamond will be within a range of 0.002 to 0.006 inch (0.05 to 0.15 mm) per wheel revolution.

The following example should illustrate the application of the preceding values:

Time Values for Checking Work Diameters

Accuracy Grade	Feature Size Range, Diam, in in. (mm)	Time for Each Gaging Position, in mins.
7 (equivalent to IT7)	0.400 to 2 (10 to 50)	0.2
	over 2 to 8 (over 50 to 200)	0.2 to 0.4
5 (equivalent to IT5)	0.400 to 2 (10 to 50)	0.3
	over 2 to 8 (over 50 to 200)	0.3 to 0.7

Wheel size: 20 inches in diam. $\times$ 2 inches wide.
Wheel speed: 6500 fpm, equivalent to 1240 rpm.
Depth of the removed layer: 0.0025 inch (0.064 mm). Depth-of-cut: 0.0005 inch (0.0127 mm). Number of passes: 5, plus a dead pass, resulting in 6 passes.

The theoretical length of the total diamond travel will be $6 \times 2 = 12$ inches.

Selecting a diamond advance (traverse feed rate) of 0.003 inch (0.0076 mm) per wheel revolution, produces at 1240 rpm wheel speed, a diamond travel rate of 3.720 inches per minute (0.094 meter per minute)

Consequently, the net truing time based on the diamond travel along the wheel face will be $12/3.720 = 3.22$ minutes.

In this example, should the grinding wheel require truing after the grinding of 20 parts, the time spent on truing will add $3.22/20 = 0.161$ minutes to the calculated piece time.

Setup Time

For time-estimating purposes the setup time should be distributed over the number of pieces in a lot for which a new setup is needed. Carrying out a new setup may involve assistance on the part of a machinist, the toolroom, the gage room, etc. However, the time spent by these service functions is generally contained in the operating burden of the shop, consequently it is not charged to the individual work order.

The following approximate values, intended purely for informative purposes, refer to the time spent by the machine operator of a new setup for general cylindrical grinding work. A grinding wheel change may be involved, but it is assumed that an exchange wheel is available, installed on a mount and properly balanced.

Grinding wheel diameter range, in. (mm)	Time spent on a new setup, in minutes
to 12 (300)	10 to 15
over 12 to 20 (300 to 500)	20 to 30
over 20 to 30 (500 to 750)	30 to 45

Setup should include the installation of the appropriate work-holding devices, the setting of speed and feed rates, as well as the limit positions for rapid approach and power infeed and, in the case of traverse grinding, the adjusting of the table reversal dog positions. Also included are the initial truing of a newly installed grinding wheel and, perhaps, the grinding of a first part for checking the correctness of the setup.

The above quoted informative values do not apply to the setting up of cylindrical grinding machines for semi- or fully automatic operations. Such tasks are usually assigned to specialized setup men, and the time involved may vary from one to several hours.

CENTERLESS GRINDING

The Principles and Characteristics

In centerless grinding the workpiece is introduced between two wheels, the grinding wheel and the regulating wheel, both rotating in the same direction but at different speeds. At the same time the workpiece is supported from below by a fixed work-rest blade. A component of the grinding force, acting in a direction toward the work-rest blade, tends to impart a rotating motion to the workpiece while it is also in contact with the much slower rotating regulating wheel. The regulating wheel acts as a brake which prevents the too rapid rotation of the workpiece and, at the same time, causes the work to rotate at a peripheral speed to match its own.

In addition to the slower peripheral speed of the regulating wheel, other factors will, in the majority of centerless grinding setups, also contribute to the frictional force acting upon the workpiece, thus further controlling its rotational speed:

(a) The bond of the regulating wheel which is usually more resilient (e.g., rubber bond) than that of the grinding wheel, thus extending the area of contact between the work and the regulating wheel.

(b) The top angle of the work-rest blade, sloping toward the regulating wheel, introduces a force component which increases the friction between the work and the regulating wheel.

By gradually decreasing the gap between the two wheels the diameter of the rotating workpiece will be reduced through the abrading action of the grinding wheel, producing an essentially round shape of uniform diameter around its entire periphery. The fundamental concepts of centerless grinding are illustrated diagrammatically in Fig. 2-1.1, presenting a simplified, and, for purposes of illustration, a somewhat out-of-scale arrangement of the principal elements.

In centerless grinding the location of the work axis is not constant in relation to the fixed elements of the grinding machine. The work is supported on its outer surface, concurrently with the machining of that same surface for reducing its diameter. The changing diameter of the supporting work surface causes a continual change in the position of the work axis.

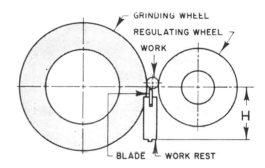

Fig. 2-1.1 The basic elements of the centerless grinding process. The diagram is drawn to a scale which shows an example of the relative arrangement of the principal participating members.

The difference is obvious between the conventional center type cylindrical grinding with the work rotating around a fixed axis, generally related to, and often actually represented by two centers, and the centerless grinding in which the concept of a fixed axis is replaced by that of a controlled diameter. While for producing round surfaces the conventional methods, such as lathe turning and cylindrical grinding maintain a fixed work axis, the centerless grinding method may be characterized as operating with a *floating work axis*.

The Basic Systems of Centerless Grinding

A traditional and informative way of distinguishing between the more frequently used operational systems of centerless grinding is related to the translational movement of the workpiece between the wheels during the process.

That movement may be a continuous traverse, the work passing in an axial direction through the area bounded by the peripheries of the grinding wheel and the regulating wheel; a method known as *thrufeed grinding.*

In another system the work performs no traverse movement at all. At the start of the operation the workpiece is introduced to a fixed lateral position between the wheels while one of them is temporarily retracted. After the return of the retracted wheel to the operating position, the gap between the wheels is gradually reduced by a feed movement directed radially into the work, hence the designation *infeed grinding.*

In the two other basic methods of centerless grinding, the work performs a limited traverse movement:

(1) The work traverses from the start of the grinding until it reaches a fixed lateral position, a method called *endfeed grinding.*

(2) The operation starts like an infeed process which, after having reached the full depth of the intended wheel penetration, changes over to a thrufeed process till the work totally emerges from between the wheels, a method distinguished as *combination infeed and thrufeed grinding.*

Table 2-1.1 contains a comparative survey of these basic methods .

Actually, there is still another method of centerless grinding which differs from the more conventional types in the equipment it uses and in some other respects as well. However, because of the fixed axial location of the workpiece during the grinding, this method conforms to the principles of infeed grinding. Designated as *shoe type centerless grinding*, it will be discussed separately in Chapter 2-5 of this section.

Thrufeed Grinding

Thrufeed grinding is that centerless grinding method which, when applicable, may present the most substantial advantages over alternate processes, primarily because it is continuous, a characteristic seldom found in other types of metalworking methods. Centerless thrufeed grinding also provides, without the need for auxiliary devices, the translational movement of the work at a preselected traverse rate, which is generally maintained throughout the process

with only insignificant fluctuations. Thrufeed centerless grinding also utilizes, in an essentially uniform manner, the entire periphery of the grinding wheel, a condition beneficial to wheel life, size control, and grinding efficiency. It also reduces the frequency of required wheel truing. One of the limitations of thrufeed centerless grinding is that, in principle, it is only adaptable for cylindrical parts or, when these are not uniformly cylindrical, the offset surface sections have to be recessed with respect to the basic cylindrical envelope surface.

Such limitations can, in some cases, be bypassed by means of special tooling which causes the workpiece to act as if it were of cylindrical form.

Infeed Grinding

The second most extensively used centerless grinding method is infeed grinding which, however, is not entirely continuous and, unless operated manually, requires mechanized loading and feed-actuating devices. On the other hand, infeed centerless grinding offers great flexibility of adaptation to workpiece configurations and work surface profiles, besides providing most of the operational advantages which characterize the centerless grinding methods in general. While the surfaces ground by this method must also have circular cross section, rotational surfaces of widely varying profiles and diameters, within the capacity of the individual machines, can be ground. Also parts of nonsymmetrical or unbalanced configurations can be ground on their rotational surfaces by infeed centerless grinding when using appropriate fixturing. The tooling of centerless grinders for parts of non-cylindrical configuration, particularly when ground in a mechanized process, can be quite expensive. However, when warranted by the number of parts to be produced, centerless infeed grinding often proves to be the most economical and dependable process for the required grinding operation.

Infeed grinding is also the proper method for cylindrical parts which, with respect to their general shape, are well suited for thrufeed grinding, but, the specified stock removal would require several thrufeed passes. Multiple passes on a single machine would call for repeated reprocessing of batches to perform, e.g., roughing, semifinishing and finishing, or the use of a series of grinding machines, interlocked or arranged in line. Repeated reprocessing is often uneconomical and using a battery of centerless grinders for a specific type of part is only advised in the case of high-volume production. Hence, a frequently selected alternative in such cases is infeed centerless grinding, which permits substantial stock removal by

Table 2-1.1 Basic Systems of Centerless Grinding

DESIG-NATION	DIAGRAM	PRINCIPLES	APPLICATIONS
THRU-FEED CENTER-LESS GRIND-ING	GRINDING WHEEL WORK REST GUIDES WORK REGULATING WHEEL	The work passes in a continual traverse movement through the gradually narrowing gap between the grinding wheel and the regulating wheel, advanced by the axial force component on the surface of the rotating regulating wheel whose axis is set at a specific angle of inclination in relation to the grinding wheel axis.	Used primarily for parts with straight cylindrical surface, which may be interrupted by recessed areas of annular shape, but axial grooves or protruding surface sections cannot be tolerated. The length of the part can exceed the width of the grinding wheel, such as in bars. Short cylindrical parts are ground consecutively in a continuous stream, several parts being in simultaneous engagement with the wheels. The system providing the highest degree of productivity, best wheel utilization and excellent size control, also well suited to automation. Sometimes several passes are needed to remove large stock allowance or to separate roughing from finishing.
INFEED CENTER-LESS GRIND-ING	GRINDING WHEEL END STOP AND EJECTOR WORK REGULA-TING WHEEL	The work is placed on the rest blade in the area between the two wheels while one of the wheel slides is in a retracted position. A rapid approach of the wheel slide precedes the actual stock removal which then is continued by a gradual infeed and final sparkout in a process comparable to plunge grinding on a center type cylindrical grinding machine, until the finished work size is attained. The infeed movement is actuated by hand (lever) or by power in automatic cycle which may include work handling too.	Applied for parts which are not suitable for thrufeed grinding because of portions larger than the diameter being ground, such as heads, shoulders, etc. Also adaptable to the simultaneous grinding of several different diameters or of various profiles on parts whose length does not exceed the width of the grinding wheel. The axis of the regulating wheel is often inclined by a small amount to impart to the work a light axial force holding it in steady contact with the end stop, which may also function as an ejector at the end of the cycle Complex work profiles or tight finish tolerances may require consecutive roughing and finishing infeed grinding operations.

(Continued on next page.)

Table 2-1.1 *(Cont.)* **Basic Systems of Centerless Grinding**

DESIG-NATION	DIAGRAM	PRINCIPLES	APPLICATIONS
END-FEED CENTER-LESS GRIND-ING	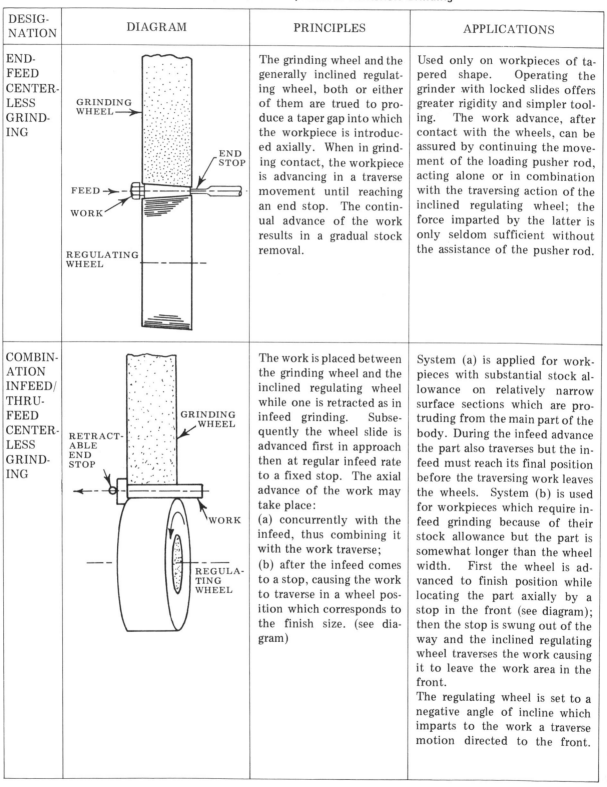	The grinding wheel and the generally inclined regulating wheel, both or either of them are trued to produce a taper gap into which the workpiece is introduced axially. When in grinding contact, the workpiece is advancing in a traverse movement until reaching an end stop. The continual advance of the work results in a gradual stock removal.	Used only on workpieces of tapered shape. Operating the grinder with locked slides offers greater rigidity and simpler tooling. The work advance, after contact with the wheels, can be assured by continuing the movement of the loading pusher rod, acting alone or in combination with the traversing action of the inclined regulating wheel; the force imparted by the latter is only seldom sufficient without the assistance of the pusher rod.
COMBIN-ATION INFEED/ THRU-FEED CENTER-LESS GRIND-ING		The work is placed between the grinding wheel and the inclined regulating wheel while one is retracted as in infeed grinding. Subsequently the wheel slide is advanced first in approach then at regular infeed rate to a fixed stop. The axial advance of the work may take place: (a) concurrently with the infeed, thus combining it with the work traverse; (b) after the infeed comes to a stop, causing the work to traverse in a wheel position which corresponds to the finish size. (see diagram)	System (a) is applied for workpieces with substantial stock allowance on relatively narrow surface sections which are protruding from the main part of the body. During the infeed advance the part also traverses but the infeed must reach its final position before the traversing work leaves the wheels. System (b) is used for workpieces which require infeed grinding because of their stock allowance but the part is somewhat longer than the wheel width. First the wheel is advanced to finish position while locating the part axially by a stop in the front (see diagram); then the stop is swung out of the way and the inclined regulating wheel traverses the work causing it to leave the work area in the front. The regulating wheel is set to a negative angle of incline which imparts to the work a traverse motion directed to the front.

the gradual infeed of one of the wheel slides; the whole operation often being finished in a single stage.

Operational Characteristics of Centerless Grinding

In appraising centerless grinding as an alternative for comparable processes, a review of the principal characteristics of this method can provide a useful background.

Many of these characteristics can have such favorable effects on the performance and economics of the planned operation as improved productivity due to increased stock removal rates and the reduction of nonproductive time, upgraded accuracy, and other quality-related properties of the workpiece, simplification of the setup and the avoidance of the need for special tooling.

There are various other unique characteristics which make centerless grinding the unquestionably superior, and even the only technically applicable method for certain operations.

To assist in appraising the benefits which can be derived from the properly engineered application of centerless grinding, a list of potential advantages has been compiled in Table 2-1.2. While there is not a single type of operation in which all these characteristics can be utilized, a comprehensive review of this kind should contribute to the better appraisal and the judicious application of the centerless grinding method, in general, and its most suitable operational system in a more particular sense.

The Rounding Effect of Centerless Grinding

The uniform work diameter, to which reference was made in the preceding discussions as one of the performance characteristics of centerless grinding, is not, however, equivalent with roundness. A body of uniform diameter may have the shape of a polygon with rounded corners which will, in every orientation, fill-in the space between two parallel surfaces, thus satisfying the concept of equal diameter, yet without being round (Fig. 2-1.2). Roundness expresses a geometric condition in which the cross sectional contour of the body is represented by a circle, that is, a figure having every single point on its periphery at an equal distance from a common imaginary point, called the center.

Assume that an out-of-round workpiece is introduced into a centerless grinder where it is supported on the flat top of a blade at a level in which its center is on the centerline of the two wheels. The diagram in Fig. 2-1.3 shows that simplified, but generally im-

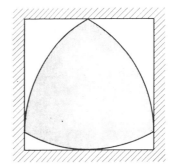

Fig. 2-1.2 Cross-sectional diagram of an imaginary body which has equal diameters in every orientation, but is still out-of-round.

practical, setup in which the development of the following conditions could be expected:

A high spot on the periphery of the out-of-round work, when contacting the regulating wheel will cause the part to be pushed toward the grinding wheel into a position, which, due to increased stock removal, will result in a depressed area opposite the high spot. Conversely, when the depressed area on the work periphery gets into contact with the regulating wheel, the fixed gap between the wheels will permit the work to recede from the grinding wheel, leaving a high spot diametrically opposite the low spot. This process will continue, repeating itself in different orientations around the work periphery, until the operation is terminated because the specified work diameter has been reached. The resulting workpiece will have the desired diameter uniformly around its entire surface, yet will not be round.

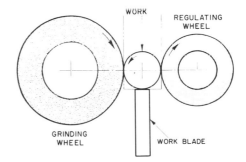

Fig. 2-1.3 Simplified centerless grinding setup with the work resting on the flat top of the blade and the work center coinciding with the plane of the wheel centers. Such a setup is generally impractical because of the poor rounding effect.

However, in many mechanical parts, it is frequently mandatory that, in addition to uniform diameter, the ground part shall also be round. Two factors in the centerless grinding process contribute to achieving a progressive rounding effect:

Table 2-1.2 Potentially Advantageous Process Characteristics of Centerless Grinding

DESIGNATION	DESCRIPTION	DISCUSSION AND POTENTIAL BENEFITS
Extended work support	A three-line support of the work, in most cases along its entire length, is provided by the two wheels and the work-rest blade, without the need of the axial thrust required when the work is held between centers.	Avoids deflections of the work and makes the process adaptable to long and slender parts, like pins, spindles, etc., which can thus be ground to equal diameter along their entire length.
Continuous operation	In thrufeed grinding the parts are passing through the grinder in a continuous stream. Even in infeed grinding the process can be carried out in a continuous sequence of repetitive operations with very short interruptions for loading and ejection.	Losses in productive time due to work loading, clamping, and unloading are entirely eliminated in thrufeed, and substantially reduced in plunge (infeed) grinding. Several parts, to an aggregate length equal to the wheel width, can be ground simultaneously.
Size control of the work diameter	For centerless grinding the change of the work diameter is equal to the distance by which one of the wheel slides is advanced; while for cylindrical grinding, where the wheel position affects the radius of the work, the advance of the wheel over a given distance causes twice as great diameter changes of the work.	The reduced sensitivity of the work diameter to the adjustments of the size-controlling machine elements contributes to the particular suitability of the centerless grinding system for work requiring precise size control.
Floating work support	The floating support of the work in centerless grinding eliminates the need for fixtures with clamping action, like those widely used in other conventional processes when grinding similar parts.	Dispensing with the initial and maintenance costs of clamping fixtures, and avoiding the time which their operation involves, may result in substantial savings and increased productivity in comparison with the fixed center type cylindrical grinding.
Simple setup	The setup variables, while critical for the proper execution of the process, are generally easy to establish, in most cases simply by the adjustment of standard machine elements.	Positioning the wheel slides and the work rest, adjusting swivel and tilt settings according to scales mounted on the machine, often represent the major functions in the setup of centerless grinders which thereafter can provide semiautomatic operation.

Table 2-1.2 *(Cont.)* **Potentially Advantageous Process Characteristics of Centerless Grinding**

DESIGNATION	DESCRIPTION	DISCUSSION AND POTENTIAL BENEFITS
Consistently uniform performance	The limited number of machine elements which directly participate in the retention, rotation, and traverse movement of the work, in combination with the basically sturdy and rigid design of the centerless grinding machines—generally operating with large size grinding wheels—are factors of the uniquely consistent size-holding capability.	Having a basic design free from the inherent inaccuracies of work holders, the runout of work spindles, and centers and the possible deflections of an insufficiently supported workpiece, and requiring only limited in-process movements of machine slides, are important characteristics which make the centerless grinding system well suited for producing tightly toleranced parts.
Dispensing with the preparation of the workpiece	The basic three-line work support assures the positive and uniformly repetitive position of the work in the machine, without the need for developing auxiliary surfaces for work location.	Drilling and grinding center holes, mounting drivers, grinding support sections for steady rests, adjusting and possibly following up during the work with the steady rest position, are examples of operational steps which are never required in centerless grinding.
Assured concentricity, even for re-positioned parts	The self-centering effect of the work positioning in centerless grinding, eliminates the need of separate work-centering operation with its potential errors, and assures both the uniform stock removal around the work surface and the co-axial location of interrelated surface sections.	The self-acting, re-positioning capacity, which assures the duplicated positioning of the work's rotational axis, is one of the favorable characteristics of centerless grinding in comparison to the fixed-center type grinding methods, particularly for parts which must be ground concentric to previously produced surfaces.
Mechanized loading easily added	By avoiding the need for clamping in work holders and locating in fixtures, the centerless grinding method makes the loading and ejection of parts relatively simple operational steps.	Upgrading processes from manual to automatic by completing the cycle with mechanized and unattended work handling, from magazines, hoppers, or feed lines connected with preceding work stations, is more easily accomplished in centerless grinding than in any other grinding method.
Gangs of grinders for continuous multi-pass cylindrical grinding	The continuous flow of work through the centerless grinder in the thrufeed process permits the setting up of several machines in a row for progressive stock removal from work of cylindrical shape.	Particularly advantageous in the grinding of bars, as well as of shorter cylindrical workpieces, which can be traversed continuously, but require stock removal in amounts exceeding the capacity of a single grinding pass. The in-line arrangement of a battery of centerless grinders performs a series of passes without additional work handling.

(Continued on next page.)

Table 2-1.2 *(Cont.)* **Potentially Advantageous Process Characteristics of Centerless Grinding**

DESIGNATION	DESCRIPTION	DISCUSSION AND POTENTIAL BENEFITS
Flexibility of adaptation for different part configurations	Special fixturing, frequently available as developed optional accessories, make centerless grinding adaptable for many parts of irregular shape and nonsymmetrical configuration.	Different types of truing devices for both wheels, feeding and loading fixtures, special hold-down devices and work supports for out-of-balance parts, and axial loading and grinding attachments, are a few examples of accessories available for extending the applications of centerless grinding machines.
Applicable for diverse work materials	The rigid and extended work support, the distribution of the grinding force over the entire length of most types of parts, together with the high production rates, make centerless grinding a method economically applicable for a wide range of work materials.	In addition to all types of steels and non-ferrous metals, centerless grinding is also applicable to brittle materials such as glass and ceramics, soft materials such as wood, plastics, and rubber, and extremely hard materials such as cemented carbides.
High productivity	Examples of factors contributing to high productivity are: the elimination of idle time in thrufeed grinding, the use of generally large and wide wheels often in simultaneous contact with the full length of the workpiece, rigid work support permitting substantial stock removal rates, etc.	The high productivity of centerless grinding in comparison with other methods is particularly evident in the continuous large scale production of tightly toleranced parts which utilize the full width of the centerless wheels either in thrufeed grinding or in the infeed grinding of long or multiple parts.
Adaptable for completely automatic operation	The major elements which characterize automatic operations on centerless grinders are continuous work feeding, mechanized loading and ejection, self-acting work positioning, and size control with automatic compensation for wheel wear.	Centerless grinders either provide these elements of complete automation as inherent in the system, or are easily adaptable to accept as complementary attachments readily available accessories which are suitable for a wide variety of workpieces.
Form grinding with simple tooling	The wide grinding wheels for which centerless grinders are generally designed, together with the extended support of the work by both the regulating wheel and the work rest, are conducive to applying plunge type form grinding on standard types of centerless grinding machines.	The truing of the grinding wheel to the desired profile can be carried out by means of a single diamond guided by a profile bar, by means of crush dressing, or by applying rotary diamond dressing, with the aid of fixtures supplied as readily available optional accessories. The regulating wheels are trued to provide full or selective work support.

Table 2-1.2 *(Cont.)* **Potentially Advantageous Process Characteristics of Centerless Grinding**

DESIGNATION	DESCRIPTION	DISCUSSION AND POTENTIAL BENEFITS
Rigidity of the interacting machine elements and setup components	The basic design of the centerless grinding machine, particularly when operating with generally locked slide movements in thrufeed grinding, provides an inherent rigidity of the machine members and work support which is seldom approached by any other grinding machine system.	The rigidity of the functional elements permits greater stock removal rates, thus raising productivity and, at the same time, assures consistent control of the ground work with respect to geometric form, dimensional accuracy, and surface finish.

(a) Holding the work center above the centerline of the wheels

(b) Using a work-rest blade with a top angle greater than zero, commonly, 20 to 40 degrees.

The effect of these factors is illustrated in Fig. 2-1.4, and explained in the following:

When a high spot on the periphery of the rotating workpiece reaches the grinding wheel, an increased radial force will act on the work, tending to push it away. The work, while receding from the grinding wheel, will also move down on the sloping top of the work-rest blade, thereby approaching the plane of the wheel centers where the gap between the wheels is the smallest, thus the work is forced with greater pressure toward the grinding wheel. This condition increases momentarily the rate of stock removal from the high spot on the work periphery.

When a high spot is contacting the regulating wheel, the work will be forced toward the grinding wheel, but also up on the slope of the work-rest blade to a level where the gap between the wheels becomes wider. Thereby the tendency toward increased stock removal opposite the high spot is moderated, or does not take place at all.

Furthermore, because the center of the workpiece is above the wheel centerline, the two wheels are not making contact with the work at its diametrically opposite sides, a condition which causes a continual shift in the relative locations of the high and low spots on the work periphery, a process often referred to as "phasing."

The gradual cutting down of the high spots and, at the same time, counteracting the development of low spots, tends to improve the roundness of the work while progressively approaching the desired finished diameter. Although lengthy to describe, the progress of the process is very fast. Often within a few seconds of grinding time the roundness of some parts may be improved to a fraction of the starting value. As a rule of thumb it may be expected that with

properly selected setup parameters each revolution of the work will cause the out-of-roundness to diminish by about five per cent.

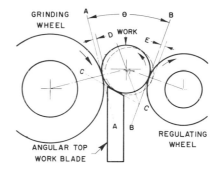

Fig. 2-1.4 Diagram showing the operating principles of the rounding process in centerless grinding. *A-A* and *B-B* are tangent planes to the wheels at the points of work contact, and *C-C* is the plane of the work-blade top. Angle θ bounded by the tangent planes (the "throat angle"), together with the blade-top angle, are major factors in achieving the rounding action. *D* and *E* are *not* in diametrically opposite positions.

As a general rule the following measures may serve to accomplish and to accelerate the improvement of the roundness conditions of an originally out-of-round workpiece:

(a) Setting the work above the plane of the wheel centers — limited by a condition where the work may tend to lift off the blade because the force component holding the work on the rest blade has been excessively reduced.

(b) Increasing the top angle of the work-rest blade—limited particularly in the case of long blades such as those used for wide wheels and thin blades needed for small work diameters, because of vibrations caused by the increased lateral force acting on the blade.

(c) Assuring a relatively high number of work revolutions during the grinding operation. This objective can be achieved by:

Increasing the work speed for faster rounding effect — a measure limited by the stability of the work, the desired degree of surface finish, etc.

Reducing the work-traverse rate in thrufeed grinding, or reducing the wheel infeed rate in infeed grinding, thus postponing the attainment of the finished size and allowing a higher number of work revolutions during the process. However, either of these measures involves a proportionate extension of the grinding time, and consequently, a reduction of the production output.

It may be concluded that the centerless grinding method provides the means for producing work surfaces of excellent roundness, by gradually diminishing the original out-of-roundness of the workpiece. The improvement of the roundness is predicated on properly selected setup parameters. (These will be discussed in greater detail in Chapter 2-8 of this section, in conjunction with the setup of centerless grinding machines.) The degree and the rate of the achievable roundness improvement are controlled by technological factors, the choice of some of which may be influenced by considerations related to productivity.

The Adaptability and Capabilities of Centerless Grinding

Typical Applications of Centerless Grinding Machines

Centerless grinding is essentially a production process, in distinction to processes serving supporting or auxiliary objectives, like tool and fixture making, repairs, replacement parts, etc. While high volume production is the most rewarding field for centerless grinding, the ease with which simple machine setups can be carried out often justifies the use of centerless grinding for relatively small lots, particularly when the achievable product or productivity advantages outweigh the time expended on setting up the grinder for a limited number of parts. The photographs of a

Courtesy of Landis Tool Co.

Fig. 2-2.1 Grinding of a bar in thrufeed centerless grinding using attached stands with guide rollers to support the ingoing and the outgoing work.

few typical centerless grinding operations are shown in Figs. 2-2.1 to 2-2.4.

For a quick appraisal of whether the work is technologically adaptable for centerless grinding, a survey of work configurations and operations for which this method is frequently applied, can prove helpful. In

Courtesy of Landis Tool Co.

Fig. 2-2.2 Centerless plunge grinding a section at one end of a long flanged shaft: an automobile rear axle. A previously ground cylindrical section at the flange end is supported on rollers installed on an extension arm, which also provides clearance for the still unfinished flange.

order to facilitate such a preliminary appraisal of adaptability, characteristic types of operations have been compiled in Table 2-2.1. The sequence of listing in this table does not represent an order of importance or frequency of usage. While characteristic

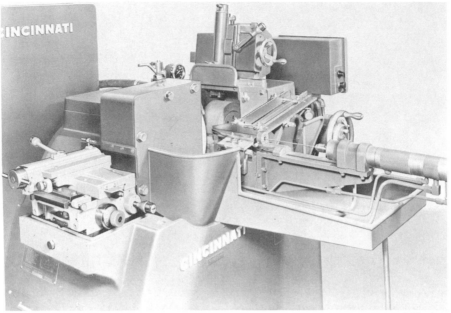

Courtesy of Cincinnati Milacron, Inc.

Fig. 2-2.3 Infeed centerless grinding of small straight shank twist drills to a profile comprising the cylindrical shaft and the cutting section with back taper. Magazine feeding and mechanical loading operate in combination with the automatic infeed of the wheel slide.

examples for many types of operations are mentioned, the list is not complete and should be considered as indicative only. Varieties of the listed applications and particularly a large number of special uses, often

Courtesy of Cincinnati Milacron, Inc.

Fig. 2-2.4 Infeed centerless grinding of groups of large roller bearing rings, combining the rough and finish grinding into the automatic cycle of a single operation. Special hand-operated fixture, at the left, removes the group when finished and will then load another group of workpieces into the grinding position.

with the aid of special tooling, are frequently applied, particularly in industries with substantial production runs. Only a very limited number of examples of such special workpieces and setups for which centerless grinding is successfully used, are indicated in the table.

In essence, centerless grinding should always be considered as a possibly advantageous method for the machining, and particularly finishing, of surfaces of revolution, when one or several of the following conditions are present:

1. High volume production
2. Parts are of generally cylindrical shape and suitable for thrufeed grinding
3. Parts are without centers or corresponding holding surfaces
4. Concentricity control on parts with several diameters is critical
5. High diametrical accuracy, combined with rigorously controlled roundness of the worked surface is required
6. Need for advanced automation, including feeding, loading and gaging with feedback for automatic size control
7. A series of operations or several passes are to be performed in a continual process on several grinders arranged in tandem, i.e., several in-line or interlocked by feed lines.

Table 2-2.1 Examples of Typical Centerless Grinding Operations

DESIG-NATION	DIAGRAM	DESCRIPTION	APPLICATION
THRUFEED GRINDING OF BARS		The work is supported on the work rest blade and restrained by the work guides while being introduced axially between the wheels. The inclined regulating wheel controls the rotation and, concurrently, imparts the traverse movement to the work.	Centerless thrufeed grinding is the best known method for the grinding of long bars and tubes in a continuous process. When the amount of stock allowance exceeds the capacity of a single pass, savings in material handling are achieved by operating two or more centerless grinders arranged in line, the work passing through successive units.
THRUFEED GRINDING OF SHORT CYLINDRICAL PARTS		Essentially identical with the bar grinding, except that a succession of parts, possibly as an uninterrupted stream, is ground instead of a solid bar. An appropriate loading device, such as an inclined chute, is needed to assure the continuity of the work feed.	Offers high productivity due to loading while grinding and by utilizing the entire wheel face in continual engagement with the work. The rigidity of the setup and the generally slow wheel wear assure excellent size control. Work is ground in a single pass, or in several passes, subsequently or in succession when passing continually through several machines.
INFEED GRINDING OF PLAIN CYLINDRICAL PARTS		One of the wheel slides is retracted for loading to widen the grinding area into which the work is introduced either axially (pusher rod) or from above (work loader) to bear against an end stop. An approach, followed by the gradual infeed of the wheel slide to a fixed stop, with dwell for sparkout, produces the finish work size.	Infeed grinding is used for plain cylindrical parts when the stock allowance exceeds the single pass thrufeed capacity. A substantial amount of stock can be removed in a single infeed operation which, in the case of automatic control, may comprise the roughing and finishing phases, each at different feed rates, followed by sparkout and slide retraction.

(Continued on next page.)

Table 2-2.1 *(Cont.)* Examples of Typical Centerless Grinding Operations

DESIG-NATION	DIAGRAM	DESCRIPTION	APPLICATION
INFEED GRINDING OF THE CYLINDRICAL SECTION OF PARTS WITH A HEAD.		Process similar to the preceding, however, special work rest is needed and the work loading is generally limited in choice, relying on work loaders with grippers.	The configuration of the part requires infeed grinding independently of the amount of stock allowance. When the cylindrical section is several times longer than the overhanging head, the restraining forces acting on the worked section are sufficient for keeping the part in a stable position.
INFEED GRINDING OF SEVERAL CYLINDRICAL SECTIONS WITH DIFFERENT DIAMETERS		Operated by using a single grinding wheel of proper width, or several wheels on a common spindle; this latter mode is preferred when substantial differences in diameter have to be balanced or the worked sections are far apart. Wheel truing with the use of profile bar (cam).	Time saving in production and consistently high accuracy with respect to concentricity and diametric relationship of the ground surfaces can be achieved by this method which is adaptable for parts whose length is within the width of the grinding wheel and with relieved transitions between the sections having different diameters, to permit truing by cam and follower.
INFEED GRINDING OF SECTIONS STRADDLING A PROTRUDING CENTER PORTION		Two pairs of grinding and regulating wheels are used, mounted with spacers to permit the unobstructed rotation of the hub separating the sections being ground, which may also comprise adjacent fillets. A special work rest is needed to avoid interference.	A typical example is the spider of universal joints, centerless ground in two operations, around mutually perpendicular axes. Other parts are, e.g., small crankshafts, with the cylindrical sections not necessarily symmetrical. The avoidance of using a work driver is the major advantage of grinding these parts by the centerless method.
INFEED GRINDING OF PROFILED WORK SURFACES		The grinding wheel is trued to the inverse contour of the required work shape, while the regulating wheel may have a partial profile for a selective work supporting contact. The shape of the work rest, too, must be adapted to provide adequate support.	The grinding of the profile along the entire work length is advantageous with respect to both productivity and product accuracy. Modern centerless grinders with wide wheels extend this method's field of application to single as well as to multiple part grinding.

Table 2-2.1 *(Cont.)* **Examples of Typical Centerless Grinding Operations**

DESIG-NATION	DIAGRAM	DESCRIPTION	APPLICATION
INFEED FORM GRINDING WITH CRUSH TRUED OR ROTARY DIAMOND TRUED WHEEL		The shape of the grinding wheel is produced by a freely rotating crush roll made of hardened alloy steel or cemented carbide, while the wheel is running at reduced speed. The diamond impregnated rolls are rotated by a drive motor and true the grinding wheel running at operational speed.	Intricate profiles with sudden transitions, such as multiple narrow lands, cannot be produced by conventional profile truing using followers guided by cams. Crush truing, or the more recently developed rotary diamond truing, are adaptable to complex profiles and also substantially reduce the truing time.
INFEED GRINDING THE CYLINDRICAL SECTIONS OF NONSYMMETRICAL PARTS		The volume of the part section being ground should safely exceed that of the overhanging portion. Occasionally an upper guide, incorporated into the work loader, is used to assure the correct seating of the ground portion on the work-rest blade.	Another uncommon, yet successful application, presented as an example of the adaptational flexibility of the centerless grinding method. Small crankshafts of cantilever design are other examples of workpieces in this category of centerless grinding operations.
INFEED GRINDING OF RELATIVELY SHORT CYLINDRICAL SECTIONS ON LONG PARTS		The non-ground portion of the part is supported in precise alignment with the operating position in a manner permitting free rotation of the workpiece. Support rolls are in a bracket mounted on the grinder or a separate stand is attached to the machine, with individual drive for rotating heavy parts, at the surface speed of the regulating wheel.	Centerless grinding is often preferred for parts which are also suited for cylindrical grinding, such as freight car axles. That choice is based on any or several of the following reasons: higher productivity, simpler work loading, no need for center holes, avoiding the interference of work drivers, etc.

(Continued on next page.)

Table 2-2.1 *(Cont.)* **Examples of Typical Centerless Grinding Operations**

DESIG-NATION	DIAGRAM	DESCRIPTION	APPLICATION
INFEED GRINDING OF MULTIPLE PARTS		A single wheel, or several wheels on a common spindle, adjacent or with spacers, are fed into two or more relatively short parts, aligned in the operating area along a common axis of rotation. Wheel truing is by profile cam. The system may be designated as multi-station infeed grinding.	This centerless grinding system is adaptable to two different processes: (a) the parts are loaded in batches and have the same operation performed simultaneously; (b) a step-by-step development of the part profile is carried out in consecutive positions using a "walking beam" type of work-handling device, thus operating as a transfer grinding method.
ENDFEED GRINDING OF TAPERED PARTS		Functionally similar to infeed grinding, however the infeed is produced as the resultant of an axial advance into a tapered throat bounded by the two wheels, either or preferably both of which are trued to provide that taper. The diameter of the workpiece is controlled by the position of the end stop at the small end of the tapered work area.	The workpiece, introduced from the front, is advanced manually or by a mechanical device in an axial direction up to the end stop at a speed selected to control the rate of stock removal. The ratio of stock removal to axial advance is defined by the design angle of the tapered workpiece. Applications limited to slender tapers and the work face at the small end must be usable for referencing the size.
COMBINED INFEED AND THRUFEED GRINDING OF PARTS WITH NARROW WORKED SURFACES		The radial infeed movement is superimposed on the concurrently operating thrufeed traverse of the part between two parallel wheel contact lines, thereby distributing the stock removal load over most of the grinding wheel periphery. The work traversing force is imparted by the inclination of the regulating wheel.	Thrufeed grinding is preferable for parts with narrow worked surfaces to distribute the grinding action over the entire wheel face, for better wheel utilization and work size control. When the stock allowance exceeds the capacity of a thrufeed pass, the combined application of both methods can be the preferable process for workpieces with suitable dimensions and configurations.

(Continued on next page.)

Table 2-2.1 *(Cont.)* **Examples of Typical Centerless Grinding Operations**

DESIG-NATION	DIAGRAM	DESCRIPTION	APPLICATION
COMBINED INFEED AND THRUFEED GRINDING OF SURFACES LONGER THAN THE WHEEL WIDTH		The operation starts as a regular infeed grinding applied to one section of the part while an end stop restrains the traversing force of the inclined regulating wheel. After removal of the stop the work traverses along the wheel face exposing the unground section to the action of the grinding wheel.	Applied for parts which due to configuration require infeed grinding, yet the surface being ground is longer than the width of the wheel. Application is limited to parts with stock allowances within the capacity of the thrufeed pass which is the only one used for grinding the work section left untouched by the preceding infeed phase of the operation.
INFEED GRINDING FOLLOWED BY AXIAL WHEEL ADVANCE		A special device permits combining the regular infeed grinding, having a radial wheel advance, with feeding the wheel also parallel with the work axis. An axially movable wheel mount, replacing or complementing the fixed wheel mount is installed on the wheel spindle, with movement controlled by handwheel or hydraulic plunger.	Shallow shoulders, square or tapered, in precisely controlled relation to the centerless ground cylindrical section, can be face-ground by this device, without the need for an additional operation. For shoulder grinding, wheels with one end face trued to a saucer shape are generally used, and a special truing device is needed.
THRUFEED GRINDING OF TAPERED PARTS		Short tapered parts can be thrufeed ground by presenting them to the grinding wheel in a manner functionally equivalent to cylinders, thus offering a single line of contact. A steel drum with helical track of special configuration replaces the conventional regulating wheel.	The helical track drum which replaces the regulating wheel for providing a continually identical supporting base for the workpiece while compensating for the deviation from the cylindrical shape, also providing for the rotation and the translation of the work, is an essential equipment member for the high volume centerless grinding of tapered bearing rollers.

(Continued on next page.)

Table 2-2.1 *(Cont.)* **Examples of Typical Centerless Grinding Operations**

DESIG-NATION	DIAGRAM	DESCRIPTION	APPLICATION
THRUFEED GRINDING OF CROWNED CYLIN-DRICAL ROLLERS		All sections of a circular arc have identical curvature, consequently a workpiece passing, in an axial plane, along the face of a grinding wheel which is trued to a circular arc contour will be ground with a corresponding profile. A concave arc on the wheel face will result in a convex (crowned) work profile.	Workpieces with shallow (large radius) convexity, known as "crowned profile," are required for controlled stress distribution under load. Typical examples are bearing rollers, both cylindrical and tapered, having the crowning superimposed on the basic contour, and located symmetrically along the work surface, with the peak at equal distance from the opposite ends.
SHOE TYPE CENTERLESS GRINDING OF OUTSIDE PROFILES		The work is held on its face by a rotating magnetic backing plate and is radially supported by two shoes on which the rotating part is firmly seated by the wind-in action from a small offset between the axes of the plate and the work. The spindles of the wheel head and workhead are mutually parallel.	The basic version of the shoe type centerless grinding is preferred for disc shaped parts whose periphery must be ground precisely perpendicular to the face. It is of particular advantage for parts with profiled OD, such as ball bearing inner rings, and is applicable as long as a dependable peripheral surface is available for shoe contact.
SHOE TYPE CENTER-LESS GRINDING OF OD AND FACE OR SHOULDER		Setup similar to the preceding, except for the relative orientation of the work and wheel axes, at an angle such as in the angular approach cylindrical grinding. Commonly, the wheel head is swiveled to present the workpiece at an angle to the wheel, which is trued on two or more adjacent operating surfaces.	The benefits from combining two operations and producing mutually perpendicular surfaces with excellent squareness, are additional to the basic advantages of shoe type centerless grinding. Typical work examples are the raceways and shoulders of roller bearing inner rings. The truing device uses profile cams and frequently, multiple diamond heads for optimum diamond point orientation.

(Continued on next page.)

Table 2-2.1 (*Cont.*) **Examples of Typical Centerless Grinding Operations**

DESIG- NATION	DIAGRAM	DESCRIPTION	APPLICATION
SHOE TYPE CONJUGATE CENTER- LESS GRINDING		Joining together, as the designation term implies, the grinding of two or more coaxial surfaces by selecting an approach angle which will provide the optimum balance between work surfaces constituting functionally different work contact areas. For shoulders inclined by less than $90°$ the swivel angle of the wheel head will also control the wheel diameter.	Composite surfaces consisting of mutually inclined straight element sections, combined with adjoining corner radii, such as the entire operative outer surface of tapered roller bearing cones, are ground by this method. Complex wheel profiles can be trued by copying, or very efficiently with rotary diamond rolls.

Although the examples listed in the table are primarily products of metalworking industries, the review of possible or preferred applications would be incomplete without mentioning the wide areas of advantageous uses of centerless grinding, for workpieces made of nonmetallic materials such as glass, ceramics, stone, wood, plastics, rubber, etc. Whenever one or more of the above listed conditions apply to parts the machined portions of which are surfaces of revolution, centerless grinding may prove to be superior to alternate processes, nearly independently of the work material.

Special Applications of the Centerless Grinding Method

The basic applications, comprising the majority of uses for centerless grinding, operate by a system in which the workpiece, while being ground on its periphery, is supported on its entire length along three lines of contact with the grinding wheel, the regulating wheel, and the work-rest blade, respectively. In the thrufeed grinding of long bars, it is the work section being ground momentarily to which this definition applies.

There are also workpiece configurations which are not adaptable to the strict application of that basic system but, nevertheless, possess surface sections whose grinding can most effectively be done by the centerless method. The desirability of applying the efficient centerless grinding method is particularly pronounced for workpieces which are manufactured in large volumes, consequently they offer additional incentives for developing means permitting the use of centerless grinding.

Engineering efforts prompted by the potential advantages of applying centerless grinding also to workpieces of other than simple configurations, led to the development of many different machine setup systems and machine accessories. Some of these centerless grinding systems actually represent the predominant or only method by which certain types of widely used machine elements are being ground in industrial production.

A limited number of examples of special applications of centerless grinding will be discussed in the following with the purpose of conveying the principles of some well-proven approaches for utilizing the advantages of the basic method without being hampered by the partial absence of the fundamental conditions.

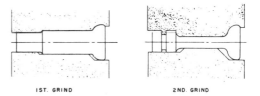

IST. GRIND   2ND. GRIND

Fig. 2-2.5 Diagram showing a workpiece with complex profile which is centerless infeed ground in two operations, developing the shape of the part to different degrees of detail. In each operation a regulating wheel, which supports the part only along selected sections of its surface, is used.

Partial support of the workpiece by the regulating wheel (Fig. 2-2.5). Workpieces representing bodies of revolution may have profiles comprising several sections of varying diameters and inclinations. Often, such parts can be ground successfully by centerless infeed grinding in which the grinding wheel only, is trued to the complete profile of the finished work, while the contact with the regulating wheel is limited to major sections of the part contour. When the workpiece has cylindrical surface sections of sufficient lengths and preferably in balanced locations, then a partial support of the work, permitting simple regulating wheel truing, is functionally adequate.

Unbalanced location of the surface being ground (Fig. 2-2.6). Workpieces with cylindrical or profiled sections at one end, which must be ground concentric with another previously finished section located at, or near, the opposite end of the part, can receive supplementary guidance outside the grinding area, by means of supporting rolls held in brackets which are mounted on the regulating wheel housing.

Consecutively ground sections on both ends of a long and heavy workpiece (Fig. 2-2.7). Typical examples for such applications of centerless grinding are freight car axles, which require an outboard support mounted apart from, but precisely aligned with the position of the work held in the centerless grinder. Such out-

Courtesy of Cincinnati Milacron, Inc.

Fig. 2-2.6 Centerless infeed grinding of a combined cylindrical and tapered surface section connected by a curved transition, all located along the surface of a part, the end section of which has been ground in a preceding operation. Concentricity between the successively finished surfaces is assured by holding the ground end section in a well-aligned outboard support.

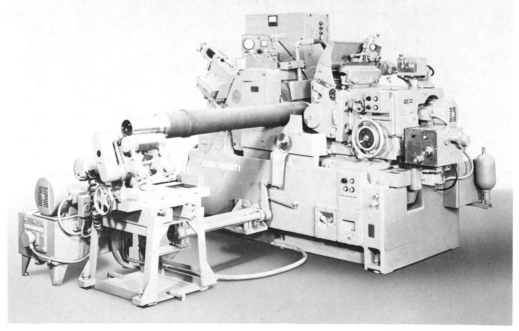

Courtesy of Cincinnati Milacron, Inc.

Fig. 2-2.7 The journals at both ends of railroad car axles are centerless infeed ground consecutively while holding the opposite end journal in a detached motor-driven support stand.

board supports consist of a properly dimensioned pair of rolls arranged to duplicate the effect of a V-block. One of the rolls is driven from a motor to provide a peripheral speed essentially equal to that of the regulating wheel.

Centerless infeed grinding of profiled sections straddling a central hub (Fig. 2-2.8). Parts of such configurations may have the sections to be ground located symmetrically, as in the spiders of universal joints, or the sections may be of different lengths and profiles. In either case the setup will comprise two grinding wheels and two regulating wheels mounted on common spindles, yet separated by spacers, to avoid interference with the intermediate section having, generally, a larger diameter than the surfaces being ground. The latter can have plain cylindrical forms, or combined profiles, comprising also tapers, radii, and fillets which are produced by the appropriate form truing of the grinding wheel.

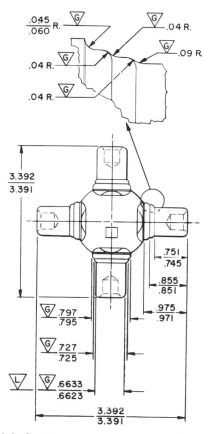

Fig. 2-2.8 Contour drawing of a universal joint spider showing the complex configurations which can be ground in consecutive operations by the centerless infeed method, producing one grind at each revolution of the regulating wheel, which has two gaps: one for loading and the second for unloading.

Assymmetrical part element protruding from a body of revolution (Fig. 2-2.9). The part illustrating that condition, a knitting finger, is also another example of a partially contacting regulating wheel. The grinding wheel has a crush-trued form for grinding the entire profile of the coaxially round elements, including a tapered section. The successful processing of such parts in centerless grinding is predicated on the sufficient length of the supported section in relation to the overhanging one.

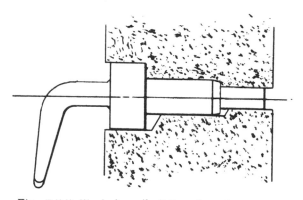

Fig. 2-2.9 Workpiece (knitting finger) with overhanging section of irregular shape which will not exclude the centerless grinding of the round section having a length sufficient to assure controlled work position during the grinding process.

Square shoulders ground in centerless grinding (Fig. 2-2.10). Grinding the square shoulder in the same operation in which the cylindrical sections of the part are being ground offers the double advantage of time saving by avoiding a second operation, and of

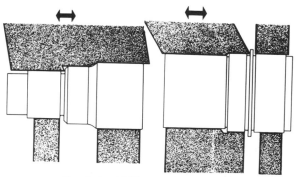

Courtesy of Lidkoeping A.B. — Hirshmann Corp.

Fig. 2-2.10 Combining the grinding of square shoulders with the regular centerless grinding of cylindrical and tapered surfaces in infeed operation requires a supplementary lateral advance of the grinding wheel with recessed (saucer shaped) side. (*Left*) Single grinding wheel with supplementary axial movement; (*Right*) Two wheels on separate mounts, one with axial movement, the other laterally fixed.

assuring excellent geometric accuracy in the relation of the surfaces which are nominally at right angles to each other. As was pointed out in the preceding chapter, the accurate grinding of shoulders perpendicular to the axis of work rotation requires the use of a grinding wheel with recessed side and a relative axial movement between the work and the wheel for producing the infeed advance. The fixed axial relationship of the work and the grinding wheel in centerless infeed grinding excludes the meeting of this requirement. Such centerless grinding problems can be resolved by means of a special system, mounting two grinding wheels on the main spindle, one directly and the other on an intermediate sleeve. This latter can be moved axially by hand, after the readial infeed advance reached its terminal point, or hydraulically, combining the two mutually perpendicular advance movements. The use of such special equipment also involves the application of an appropriate truing device by which the recessed side of the grinding wheeel can be formed.

Grinding of balls on centerless grinder (Fig. 2-2.11). The ball which is a perfect body of revolution differs from the basic cylindrical form for which centerless grinding is primarily adapted, by having a round shape in an infinite number of orientations. Combining a rotating shift of the momentary part axis with the basic rotation of the part, accomplished by a skewed regulating wheel surface acting on an axially confined workpiece, the theoretically defined condition of a continuous shift in the orientation of the axis of rotation can be satisfied. Typical workpieces are bowling balls which, made of hard vulcanized rubber, are also examples for the adaptability of centerless grinding to a wide range of work materials.

Thrufeed centerless grinding of tapered parts (Fig. 2-2.12). In principle, the application of the thrufeed centerless method is predicated on the essentially cylindrical form of the workpiece. Exceptions to that rule can be made by means of special fixturing, which will make the part act as if it were cylindrical during its contact with the grinding wheel. Tapered bearing rollers are typical examples of a widely applied method which uses a steel regulating wheel with a helical track having an incline essentially equal to the included angle of the workpiece. By proper design and dimensioning of the track profile, a continuous helicoidal nest is provided for the tapered workpiece; the nest holds the part in a position so that a side element (generatrix) of the work is presented for contact with the grinding wheel.

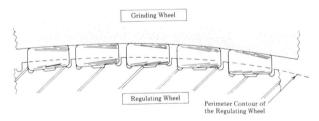

Fig. 2-2.12 The principles of regulating wheel design for the centerless thrufeed grinding of tapered bearing rollers. The steel regulating wheel has a helical groove with inclined track base for presenting the rollers to the grinding wheel along a single line of contact. That line may be straight or curved, as shown, to produce tapered rollers with "crowned" surface.

This element is contained in a common line for all the parts being ground concurrently. In this application the axis of the regulating wheel does not have to be tilted as in common thrufeed grinding, because the shoulder of the helical track imparts the needed traverse movement to the workpiece.

Economic Aspects in the Selection of the Centerless Grinding Process

Centerless grinding may be the only process by which certain operations can be carried out, or the technological advantages over alternative processes,

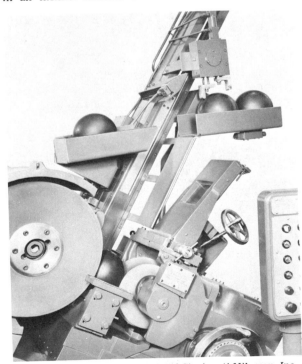

Courtesy of Cincinnati Milacron, Inc.

Fig. 2-2.11 By combining the rotation of the workpiece with a continual shift in the position of its axis of rotation (achieved by a skewed regulating wheel mounting) spherical shapes, such as bowling balls, can be produced in centerless infeed grinding.

primarily in high volume production, may be so obvious that the choice of the process is evident. The process selection becomes less clear, however, when other methods, particularly the regular center type cylindrical grinding, are also adaptable for the operation and will have to be considered as alternatives.

In such cases of alternative processes two major cost factors will often have to be considered, namely:

(a) The machine setup, which is generally simpler and less time-consuming in the regular cylindrical grinding

(b) The piece time, having generally as a main factor the rate of stock removal, which, as a rule, is faster in centerless grinding where the work is more rigidly supported.

When the sum of the setup and grinding times required on a specific lot are the only elements to be considered, it will be the number of pieces per lot which determines the position of the break-even point of cost for the compared processes; beneath that point it will be the regular cylindrical grinding, while above that point it is centerless grinding, which may be considered the more economical.

For finding the lot size (L) for which the total time expended by either of these processes is essentially equal, the following simple formula may be used:

$$L = \frac{\text{(Setup time centerless)} - \text{(Setup time center type)}}{\text{(Piece time center type)} - \text{(Piece time centerless)}}$$

The piece time comprises, in addition to the time used for the actual grinding, also the work-handling time, particularly loading and unloading, as well as the time needed for slide retraction and advance, etc. ("floor-to-floor time"). Because of the self-locating properties of the workpiece in centerless grinders the time spent on these auxiliary operational elements will often tip the balance in favor of centerless grinding.

Other factors which may be considered for making the cost comparison more dependable are the following:

(a) Gaging and size control time

(b) The need for operator's attendance (whether, e.g., the operator can handle several machines concurrently)

(c) Work preparation, such as drilling and lapping of centerholes and mounting driver dogs for center type cylindrical grinding

(d) The frequency of wheel truing, which is the principal factor in grinding wheel and diamond usage, and can also represent a significant time factor

(e) Tooling (work-holding) cost, both general and special.

With regard to the setup time of centerless grinders for small lot sizes, as in toolroom and jobbing shop operations, it is well to remember that these can be reduced substantially when arranging the workpiece lots to be ground in a sequence in which the setup changes between consecutive lots will require only minor adjustments. In this respect it is particularly important that the frequency of changeover from thrufeed to infeed, or vice versa, should be kept as low as is compatible with the work schedules.

The Capabilities of Centerless Grinding

Centerless grinding is essentially a production type manufacturing process which is particularly well adapted to the grinding of large quantities of uniform workpieces. In such applications the most valuable single characteristic of the process is the *consistency of performance*. It is particularly with respect to consistent size and form control that centerless grinding offers a uniquely high degree of performance capabilities. There are several reasons for this characteristic of the method, a few of which will be discussed in the following:

(a) In thrufeed centerless grinding the fixed position of the grinding wheel axis with respect to the work is constantly maintained, with very small changes in the wheel surface only, due to wear which is generally predictable and can be compensated for.

(b) The relatively minor effect of wheel wear on size control is due to the generally large ratio of wheel diameter to work diameter and, in thrufeed grinding, also to the fact that the width of the wheel is often several times larger than the length of the workpiece.

(c) Rigidity of the setup with, generally, full-length support of the workpiece, thereby avoiding deflections and inaccuracies caused by lack of work stability.

(d) Obviating the reliance on rotating work-holding devices and on the condition of auxiliary work surfaces, such as center holes and clamping sections.

Capability values for a process can never be considered as generally valid because of the dependence on a large number of variables in the operational conditions. Still, a few *examples of performance accuracy* which are accomplished, or may even be exceeded in regular production runs, will indicate the capabilities of the centerless grinding method, predicated on well controlled, but definitely feasible and consistently maintainable conditions.

Work accuracy data for precision type centerless grinding:

Roundness
(radial separation of concentric envelope circles)
.000 0025 in. (0.64μm) for diameters of

about $\frac{1}{2}$ in. (13 mm) or less,

.000 050 in. (1.27 μm) for larger diameters

Straightness
0.0001 in./in. (0.0025 mm/25mm)

Concentricity
[of simultaneously ground surfaces, each with an individual size tolerance of 0.0001 in. (0.0025 mm) max]: 0.0005 in (0.013 mm) T.I.R.

Size (diameter)
Attainable setup accuracy ±0.000 050 in (1.27 μm)

Consistency of size holding:
(total range of variations over extended production runs with appropriate sizing controls):

thrufeed	0.000 050 in. (1.27 μm)
infeed	0.000 075 in. (1.91 μm)

Surface finish
6 to 8 microinches AA (0.15 to 0.20 micrometer R_a), exceptionally as fine as 2 microinches AA (0.05 micrometer R_a) for hardened workpieces.

Regarding the *production capability* of centerless thrufeed grinding, a few examples are listed in the following table, for maximum performances attainable under optimum conditions:

Attainable Thrufeed Values under Favorable Conditions

Part diameter	Thrufeed advance rate of work
1 in. (25 mm) (rod)	200 in. (5 m)/min
0.500 in. (13mm) (cylindrical roller)	400 in. (10 m)/min
0.200 in. (5 mm) (needle roller)	600 in. (15 m)/min

Conditions less than optimum with respect to equipment, work preparation, and setup will each necessitate reduction in thrufeed rates which, due to the cumulative effect of such limiting factors, may result in a fraction of the listed maximum performances. Guidelines for calculating production rates under average conditions are indicated in Chapter 2-9 of this section, discussing the operational data of centerless grinding.

As a general rule, the accuracy of the centerless grinding process will improve, when:

The operation requires only a small amount of stock removal

The workpieces to be ground have well controlled geometry (roundness and straightness) resulting from the preceding operation

The selected operational data call for a number of revolutions of the part sufficient to produce the desired degree of roundness improvement.

The Principal Machine Elements

General Design Characteristics of Centerless Grinding Machines

The basic principles of the centerless grinding method determine the essential design characteristics of these grinding machines. Independently of the size of the work which they are expected to handle, the centerless grinding machines comprise as major elements:

(a) The bed
(b) The grinding wheel head (optionally mounted on an individual slide)
(c) The regulating wheel head (on its slide)
(d) The work rest.

While the basic machine design objectives are identical, there are certain differences in the concepts of various machine tool manufacturers regarding the ways in which these objectives can best be achieved. A few essential design alternatives will be pointed out without expressing any preference. As a matter of fact, most of the currently applied design alternatives have their own merits on the basis of which the selections were probably made by the respective manufacturers.

One of the fundamental differences in the design principles as applied by various prominent grinding machine manufacturers, affects the mutual positional adjustments of the three basic machine elements: the grinding wheel, the regulating wheel, and the work rest. The following are the two most widely used design systems:

(1) *Fixed wheel head and two-level slide for the regulating wheel head, the lower slide supporting the work rest* (Fig. 2-3.1). The upper slide of the regulating wheel head serves to establish the distance between the work-rest blade and the regulating wheel during the original machine setup, and is used thereafter only following the infrequently needed dressing of the regulating wheel. Otherwise, the upper slide is locked to the lower one and moves together with it when the lower slide is advanced or retracted.

The movement of the lower slide sets the location of the work rest with respect to the surface of the grinding wheel, and also the distance between the two wheels, the regulating wheel being at a pre-set position from the work rest. Such adjustments are

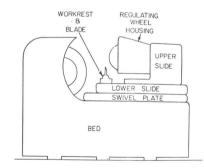

Fig. 2-3.1 Diagram of the design principles which determine the operational movement of a centerless grinder with two-level regulating wheel slide and slide mounted work rest. The housing of the grinding wheel spindle is integral with the rigid bed. The swivel plate permits the sensitive alignment adjustment of the regulating wheel. During the grinding operations the two slides are mutually locked and the lower slide is advanced for feed and wheel wear compensation. In plunge grinding operations it is the lower slide which is generally retracted for loading and unloading of the work. Plunge grinding by moving the upper slide while the lower slide is locked to the swivel plate is also possible, but only rarely used.

needed, during the operation of the grinder, for correcting size variations of the work in thrufeed grinding, for approaching and then gradually advancing the work in infeed grinding and also for size compensation following each truing of the grinding wheel.

In this design there is no need for moving a relatively heavy grinding wheel slide with its drive motor, these elements being integral with, or rigidly bolted to the machine bed, instead of being mounted on a movable slide.

One condition related to this grinding machine concept may be mentioned: in infeed grinding it is sometimes the upper regulating wheel slide which is retracted for changing the work. The compensation for grinding wheel wear and dress-off can be a part of the automatically controlled upper slide retraction of advance movement. However, an occasional resetting of the lower slide is still needed for avoiding too-wide variations in the position of the work-rest blade relative to the grinding wheel.

(2) *Fixed work rest and adjustable slides for both the grinding wheel and the regulating wheel* (Fig. 2-3.2). Both wheel slides must be adjusted in the setup of these machines for establishing the proper positions of the wheels with respect to each other as well as to the work-rest blade. During the grinding process, slide adjustments are needed to compensate for the

wear and the dress-off of both the grinding wheel and the regulating wheel, less frequently for the latter.

In infeed grinding the retraction movement is generally carried out by moving the grinding wheel slide, unless the process requires the release of the ground part by letting it roll down toward the machine bed; in that case the retraction is effected by the regulating wheel slide.

This system offers advantages for the in-line type arrangement of a battery of several centerless grinders through which long bars are processed in consecutive passes. The fixed work-rest position will assure consistent alignment of the work during its traverse through several grinding machines.

The fixed work rest, solidly bolted to the machine bed provides a high degree of rigidity in supporting the work and permits the use of a built-in adjusting device for a sensitive raising or lowering of the work-rest-blade position.

The Grinding Wheel Head

Because centerless grinding machines are generally built to accept grinding wheels with large diameters and widths for carrying out operations which often involve substantial grinding forces, powerful motors drive the spindles of the wheel heads which character-

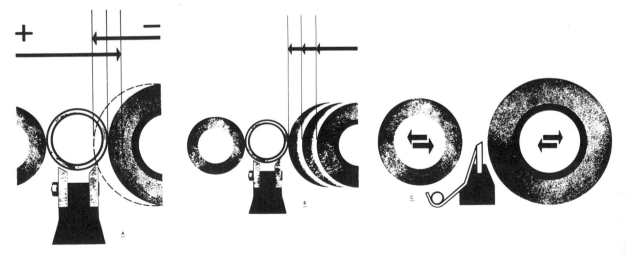

Courtesy of Lidkoeping A.B. — Hirschmann Corp.

Fig. 2-3.2 Diagram A showing the operational principles of centerless grinding machines with laterally fixed work rest, while the headstocks of both the grinding wheel (at right) and the regulating wheel are mounted on individual slides, permitting infeed and retraction. (NOTE: The blade on the left side of the rest is an optional feature for supporting the work during loading but has no function during actual grinding. The angle of the right-hand side blade is exaggerated for illustration purposes.
Alternative operations:
B. Wheel slide movement during regular infeed grinding.
C. Movements for discharging the work by gravity when the regulating wheel slide retracts, while the movement of the grinding wheel slide assures size control.

istically are of generous dimensioning and sturdy construction.

There are two different principles applied in the design of the grinding wheel spindle, which are found in different types and makes of centerless grinding machines: (a) the cantilever, and (b) the straddling support design.

In the cantilever design the grinding wheel spindle is supported in two adjacent bearings, leaving one end of the spindle free, to permit the unrestricted mounting or the removal of the grinding wheel from the firmly supported spindle (see Fig. 2-3.3). This design is of advantage in toolroom and job shop practice where frequent wheel changes during the useful service life of the grinding wheel may be needed to suit varying work materials and process requirements. The easy wheel change, requiring no particular equipment, is also preferred in general shops even when the wheels are removed only after they have been used up.

Courtesy of Cincinnati Milacron, Inc.

Fig. 2-3.4 Changing the grinding wheel of a centerless grinding machine, with straddling bearing arrangement for the wheel spindle, requires the removal of the entire wheel spindle together with its cartridge type bearings.

Courtesy of Cincinnati Milacron, Inc.

Fig. 2-3.3 The cantilever support of the grinding wheel spindle offers the convenience of easy wheel change.

The straddling bearing support requires the entire spindle to be withdrawn from the wheel head, usually by removing the caps of the bearing housings and then lifting out the spindle with its bearings and the mounted grinding wheel (see Fig. 2-3.4). The bearings are usually enclosed in a cartridge whose outer surface is supported in the housing, while the sensitive bearing surfaces remain encapsulated and well protected. The wheel itself is usually mounted on a sleeve which is slipped over the spindle.

The straddling bearing design offers a high degree of rigidity, and reduces the deflections of the spindle

even when substantial grinding forces are acting on the wheel. This design is preferred for machines intended for heavy work and high performance, or extreme accuracy, when even the slightest "orbiting," experienced in spindles supported at one end only, cannot be tolerated. Some manufacturers use the straddling support on all their models, even though it requires a more difficult grinding wheel changing procedure than the cantilever design.

The bearings used for supporting the grinding wheel spindles of centerless grinding machines are, with few exceptions, either of the sliding or of the rolling type. The sliding bearings operate by the hydrodynamic principle, four or more circumferentially arranged shoes producing gradually narrowing gaps around the bearing surface of the spindle. The rotating spindle forces the lubricating oil toward the tight ends of these wedge shaped gaps, where the confined oil produces a high pressure providing a very stiff support for the spindle. The circulation of the lubricating oil into the bearings is assured by an oil pump; an interruption of the oil supply will actuate a switch which automatically disconnects the current of the spindle drive motor.

For rolling bearing types of spindle supports, preloaded roller bearings of special design and accuracy are generally used, complemented with ball thrust bearings for taking up the axial forces.

A few machine tool manufacturers also supply hydrostatic spindle bearings, although that system has, so far, not attained wide acceptance for centerless grinders.

The Regulating Wheel Head and Its Slides

The regulating wheel unit of the centerless grinders has several kinds of adjustments which are needed to assure the proper execution of the various functions which these elements have to provide. The following listing of these functions will be accompanied by a brief description of the pertinent elements and their adjustments.

(a) Controlling the rotation of the work at the speed best suited for the operation. The regulating wheel drives of centerless grinding machines are generally designed to provide a wide range of different rotational speeds, usually infinitely variable by mechanical (see Fig. 2-3.5) or electrical means. Commonly, a tachometer is applied, with its dial mounted on the front of the regulating wheel head, to indicate the actual rotational speed of the regulating wheel spindle. A typical range is 15 to 300 rpm, the highest speed to be used for the diamond dressing of the regulating wheel.

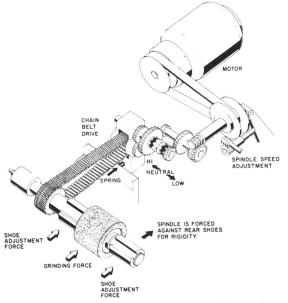

Courtesy of Cincinnati Milacron, Inc.

Fig. 2-3.5 Diagram of a typical drive mechanism for a centerless grinder regulating wheel with infinitely variable speed which can be applied in either a high or low range.

(b) Setting up the centerless grinder for a specific work diameter, correcting the variations of the work size during grinding (optionally), compensating for wheel wear and dress-off, retraction and advance for

infeed grinding (optionally). These functions are carried out by the movement of the regulating wheel slide, or slides, depending on the basic design principles of the grinder, as discussed before. The slide movements are operated by a handwheel, generally providing both coarse and fine adjustment. When infeed grinding by moving the slide of the regulating wheel, a quick-acting hand lever for a fixed amount of retraction, e.g., 0.050 in. (1.27 mm), or power actuated retracting devices (the latter often tied in with an automated operational cycle), are frequently used.

(c) Imparting a traversing motion to the work in thrufeed and endfeed grinding, exceptionally in infeed grinding, for assuring a dependable positioning against a fixed stop. This axial motion of the rotating workpiece is brought about by the tilted position of the regulating wheel axis with respect to the plane containing the grinding wheel axis. When so tilted each single element on the surface of the regulating wheel will rotate in an inclined plane, describing an orbit during which in one half of its rotation it will move away from, and in the other half return to, an imaginary vertical plane. By adjusting the tilt of the regulating wheel axis, a combined rotational and axial movement can be imparted to the work. Usually an advancing movement of the work with respect to the point of loading in front of the machine is required and the pertinent regulating wheel tilt is termed positive. In particular cases of thrufeed grinding, for reasons of fixturing, the work is introduced from the rear of the machine, tilting the regulating wheel spindle in the negative sense. The amount of tilt, displayed on a graduated sector on the regulating wheel head will determine the rate of the axial work movement in relation to its peripheral speed. Details regarding the applicable values, as well as of the required special wheel truing procedure, will be discussed in Chapter 2-8 of this section, explaining the setup of centerless grinding machines. A typical range for the tilt adjustment of the regulating wheel head is plus 8 and minus 2 to 4 degrees.

(d) The angular adjustment of the work support unit, containing the regulating wheel with the work guides, often the work rest, too. This adjustment is used for correcting the taper of the work in infeed grinding and also for distributing the cut uniformly across the face of the grinding wheel in thrufeed grinding. It can be carried out conveniently in most types of centerless grinders which have a *swivel plate* supporting the entire regulating wheel unit, including the slideways and the slides. That plate is mounted directly on the machine bed, with its pivot point located at a fixed position which, however, is not

identical in grinders of different makes. Some models have the pivot point at one end, in the center, or at a location closer to the rear end of the grinding wheel face (see Fig. 2-3.6), depending on the design concept of the respective manufacturer. Two extended screws, acting in opposite directions and accessible in the front of the machine, serve to carry out the swivel adjustment of the plate in the required direction. Some refined types of centerless grinders are also equipped with means to guide the swivel setting by an indicator and to establish its reference position by inserting a gage block.

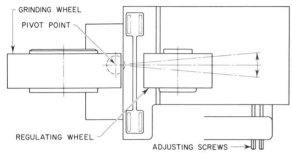

Courtesy of Cincinnati Milacron, Inc.

Fig. 2-3.6 Diagram showing the arrangement and operation of a swivel plate which is supporting the entire regulating wheel unit with its slide and the work rest.

The Work Rest

The work rest, consisting of a base retaining the blade, is one of the three elements which support and restrain the workpiece during the centerless grinding process. While the relative positions of the two other elements, the grinding wheel and the regulating wheel, are the controlling factors of the work size and form, the general configuration and the position of the work-rest blade will provide the staging of the work in agreement with the requirements of an effective centerless grinding process.

The work-rest blade must also absorb a substantial part of the forces arising in centerless grinding, a method which substitutes the balancing of interacting forces for the fixed position type holding of the workpiece. The work-rest blade should hold the work at a specific level with respect to the center line of the wheel; variations in that level will affect the depth of penetration of the grinding wheel into the work surface when the distance between the two wheels remains constant.

One of the functions of the work-rest blade is to divert, by means of its inclined top surface, a part of the grinding force acting on the work in a direc-

tion causing increased contact pressure of the work against the regulating wheel.

Finally, because the dimensions of the work-rest blade must be adapted to the work and the operational conditions, small-diameter parts will require very thin blades.

This combination of essential roles in the centerless grinding process, the heavy burden which it must carry, and the often severe dimensional limitations make the work rest, and more particularly, the rest blade, a critical element in the setup and operation of centerless grinding machines.

The Functional Dimensions of Work-Rest Blades

The general configuration and external dimensions of the blades to be used for any specific model of centerless grinder are determined by the design of the work-rest base and are specified in the pertinent publications of grinding machine manufacturers. Special blade configurations are needed for certain types of workpieces which also have noncylindrical sections or surfaces with different diameters.

The length of the work blade is commonly equal to the width of the grinding wheel, but special setups or work shapes may require exceptions. The thickness of the blade must always be less than the diameter of the workpiece and will seldom exceed $\frac{1}{2}$ to $\frac{5}{8}$ inch (13 to 16 mm).

The top angle of the work blade has a double role in centerless grinding:

(a) It diverts the vertical component of the grinding force which acts on the workpiece into an essentially horizontal direction, thereby assuring a more powerful contact between the work and the regulating wheel.

(b) It provides the geometric conditions of a shifting work center when grinding out-of-round parts, an important factor in achieving the rounding effect of the centerless grinding process (see Fig. 2-1.4).

The most commonly selected top angle of the work-rest blade is 30 degrees (inclined from the horizontal in the direction of the regulating wheel), although quite frequently, other top angles are used, usually within a range of from 10 to 40 degrees, because they are better suited to the work conditions. As a general rule, a greater top angle will accelerate the rounding effect and increase the horizontal force component. That latter effect can be of advantage or, under different conditions, undesirable because of the greater friction between the blade and the work and the tendency to deflect the blade, particularly in the case of very thin or long blades. Deflections of the work blade during the process must be avoided because of harmful effects (chatter) on the ground work surface.

The Mounting and Setting of the Work-Rest Blade

The work-rest blades, which are interchangeable and replaceable elements, are usually bolted or clamped to the base, sometimes designated as the cradle. The bolts are applied through elongated holes in the blade, or have nuts which can slide in appropriate grooves in the base, to permit the adjustment of the blade height. Such adjustment is assisted at times by placing shims under the blade. Certain makes of centerless grinding machines with work rest mounted on the machine bed, are equipped with crank-actuated blade height adjusting devices; this makes the accurate setting of the required blade height easier to carry out (see Fig. 2-3.7), preferably with the aid of a direct contacting height gage for the accurate measurement of the actual work position with respect to the wheel center line.

The height setting of the blade is a major factor in obtaining the required rounding effect in the centerless grinding process. For general operations the work rest should be set to support the work in a position where its center is at about one-half of the work diameter, but by not more than $\frac{1}{2}$ inch, (about 13 mm), above the center line of the wheels. The rough thrufeed grinding of long bars constitutes one of the rare exceptions to this general rule; such work is usually ground by setting its center below the center line of the wheels, in order to achieve a more stable work support.

For reasons explained earlier (see "The Rounding Effect of Centerless Grinding") raising the work center above the wheel center line assists the rounding process, but also reduces the stability of the work retainment in the gap between the wheels and in continuous contact with the rest blade. Should that contact be momentarily interrupted during the grinding process due to the unstable positioning of the work (which thus may tend to rise from its support), the lack of continuity of the grinding contact will result in closely spaced circumferential waves, known as "chatter marks," on the ground surface. Conditions such as the top angle of the blade, the composition of the regulating wheel (e.g., high resilience), and the properties of the grinding fluid (good lubricity) can reduce the tendency of the work to develop chatter.

The machine tool manufacturers distinctly specify the height of the wheel center line in relation to a fixed level, commonly, the top of the bed. The setting of the blade height can be measured either indirectly, by placing a workpiece on the blade when the

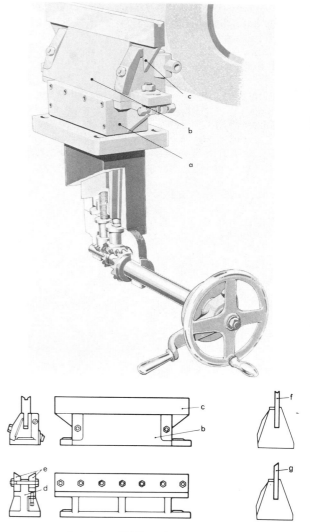

Courtesy of Lidkoeping A.B.—Hirschmann Corp.

Fig. 2-3.7 Laterally fixed work support in special design with hand wheel operated height adjustment. During grinding the work rest is solidly locked to the machine bed by means of tapered keys actuated through a lever behind the handwheel. This particular design comprises the work rest (a), carrying the work holder (b), with the blade (c). For heavy workpieces the work rest can be equipped with a special blade holder (d) to accept two blades (e), one of which provides complementary work support during loading and unloading. Similar complementary support is provided for smaller work by the V-top single blade (f), in distinction to the regular bevel top blade (g).

wheels are in operating position, a method preferred for work rests with means for vertical adjustment, or directly by measuring the height of the blade edge. In the latter case the resulting work center height can be calculated based on the pertinent setup values, including the work diameter, the blade top angle, and the distance between the blade and the grinding wheel (see Fig. 2-3.8).

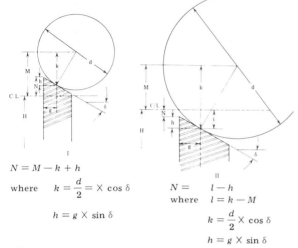

$$N = M - k + h$$

where $\quad k = \dfrac{d}{2} = \times \cos \delta$

$$h = g \times \sin \delta$$

$$N = \quad l - h$$

where $\quad l = k - M$

$$k = \dfrac{d}{2} \times \cos \delta$$

$$h = g \times \sin \delta$$

Fig. 2-3.8 Geometric conditions involved in calculating the height setting (N) of the edge of the work-rest blade in relation to the height of the wheel centerline (C/L) for holding the work center by a selected M distance above the C/L (at H height from the machine reference plane).

Diag. I - For workpieces with diameter (d) of one inch or less (N to be added to H)

Diag. II - For workpieces with larger diameter (N to be subtracted from H)

$\delta \;=\;$ top angle of the work-rest blade

$g \;=\;$ distance of the work contact from the side of the work blade.

Material and Maintenance of Work-Rest Blades

The selection of the work blade material will usually represent a compromise of objectives which, in some respects, may be conflicting:

(a) Hardness of the blade is desirable to obtain excellent finish in making and in reconditioning the blades; and particularly for improving the wear resistance in use.

(b) The blade must not damage the work material such as by showing a tendency for "pick-up" and, for that reason, the hardness of the blade may have to be selected to suit the characteristics of the work material.

(c) Toughness of the blade material may be a major consideration when the workpieces or the grinding conditions are prone to cause chipping of the blade edges.

(d) A flexible blade material may be needed in such relatively rare cases where the blade must be bent in its rest position to comply with the profile of the grinding wheel, as in the case of a concave wheel face needed for "crowning" the workpiece.

(e) The need for frequent regrinding may call for a blade material well adapted for that kind of regular maintenance.

(f) The size of the work, such as a very small diameter requiring a particularly thin blade, may be a limiting factor in selecting the blade material which could provide the best wear resistance.

(g) Economic considerations are often important factors, particularly in the case of short run work.

These aspects will be considered when selecting the work-rest material from the following most frequently used alternatives:

(a) Tool steel, generally an oil-hardening type, may be the least expensive, it is easy to work and recondition, and often provides satisfactory service under ordinary conditions.

(b) High speed steel is recommended for nonferrous metals and for parts which are heat treated to hardness values which differ for various sections being ground concurrently.

(c) Sintered carbide, generally used in the form of strips brazed on the edge section of a blade body made of steel, can provide the best wear resistance for many different work materials. However, carbide is the most expensive, it is sensitive to shocks which may cause chipping, and its reconditioning requires special care.

(d) Cast iron, preferably with a dense structure, may be the best choice for the finish grinding of soft steel parts, which are sensitive to scoring.

(e) Hard bronze may, in some cases, prove to be even superior to cast iron, when the finish of the sensitive work material is a major consideration.

When normal wear in usage has changed the original shape of the blade edge to an extent which affects the quality of the work, the blade should be reconditioned. That consists in grinding off the worn surface layer and producing a new edge surface, precisely duplicating the original shape, particularly the top angle.

The regrinding of the blades is generally carried out in the toolroom on peripheral surface grinders, mounting the blade in an appropriate holding device which assures the reproduction of the correct top incline angle. The applied grinding wheels and grinding data, particularly the depth of cut, must be selected in compliance with the blade material, e.g., using diamond wheels for carbide, in order to prevent damaging the blade material and to obtain a good surface finish for reducing friction during the operation of the blade.

The Basic Types of Centerless Grinding Machines

A survey of the different models of centerless grinding machines can provide useful information in several respects, particularly for comparing the available work capacities and other significant characteristics. These data may serve as the basis for the preliminary selection of that general type of centerless grinder which is suitable for carrying out the planned production process.

For a rapid survey Table 2-4.1 lists five major categories of centerless grinding machines. These categories are not based on standards and have been defined with the purpose of facilitating the review of that rather wide range of centerless grinding machines which are in general use in many branches of manufacturing. The major characteristic by which individual models were assigned to these categories is their work capacity.

While the range of acceptable workpiece diameters is a prime indicator for the general dimensions of several other major machine elements, the work capacity is not a precisely defined characteristic of any particular category. Thus, centerless grinding machines in the same category may have similar capacity ranges but with different limits for acceptable workpiece diameters. Also, the ranges of acceptable workpiece diameters usually overlap with those of the adjoining categories.

The upper limit of the workpiece diameter range of any specific centerless grinding machine is generally determined by the widest opening between the grinding wheel and the regulating wheel when both are new and have the maximum diameter. The upper limit of the capacity range consequently can be extended by using wheels of less than maximum diameter.

Although the upper work size limit, often referred to as the nominal grinder capacity, represents the work diameter which the grinding machine can accept and is capable of grinding, for high volume production machine models are usually selected which offer *greater* nominal capacity than the diameter of the largest workpiece to be processed. General exper-

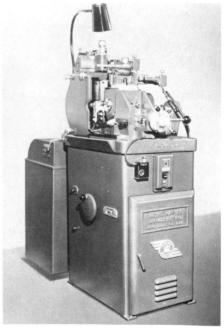

Courtesy of Royal Master Grinders, Inc.

Fig. 2-4.1 Small-size centerless grinding machine designed for grinding very small and miniature size workpieces. The coolant tank is installed behind the machine. These small centerless grinding machines are also produced in a model equipped for crush dressing.

Table 2-4.1 Typical Dimensions of Major Categories of Centerless Grinding Machines
(Grouped according to work size capacity)

MACHINE CATEGORY	NOMINAL INFEED WORK CAPACITY	THRUFEED CAPACITY FOR SOLID BARS	GRINDING WHEEL DIAMETER × WIDTH (MAX.)	REGULATING WHEEL DIAMETER × WIDTH (MAX.)	SPEED RANGE OF REGULATING WHEEL	GRINDING WHEEL DRIVE POWER	APPROX. MACHINE WEIGHT
Light, for small workpieces	1½ in. (38 mm)	1 in. (25 mm)	12 × 3 in. (300 × 75 mm)	6 × 3 in. (150 × 75 mm)	30 to 480 rpm	5 hp	2,200 lbs (1000 kg)
General purpose	3 in. (75 mm)	1½ in. (38 mm)	20 × 8 in. (500 × 200 mm)	12 × 8 in. (300 × 200 mm)	15 to 300 rpm	15 hp	7,700 lbs (3,500 kg)
Production type	6 in. (150 mm)	4 in. (100 mm)	20 × 12 in. (500 × 300 mm)	14 × 12 in. (350 × 300 mm)	10 to 300 rpm	25 hp	15,400 lbs (7,000 kg)
Heavy duty	12 in. (300 mm)	8 in. (200 mm)	24 × 12 in. (600 × 300 mm)	16 × 12 in. (400 × 300 mm)	0 to 120 rpm	40 to 100 hp	20,000 to 26,000 lbs (9,000 to 12,000 kg)
Limited purpose wide wheel	10 in. (250 mm)	Not used	20 × 24 in. (500 × 600 mm)	14 × 24 in. (350 × 600 mm)	0 to 150 rpm	75 hp	22,000 lbs (10,000 kg)

NOTE: The listed values express orders of magnitude only and do not represent the dimensional data of particular models.

ience indicates that such extra capacity is less important for infeed grinding than for thrufeed grinding, and is particularly desirable for bar grinding.

The listing in Table 2-4.1 does not include the shoe type centerless grinders, although these operate on essentially the same principles as the regular centerless grinding machines. However, in view of their several original characteristics the shoe type centerless grinders are considered as representing an independent category and will be discussed separately in Chapter 2-5 of this section.

The values listed in Table 2-4.1 are strictly for general information and do not represent the data of any particular model. The intention was to present average values which are close to the actual data of most makes and models considered as belonging to a particular category.

In the following a few typical models of centerless grinding machines will be reviewed as characteristic representatives of the categories of Table 2-4.1.

Precision Type Centerless Grinding Machines for Small Parts

Centerless grinding, which is generally considered as an efficient method of high-volume production, can be definitely economical for the grinding of small or even very small parts in moderate quantities as long as grinding machines specially adapted for such operations are used.

Effectiveness and economic advantages are, of course, not the only reasons for giving preference to the centerless grinding method. The decisive reasons may be technological; some part configurations, dimensions, or material may make centerless grinding the most dependable or accurate, or actually the only applicable process. Examples of such workpieces are parts of very small diameter; of length-to-diameter ratios requiring extended support in grinding; parts made of brittle materials such as glass, ceramics, carbides, etc.; workpieces which have to be ground along the entire length, or could not accept center holes; and finally, round parts with profile surfaces which must be ground simultaneously for assuring controlled concentricity of all sections.

The technological advantages which grinding by the centerless method can provide for many types of very small or even miniature parts has led to the development of a special category of small-size centerless grinding machines with design characteristics adapted to this particular field of application.

Figure 2-4.1 illustrates a small centerless grinding machine specially adapted to the above outlined types of workpieces. It has the following principal characteristics: the grinder is adaptable for both infeed and thrufeed grinding on workpiece diameters up to $1\frac{1}{2}$ inches (38 millimeters), although for thrufeed grinding of long bars, the upper limit of the work diameter should not exceed 1 in. (25 mm) or $\frac{1}{2}$ in. (13 mm) respectively, depending on the length of the bar. On the other hand, small diameter workpieces, down to a limit of 0.004 in. (about 0.1 mm) can be ground successfully.

As a special feature, the illustrated model is also equipped for the crush-truing of the grinding wheel, thereby providing an efficient method for the centerless infeed grinding of round parts with complex profiles. For this purpose the grinding wheel spindle has a special support incorporating an additional set of bearings in order to effectively resist the high crushing loads during the truing of the grinding wheel. Furthermore, a separate motor serves for diving the wheel spindle at a special low speed during the crushing process.

To make this precision type of centerless grinding machine adaptable for automatic operation, i.e., grinding small size parts in a continuous process, special feeding devices designed for such workpieces are also available. These include, for example, magazine feeders for infeed grinding and chute or hopper feeds for thrufeed grinding.

General Purpose Centerless Grinding Machines

The original field of application for centerless grinding machines was the production of small and medium size lots, which required the frequent changeover of the setup. Such setup changes may have involved only size and regulating wheel speed adjustment, perhaps also the exchange of the work-rest blade, but quite often more comprehensive changes, requiring grinding and regulating wheels with different specifications, special truing cams, or even the conversion from thrufeed to infeed grinding, or conversely, had also to be carried out.

In the decades following the introduction of the first centerless grinding machines, this method rapidly became accepted for high volume production, where the largest percentage of centerless grinding machines is in use today. There still are many areas of small and medium lot production, however, where centerless grinding offers potential advantages, as long as adequate equipment is available. Adequacy in this respect means flexible adaptation to different centerless grinding processes and work configurations, as well as to easy setup and operation. On the other hand, substantial stock removal rates or a high degree of automation are not major considerations for the indicated kind of small lot applications.

Recognizing the continued existence of uses outside of high volume production, most prominent manufacturers of centerless grinding machines are still building such adaptable types of grinders, which are sometimes designated as "universal" types of centerless grinding machines.

A typical example of such an adaptable, general-purpose centerless grinding machine is shown in Fig. 2-4.2, illustrating an updated version of one of the most widely used basic models of centerless grinders. That updating resulted in larger wheel sizes, (20 × 8 inches [500 × 200 mm], max, grinding wheel), a more powerful drive motor (15 hp) and

Courtesy of Cincinnati Milacron, Inc.

Fig. 2-4.2 General-purpose centerless grinding machine designed for quick setup, varied applications, and easy operation.

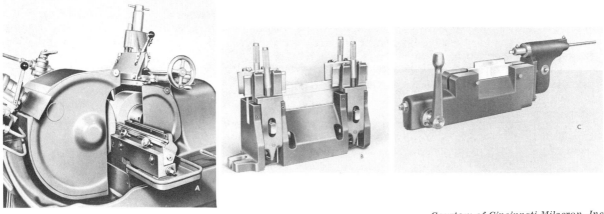

Courtesy of Cincinnati Milacron, Inc.

Fig. 2-4.3 Interchangeable work rests facilitate the conversion of general purpose centerless grinders, the work area of which is shown in (A). Alternate work rests can be used, e.g., for: (B) Thrufeed grinding; or for (C) Infeed grinding with lever-controlled manual ejection.

sturdier general execution (net weight about 7,800 lbs or 3,500 kg), all contributing to more effective stock removal, making this machine suitable also for substantial production runs.

On the other hand, several well proven features which assured flexible adaptation have been retained from earlier models.   The location of the wheels closer to the front of the machine and the concentration of the control elements at the operator's normal working position, contribute to the ease of operating the machine. The setup is made simpler by better accessibility for wheel change resulting from the open-end wheel spindle, the convenient location and the clearly visible graduations of the adjustable machine elements for feed angle, truing angle and diamond offset, and the interchangeable work rests (see Fig. 2-4.3).   Another feature assisting setup is the taper correction adjustment for the complete regulating wheel unit, including the regulating wheel head, the slides, the feed screw and the work rest.

While an infeed lever for hand operated infeed grinding is part of the standard machine equipment, an electro-hydraulic infeed device can also be supplied as an optional accessory, which permits automating the infeed grinding process and provides infinitely variable feed rates and selective stock removal adjustment.

Production Type Centerless Grinding Machines

Grinding machines assigned to this category represent the bulk of centerless grinders operating in the metalworking industry, where the trend toward increased productivity is the dominating factor in the selection of new manufacturing equipment. A char-

acteristic example of production type centerless grinders is shown in Fig. 2-4.4.

Courtesy of Cincinnati Milacron, Inc.

Fig. 2-4.4 Production type centerless grinding machine, combining rigidity of design with operator convenience, and accepting a wide range of special equipment, thereby making the machine adaptable to a multitude of different grinding operations.

Productivity in metalworking processes is controlled primarily by the rate of stock removal and the proportion of the nonproductive to the total operational time. The general engineering concepts of centerless grinding machines included in this category very distinctly reflect the recognition of these important constituents of increased productivity. A few of the major design concepts of centerless grinding machines which contribute to high productivity will now be reviewed, together with the pertinent characteristics of equipment considered in this category. It should be mentioned that along with the development of machine tool characteristics serving increased output performance, meaningful advances have also been achieved with regard to satisfying another set of equally important and progressively more stringent requirements; that of the dimensional accuracy of the product.

(a) The diameters of the grinding wheel and of the regulating wheel. Grinding machines in this category are generally designed for operating with large-diameter grinding wheels as well as regulating wheels. Wheels with greater peripheries offer a larger number of abrasive grains for consecutive engagement with the work surface, resulting in slower wheel wear. This extends the periods between wheel dressing, a nonproductive operation, and also improves the size-holding capacity of the grinding machine.

(b) The widths of the grinding and regulating wheels. A wide wheel is of calculable advantage for most types of thrufeed grinding operations. Considering that a specific number of work rotations is needed for the removal of the stock allowance and for attaining the required rounding effect on a workpiece rotating with a fixed peripheral speed, the work must stay between the wheels for a certain period of time. This time value of the wheel engagement is a basic factor in selecting the thrufeed rate, the other factor being the length of the traverse, which is essentially equal to the width of the wheels. The wider the wheel, the faster the traverse rate that can be applied for obtaining a fixed wheel engagement time.

The generally wide wheels of the production type centerless grinding machines actually permit a higher traverse rate increase than the proportion of the wide to the narrow wheel indicates. In thrufeed grinding the wheel edges are usually beveled, at the feed end to permit easier entry of the work between the wheels, and at the discharge end to avoid injuring the ground work with a sharp wheel edge. These beveled sections have the same length independently of the wheel width. Consequently, a wider wheel offers a proportionately longer effective contact section than that resulting from the difference of the wheel widths.

In infeed grinding, wide grinding wheels permit the grinding of longer parts, or the simultaneous grinding of several parts to an aggregate length defined by the width of the wheel.

(c) The driving power. The higher power consumption of driving the larger wheels is only one of the factors which justify the very powerful drive motors of modern production type centerless grinding machines. A further reason may be found in recent

advances in grinding wheel technology which permit increased specific rates of stock removal. This realization of that potential source of higher productivity is predicated on sufficient driving power for the wheel.

The substantially increased power of the drive motors used for production type centerless grinding machines involved substantial changes in the design of the older types of similar machines, often even the application of fundamentally different engineering concepts. While the very substantial weight of such powerful grinding machines is only one of the results, it is indicative of the applied design concepts.

(d) Reduction of nonproductive time. A brief review of the major elements which constitute the nonproductive portion of the operational time and the measures directed at reducing or even entirely eliminating their effect on productivity shows that:

(1.) Loading and discharging time losses can be entirely eliminated in thrufeed grinding by assuring continuous succession of the workpieces, and are greatly reduced in infeed processes by highly mechanized and rapidly acting work loading devices;

(2.) Wheel dressing is accelerated by methods such as the use of cluster diamonds for achieving faster traverse rates, crush dressing for forms, and rotary diamond dressers for either straight or profiled grinding wheels;

(3.) Automatic gaging, often combined with feedback signals acting on compensating wheel slide adjustment, all carried out automatically and without the interruption of the production process reduce or eliminate downtime due to gaging.

Courtesy of Landis Tool Co.

Fig. 2-4.5 Production-type centerless grinding machines arranged in tandem for the continuous two-step thrufeed grinding of the outside surfaces of roller bearing outer rings. The fixed position work rest design of these centerless grinders is conducive to an in-line arrangement which avoids the use of a separate conveyor system for the transfer of the workpieces.

Production type centerless grinding machines are used individually or in groups performing consecutive operations on workpieces which are often transferred by integrated conveyor systems. Another alternative method for continuously effected consecutive operations is by setting up the centerless grinders in tandem arrangement, such as shown in Figs. 2-4.5 and 2-4.6.

Courtesy of Landis Tool Co.

Fig. 2-4.6 Work area of the tandem arranged centerless grinders shown in Fig. 2-.4.5, viewed from the feed end with the retractable pusher device in front. The action of this device assures that the rings stay mutually supported and pass through the grinders in an uninterrupted stream.

Heavy-duty Centerless Grinding Machines

Workpieces with substantial material allowance are often required to be brought to finish size by centerless grinding. Such operational specifications may be the result of more recent processing systems, two examples of which follow:

(a) Obviating or reducing the work done in preliminary or roughing operations, such as when using cold extruded parts without previous machining, and bringing them to final size by applying centerless grinding for the removal of the entire excess stock.

(b) Extending the application of centerless grinding to parts much larger than were worked by this method in earlier years, a trend which proportionately increases the volume of stock to be removed when the thickness of the allowance remains the same.

Because the stock-removal rate in thrufeed centerless grinding is limited by such machine characteristics as drive power, the width of the grinding wheel, the rigidity of the grinder and of its major elements, etc., a great amount of stock could usually be removed only by passing the work successively through several centerless grinding machines, either arranged in-line, or interconnected by automatic feeding devices. In order to reduce the number of individual grinding machines involved in a single centerless grinding process, a new generation of powerful centerless grinders has been developed, capable of matching the stock removal capacity of two or three centerless grinders of earlier design. These production type grinders are commonly designated as heavy-duty centerless grinding machines.

Courtesy of Cincinnati Milacron, Inc.

Fig. 2-4.7 Multiple purpose heavy-duty centerless grinding machine designed for substantial stock removal, with rigidly supported spindles which accept wheels up to 10-inch (250-mm) width and are capable of maintaining consistent work size in continuous production runs. Compact design saves floor space and its adaptability to many types of special equipment contributes to extended applications.

The boundaries, with respect to the preceding category, are far from distinct. Still, there are a few characteristics which, as a rule, can be found in all heavy-duty grinding machines, which appear occasionally, or perhaps only to a lesser degree, in the common production types of centerless grinding machines:

(a) The general use of bridge type wheel mounting for both wheels, supporting the wheel spindles in bearings which straddle the wheel and thus more effectively resist the radial components of the grinding forces which tend to deflect the wheel spindle.

(b) Drive motors in the output ranges of 50 to 100 hp, reflecting the generally required high stock removal rates, considering that a rather closely constant ratio exists between the expended horsepower and the volume of material removed in grinding.

(c) Wheel widths in the order of 20 inches (500 mm), or more.

(d) Machine weights in the order of 18,000 to 25,000 lbs (8,000 to 11,000 kg).

Two examples of heavy-duty centerless grinding machines, representing different size groups are shown in Figs. 2-4.7 and 2-4.8.

Courtesy of Cincinnati Milacron, Inc.

Fig. 2-4.8 Large size heavy-duty centerless grinding machine adaptable for both infeed and thrufeed grinding, capable of operating with wheels up to 20 inches (500-mm) wide. Automatic control of different infeed rates applied in sequential order serves to combine roughing and finishing phases into a single infeed grinding operation. The special angular bed design is intended to provide greater contact pressure between the work and the regulating wheel.

Limited-purpose, Infeed Type, Centerless Grinding Machines for Large Workpieces

As an alternative to center type cylindrical grinding, the centerless method can offer significant advantages even in applications where the more efficient thrufeed grinding is not practicable because of the substantial dimensions and/or weight of the workpiece.

Special infeed type centerless grinding machines have been developed in order to realize the benefits of centerless grinding for large workpieces which, while not requiring or not being adaptable to thrufeed grinding, do exceed the capacities of multiple-purpose centerless grinders. By avoiding the need for a tilt adjustment of the regulating wheel as a means of providing the force component for the workpiece traverse, the wheel width of a given basic model of centerless grinding machine can be increased, thus extending the grindable workpiece length and concurrently, the acceptable work diameter can also be increased.

Centerless grinding machines in this category can be considered as providing the largest work size capacity particularly with regard to length (in one model this is 605 mm or $23\frac{1}{2}$ inches), which can be extended still further by means of special accessories. This expansion of the work capacity is accomplished partly by dispensing with the thrufeed grinding capability.

A characteristic example of the application of centerless grinders discussed in this category is shown in Fig. 2-4.9.

Courtesy of Lidkoeping A.B. — Hirschmann Corp.

Fig. 2-4.9 Partial view of a special infeed type centerless grinding machine, designed for large workpieces, several of which may be ground simultaneously utilizing the 610-mm (24-inch) wide grinding wheel. Photo shows large bearing races mounted on a common arbor, ready to be transferred into the grinding position by an automatic loader.

Shoe Type Centerless Grinding

In centerless grinding the confinement of the workpiece between the grinding wheel, the regulating wheel, and the work rest is assured through extended areas of contact with these machine elements. They are nominally line contacts and the greater their length the better can the position of the work during the centerless grinding process be maintained. For shorter parts which are processed in continuous thrufeed grinding, the butting of the faces of the consecutive parts may add another locating element, thus compensating for the relative shortness of the individual line contacts.

There are certain types of essentially disk-shaped parts whose surfaces to be ground are not plain cylinders, yet centerless grinding is still the most effective process in their manufacture. Such parts, however, are not adaptable to thrufeed grinding and must be ground individually by the infeed process. Characteristic, yet by far not exclusive, examples of such type of parts are the inner rings of ball bearings, into the outer surface of which a groove of essentially circular arc profile, the ball raceway, must be ground.

The System, its Elements and Operation

A new system of centerless grinding has been developed for workpieces of disk, annular, or similar shapes, with diameters often substantially larger than the widths, and which require infeed grinding because of the profile or the recessed location of the ground surface. This system incorporates the essential principles of conventional centerless grinding, such as locating the workpiece by supporting and restraining it on the outer surface which is the one being ground or on an adjacent round surface concentric with the former and using three lines of contact, one of

which is the operating face of the grinding wheel. The other two supporting surfaces are the two shoes mounted in exactly defined positions (see Figs. 2-5.1 and 2-5.2).

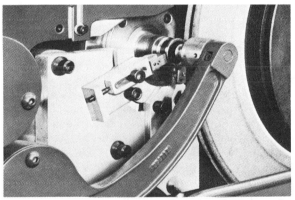

Courtesy of Cincinnati Milacron, Inc.

Fig. 2-5.1 Work area of a shoe type centerless grinding machine, showing the workpiece supported by two shoes in front of the grinding wheel. In front is the loader arm which will swing out of the work area before the grinding starts.

The workpiece is usually held by its face resting flush against a magnetic plate (see Fig. 2-5.3) mounted on the work spindle, thereby assuring the controlled location of the workpiece in a specific plane. That plate on which the magnetic force is holding the work, also provides its drive when the machine spindle is rotating. Holding the part by attaching its face to a magnetic plate does not prevent the part's radial shift when forces acting on the work are stronger than that exerted by the magnet in a direction normal to its principal pull. This balancing of forces permits the part to be ground without axial

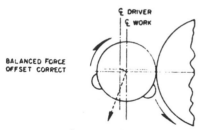

Courtesy of Cincinnati Milacron, Inc.

Fig. 2-5.2 Diagram showing the relative positions of the two shoes and of the grinding wheel around the periphery of the workpiece. The proper setup should assure a resultant force directed between the two shoes which keeps the rotating workpiece in dependable contact with these supporting elements.

movement, by applying the principles of centerless grinding in a process which operates in the following manner:

The workpiece retained by the magnetic force of the backing plate is located by the shoes in a radial position such that the center of the work and the axis of rotation of the plate on the work spindle are not coincident. The offset of the work center, as seen in Fig. 2-5.2, is toward the grinding wheel, called the front, and downward. It is easy to visualize that when the work spindle rotates, imparting a similar rotation to the workpiece it holds, that point on the plate which corresponds to the center of the work, will move in a downward direction between the two shoes. A balance of forces will result which is similar to that in the regular centerless grinder, where it is directed between the work-rest blade and the regulating wheel. The force resulting from the offset of the backing plate (known as the "wind-in"), in combina-

tion with the grinding force, keeps the work in uninterrupted contact with the two shoes while the work is continuously slipping on the surface of the backing plate.

The analogy between the regular and the shoe type centerless grinding also exists with respect to the rounding effect of the process. As can be seen in Fig. 2-5.4, the imaginary line connecting the center of the grinding wheel with the initial point of contact on the opposite shoe (functionally equivalent to the wheel center line in the regular centerless grinder), is below the center of the work. The greater that distance, the faster will be the rounding effect, however, the stability of the work retainment will concurrently be reduced, increasing the likelihood of developing chatter. These conflicting consequences of shoe adjustment will call for a compromise, aiming at a reasonably fast rounding effect, yet without impeding the stability of the work retainment. The correct position of the shoes in relation to the work can be calculated by applying formulas developed for that purpose (see Fig. 2-5.5), and in the case when instability of the work is experienced the offset will have to be reduced in the indicated manner.

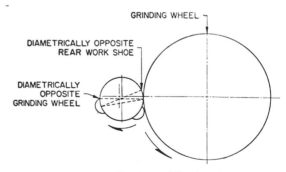

Courtesy of Cincinnati Milacron, Inc.

Fig. 2-5.4 Diagram shows that the non-diametrical work support—known as offset—tends to reduce the out-of-roundness of the work in a manner similar to holding the work above the wheel centerline in regular centerless grinding.

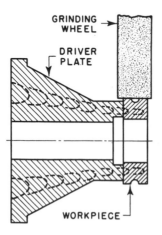

Courtesy of Cincinnati Milacron, Inc.

Fig. 2-5.3 Magnetic drive plate for holding and rotating the work in shoe type centerless grinders. The diagram indicates the lines of force when the electromagnet is energized. During the loading and unloading of the work the plate is de-energized.

Both shoes, or only one of them, are held in a pivot support, permitting minor self-adjustment of the shoe surface to that of the work. Because of the sliding contact which is similar to that on the work-rest blade, the shoes are made of wear resistant material (hardened steel, cast iron, bronze or carbide tipped) selected by taking into account the sensitivity of certain work materials to scoring.

Occasionally the shoes require readjustment to compensate for wear, and also replacement when the wear has reached an advanced state. Readjustment or exchange of the shoes is also needed when changing over to a different workpiece. In order to reduce the setup time loss in the shop and also to assure a very accurate setting of the shoes, the shoes are often mounted on interchangeable cartridge plates. In the case of this type of tooling, the setting up of the shoe cartridges can be carried out, usually with the aid of special gages, away from the grinding machine, e.g., in the toolroom. The complete shoe cartridge can then be installed into the grinding machine with a minimum of time loss, yet with well controlled accuracy.

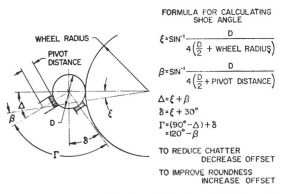

Fig. 2-5.5 Method of calculating the angles for the correct shoe positions. Offset is reduced by moving either or both shoes in a direction away from the grinding wheel.

Design Characteristics Adapted to the Principal Applications

The shoe type centerless grinding machines are used predominantly for high volume production, where the automation of the process and the reduction of the cycle time are important objectives. Loading and unloading of the work being one element of the operational cycle, efforts were made by machine tool designers to reduce the time needed for that nonproductive, yet unavoidable operational element. Figure 2-5.6 shows the principles of operation of an advanced design work loader, comprising two arms in synchronized movement for carrying out the unloading of the finished part, immediately followed by the loading of a new workpiece, the time requirement of the whole work change being in the order of about one second.

The character of the work done on shoe type centerless grinders does not, in general, call for extended wheel retraction for changing the work. This condition supports the efforts for fast grinding head movements and accurate infeed control, comprising approach, rough, and finish grinding. One system developed for satisfying the objectives of rapid retraction and approach, in combination with accurate

Fig. 2-5.6 Photo showing the operating elements of a double-arm work loader used on a shoe type centerless grinder. A-loading chute, B-rotary escapement, C-backing plate, D-unloading chute, E-loading arm, F-magnetic driver base, G-manual control for arms, H-unloading arm.

infeed control, operates by substituting a swing movement of the grinding wheel head for the conventional translation of a slide. The operating principles of the wheel head which can be tilted around a fixed trunnion by means of control elements which provide for rapid advance, two-rate infeed, pre-set tarry and accelerated retraction are shown in Fig. 2-5.7. In another model of shoe type centerless grinders the wheel head is supported on three ball slides and the traversing movement is imparted by a lead screw acting in line with the area of grinding contact.

Instead of shoes on which the workpiece is sliding, rollers can and are being used in some cases for supporting the work, particularly when its relatively heavy weight could cause scoring in sliding contact. Experience indicates, however, that a nonrotating shoe provides a more accurate work support than rollers sustained by bearings whose unavoidable inaccuracies are translated to the ground work.

Figure 2-5.8 shows a well-known model of the shoe type centerless grinding machines, adaptable to a wide range of workpiece configurations and ground surface profiles, the latter produced by means of appropriate wheel truing devices. For specific workpieces, various types of single-purpose shoe type centerless grinding machines have been developed. Some of these machines are designed for grinding simultaneously two mutually perpendicular surfaces, even in combination with adjoining radii, by applying the principles of angular approach grinding (see Chapter 1-4). Grinding machines built for such applications

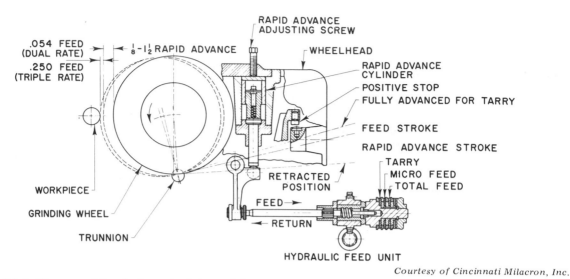

Courtesy of Cincinnati Milacron, Inc.

Fig. 2-5.7 Diagram of the design principles and feed elements of the swinging wheel head used for one model of shoe type centerless grinding machines.

have the work head swiveled at, e.g., 45 degrees, with respect to the grinding wheel spindle; that angle varying, depending on the relative widths of the concurrently ground surfaces. Typical parts for such operations are bearing races with lips or ground shoulders, e.g., the inner races of tapered roller bearings.

Multiple surface grinding operations of the mentioned type require special grinding wheel truing devices, usually having two separate slides and two diamonds. Substantial time savings can be accomplished in this process by using rotary diamond dressers for complex shapes and advancing the diamond rolls in a radial or tangential direction toward the

grinding wheel. In the latter arrangement it is possible to use more than a single diamond roll, each generating a different section of the grinding wheel profile (see Fig. 2-5.9).

Courtesy of Bryant Grinder Corp.

Fig. 2-5.9 Work area of a special shoe type centerless grinder with angled work head for grinding two essentially perpendicular surfaces on tapered roller bearing inner rings, in a process called "conjugate" grinding. In the front, at left, are the two diamond rollers for wheel truing, moved out of the operating position in which the common slide of the rolls will traverse them tangentially to the wheel face.

Advanced Automation in Shoe Type Centerless Grinding

Very impressive progress in shoe type centerless grinding has been achieved in recent years by several prominent machine tool manufacturers, with the

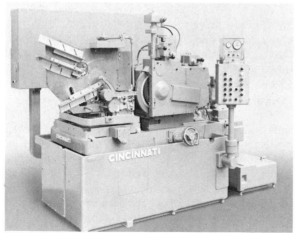

Courtesy of Cincinnati Milacron, Inc.

Fig. 2-5.8 Typical representative of a shoe type centerless grinding machine equipped with feed chutes for the automatic work loader.

application of solid state electronics for the command functions of grinding machines. The purpose of such control devices is to reduce the dependance of the production equipment on the operator's skill for accomplishing very rigorous performance objectives and, at the same time, make it possible for a battery of machines to be attended by a single trained worker. Moreover, the speed of the control functions eliminates, or substantially reduces idle times which, in combination with maintaining optimum operational values, contributes to a very high productivity.

Of course, the utilization and the economic justification of such advanced equipment is predicated on production runs which comprise large quantities of identical or closely similar parts. Typical examples of appropriate workpieces are the inner races of tapered roller bearings, whose external grinding involves the concurrent finishing of several interrelated surfaces with particular contour characteristics. These parts must be produced to very exacting dimensional tolerances with regard to diameters 0.0002 in. (0.005 mm), angles (0°02'), profile 0.0002 in. (0.005 mm), roundness 0.00008 (0.002 mm) and surface finish 10μ in (0.25 micrometers). These figures indicate typical tolerance values.

A typical representative of this new breed of highly automated shoe type centerless grinding machine will now be discussed, with particular attention given to a few of the novel control characteristics.

Figure 2-5.10 shows the work area and Fig. 2-5.11, the control panel of an automatic shoe type center-

less grinding machine with solid state control, set up for the grinding of tapered roller bearing inner races. To assure the dependable consistency of the feed rates and dwell periods, these are controlled by a 400,000 cycles per second crystal oscillator which is accurate within 1/10th of 1 per cent. The numbered thumbwheels on the control panel (see Fig. 2-5.11) permit a setting for each section of the wheel advance (coarse, fine, and micro), the advance length (in inches), the feed rate (inches/second), and the dwell time (in seconds). The advance movement for the feed is transmitted through a precision feed screw, which is actuated by a stepping motor, commanded by the digital programmed controller. The resulting slide movement, its direction, distance, and rate, are in exact agreement with the programmed commands. Also programmed are the grinding wheel truing values, including the diamond feed, the frequency of the truing, and the compensation in the wheel slide position.

For assuring an excellent size control independently of the breakdown, i.e., wear of the grinding wheel

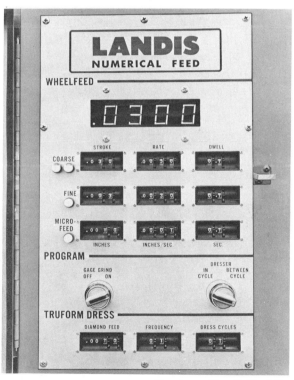

Courtesy of Landis Tool Co.

Fig. 2-5.11 Control panel of a fully automatic shoe type centerless grinding machine operating with numerically programmed wheel advances, feed rates, and dwell periods. The numbers at the top, displayed by electronic tubes, report the difference between the momentary actual work size and the attainable final size, in units of ten-thousandths of an inch.

Courtesy of Landis Tool Co.

Fig. 2-5.10 Work area of a special shoe type centerless grinder carrying out conjugate grinding on two separate but functionally interrelated and profiled sections of a bearing race. The work feeding chutes and the loader arm are on the left side, at the front. In this type of grinding machine the wheel truing device is in the rear (not visible in this photo).

and other inconsistent variables, the final wheel advance is controlled by an electronic in-process gage, whose signals guide the operation of the microfeed advance, by means of fine pulses in increments of 0.000 050 inch (0.00125 mm).

Shoe type internal grinding machines are often considered as belonging to the centerless grinder category, because the workpiece is not rotating around a fixed center. While the absence of a fixed center, in mechanical linkage with the rotating workpiece, is one of the characteristics of the centerless grinding method, the other essential condition, namely the floating position of the work's axis of rotation, is not present in the shoe type internal grinding. In this latter process the workpiece is supported on an external surface which has been produced in a preceding operation, in contrast to the centerless grinding proper, which supports the work on the very surface which is being ground in the course of the process.

In view of the significant deviation from one of the basic characteristics of the centerless grinding method, and also because the working of an internal surface is considered the dominant element of the process, shoe type internal grinding will be discussed in Section 3 on Internal Grinding.

Supplementary Process Functions and Equipment

Carrying out the centerless grinding process requires several functions which supplement the operations of the basic elements. The devices used for the implementation of these functions are essential members of the operating equipment, generally with direct effect on the productivity and the performance quality of the process. Several of the supplementary functions are required for specific types of work only, others are similar to those used in other grinding processes. There are, however, three essential functions which are needed in all types of centerless grinding processes, yet generally differ in their operation and equipment from those used for similar purposes in other types of grinding machines. These functions, which will be discussed in greater detail in the following, are (a) the truing of both the grinding wheel and the regulating wheel; (b) the work loading for both infeed and thrufeed grinding; and (c) the work gaging.

The Truing of the Grinding Wheel

Centerless grinding machines usually operate with large diameter grinding wheels, a condition which tends to reduce the frequency of the required wheel truing. In addition, in thrufeed grinding with the use of wide grinding wheels, the wear is distributed over a large wheel surface and the process commonly results in rather uniform wheel wear. By combining these process characteristics with the proper wheel selection in the interest of utilizing the self-dressing properties of the grinding wheel, the periods between the successive grinding wheel truings can be rather long, up to several hours.

On the other hand when the truing of the wheel is needed, the sheer size of the commonly used grinding wheels calls for sturdy devices and tools, providing adequate rigidity and operating accuracy in the interest of good work finish and size control.

The dressing and truing of the grinding wheel in centerless grinding is usually needed for the following purposes individually or for several concurrently:

(a) To assure true running after mounting of a new wheel
(b) To clean the glazed or loaded wheel surface in the interest of free cutting wheel action
(c) To eliminate the effects of uneven wheel wear
(d) To control size
(e) To control the produced work finish
(f) To true or maintain a specific form (profile) of the wheel face.

With regard to the type of tools used for truing the centerless grinding wheel, the following major truing systems can be distinguished, several of which will be discussed in detail:

(a) Single-point truing diamonds
(b) Multiple-point, cluster type diamonds (not applicable for form truing)
(c) Metallic (tungsten carbide) wheels (limited to coarse dressing)
(d) Crush truing (preferred for intricate profiles)
(e) Rotary diamond truing. (Because of cost, generally preferred for narrow wheels, such as used on shoe type centerless grinding machines.)

Single-point diamond truing is the most commonly used method and the only proper one when the truing of a form with the aid of a profile bar (template) has to be carried out. Usually diamond sizes in the order of one or two carats are needed for the large wheels used on most types of centerless grinding machines.

Figure 2-6.1 shows the general view and Fig. 2-6.2 the cross-sectional drawing of a characteristic single-diamond-point type grinding-wheel truing device. For reasons of space saving and for more convenient operation the device is mounted in an inclined plane on the centerless grinding wheel head. The sleeve of the device is directed radially toward the center of the grinding wheel, while the diamond holder is tilted by about 15 degrees in the direction of the wheel rotation. The resulting position of the diamond nib will present the diamond point to the grinding wheel at an angle, thus the periodic rotation of the nib in its holder will promote a uniform diamond wear with extended service life.

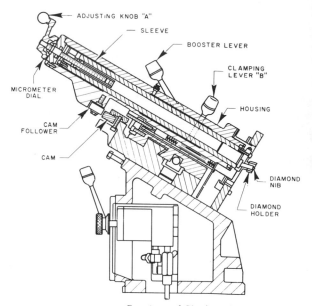

Courtesy of Cincinnati Milacron, Inc.

Fig. 2-6.2   Cross-sectional view of the grinding wheel truing unit shown in Fig. 2-6.1.

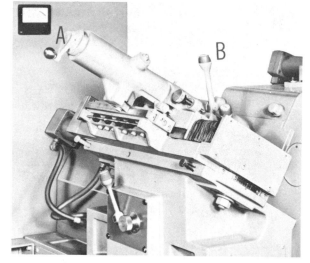

Courtesy of Cincinnati Milacron, Inc.

Fig. 2-6.1 General view of a grinding wheel truing unit in a design providing swivel adjustment. *A*. Diamond adjustment; *B*. Clamping lever.

The major functional elements of the truing device are named and indicated by arrows on these illustrations. The device can be used for truing either a straight or a profiled wheel face. For profile grinding the commonly used flat bar will be replaced by an appropriate profile bar or cam to guide the cam follower. Depending on the arrangement and the design system of the truing device, either gravity only, or some additional source of force, such as adjustable spring pressure, is used to assure that the follower bears steadily against the profile cam. For truing straight cylindrical wheel shapes the action of the cam-follower couple can be disengaged by means of the clamping lever *B* (see Fig. 2-6.2).

The truing unit slides in guideways across the face of the grinding wheel, and is traversed by a hydraulic cylinder at a rate which can be adjusted with the aid of the feed rate control knob while a lever also controls the direction of movement of the truing unit (see the vertical front panel in Fig. 2-6.1).

Recommended traverse speeds for the grinding wheel truing unit operating with a single-point diamond are:

Rough grinding   10 to 20 in. per min (250 to 500 mm per min)

Finish grinding   4 to 7 in. per min (100 to 175 mm per min)

These are general values; the optimum diamond traverse rate will also depend on the grain size of the wheel and the attainable work-surface finish.

The diamond is caused to approach or retract from the grinding wheel by the movement of the sleeve; the latter is moved by turning the adjusting knob *A* (Fig. 2-6.2) as indicated by the attached micrometer dial. The actual position of the diamond nib near to or in contact with the grinding wheel can be observed through a peephole which is normally covered (see Fig. 2-6.1). After the diamond has made a light contact with the grinding wheel, the sleeve may be advanced to produce the desired truing cut, usually in the order of 0.001 in. (0.025 mm) per pass. Because the grinding wheel tends to wear faster in use near the edges than in the center, that initial contact for establishing the starting point for the diamond advance should be made near the center of the wheel face.

The booster lever (Fig. 2-6.1) permits the cam follower to be manually assisted over steep rises in the cam which might have the tendency to jam the follower while carrying out its traverse movement.

Certain types of truing units, such as illustrated in Fig. 2-6.1, can be swiveled for the purpose of truing the grinding wheel to a slender taper shape, as needed for gradual stock removal in thrufeed grinding. To carry out such adjustment the clamping screws are first loosened, then the unit is swiveled with the aid of an adjusting screw as guided by the index plate D or by mounting a sensitive dial indicator.

Crush Truing of the Grinding Wheel

The crush truing of the centerless grinding wheel is of advantage in infeed grinding operations involving work profiles of irregular shapes, or containing small radii, narrow grooves, etc., which are difficult or even impossible to true into the wheel face by any other method. It is also applied for the grinding of slender parts which require wheel faces with particularly free cutting properties.

The crush roll, made of hardened tool steel or cemented carbide, has the profile of the ground workpiece, and is mounted on a freely rotating spindle, which is parallel with the spindle of the grinding wheel. During crush truing the grinding wheel must rotate at a slow speed of about 200 to 300 fpm (1 to 1.5 m/sec), consequently the crush truing process can only be carried out on centerless grinding machines designed to provide that low rotational speed of the grinding wheel.

The crush roll, whose spindle is mounted on a slide, is advanced to contact the grinding wheel, with which it then rotates. The infeed advance is continued at a slow rate until the complete profile of the roll is transferred to the face of the grinding wheel.

Some crush truing devices are also equipped with a drive motor for the crush roll (see Fig. 2-6.3), to be used for rotating the roll when its profile is being reground by the wheel of the centerless grinding machine. For such operation the grinding wheel has to be trued first with another roll of perfect profile, a master roll, usually retained for that purpose. Then the rolls are exchanged, the wheel is rotated at regular grinding speed while the crush roll is driven at a speed about equal to that of the crushing operation.

In crush truing the abrasive grains are not cut as in diamond truing, but dislodged from the bond of the

Courtesy of Cincinnati Milacron, Inc.

Fig. 2-6.3 Crush truing device with roll regrind motor, installed on a centerless grinding machine, behind the grinding wheel. The side cover plates of the wheel heads are removed for better visibility.

wheel, consequently the remaining grains retain their original sharp edges. Therefore, the crush trued wheel is free cutting to a degree which sensibly reduces the grinding force and increases the cutting efficiency of the wheel. The reduced grinding force is particularly advantageous when grinding slender shafts which are sensitive to deflection. Also, the higher cutting efficiency increases both the grinding rate and the size retention of the grinding wheel, resulting in less frequent truings.

The Truing of the Regulating Wheel

The profile of the regulating wheel must comply with its major functions in the centerless grinding process, which involves the following: (a) to participate in the radial restraint and support of the work, together with the grinding wheel and the work-rest blade; (b) to rotate the work at a speed essentially identical with its own peripheral speed; (c) to impart an axial movement to the rotating work in thrufeed grinding, and often an axial force as well in infeed and endfeed grinding.

These functions of the regulating wheel require that the work surface element which is momentarily in contact with the regulating wheel shall face a complying element of the regulating wheel surface, thus establishing a straight line contact. That contact usually extends over the entire length or, for complex forms, along specific sections of the work surface.

That requirement is relatively simple to satisfy as long as plain cylindrical work is ground in an infeed process and the axes of the two wheels as well as of the work are in mutually parallel positions. In the case of profiled workpieces ground in an infeed process, the truing movement must be guided by a profile bar to produce a wheel surface whose profile agrees completely, or along the selected contact sections with that of the work (see Fig. 2-6.4).

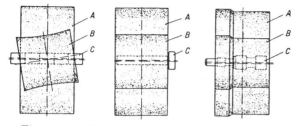

Fig. 2-6.4 The shapes of the regulating wheel in commonly used centerless grinding methods. (Left) Hyperbolic shape in thrufeed grinding, resulting in a straight line contact of the tilted regulating wheel with the cylindrical workpiece. (Center) Cylindrical shape for infeed grinding. (Right) Formed shape for profile grinding. A—Grinding wheel; B—Regulating wheel, and C—Workpiece.

However, when the setup requires a feed angle obtained by inclining the position of the regulating wheel head, the task of generating a complying regulating wheel profile which will produce a straight line contact with the work along its traverse movement becomes more complex. The reference to the complexity of the geometric conditions does not necessitate an intricate procedure in carrying out the truing of the regulating wheel. Relatively simple instructions and the use of properly designed truing devices, permit the operator of the centerless grinding machine to produce by truing, the correct regulating wheel profile.

An analysis of the conditions involved is presented in the following to explain the purpose of the specified truing adjustment procedure:

The reasons for tilting the regulating wheel head over a specific angle of inclination in thrufeed centerless grinding are discussed, together with the pertinent values, in Chapter 2-8 of this section. The inclination of the regulating wheel will produce a condition where the axis of the regulating wheel will be out-of-parallel with the axes of the grinding wheel and of the work. A straight cylindrical regulating wheel whose axis is tilted would make contact at a single point only with cylindrical work having its axis parallel with that of the grinding wheel. In order to create a line contact the regulating wheel must be trued to a concave (hyperbolic) profile. This is accomplished by traversing the truing diamond, usually through the swivel setting of the truing slide, along a line which has essentially the same angle of inclination with respect to the regulating wheel axis as the

latter is tilted in relation to the plane of the grinding wheel axis.

The truing angle setting on the regulating wheel truing device is practically identical with the incline angle of the regulating wheel axis, as long as the ratio between the diameters of the regulating wheel (D) and of the work (d) is high, e.g., 50 to 1 or more, such as in the case of small diameter workpieces.

For work of relatively large diameter, with a decreased D to d ratio, a compliance between the two contacting surfaces will, for geometric reasons, require the amount of concavity needed on the regulating wheel surface to be diminished, resulting in a reduced truing angle.

That relationship is expressed by the following formula:

$$\tan \alpha = \frac{\tan \theta}{\left(1 + \frac{r}{R}\right)^{\frac{1}{2}}}$$

Where:

$\tan \alpha$ = Tangent of the truing angle
$\tan \theta$ = Tangent of the incline angle of the regulating wheel axis
r = Radius of the workpiece
R = Radius of the regulating wheel

Based on the above formula, Table 2-6.1 has been developed for shop use, with rounded values expressed in fractional inches.

For the purpose of improving the rounding effect in centerless grinding the workpiece is generally held with its center above the plane containing the axes of the wheels, which latter are often referred to as the centerlines (C/L) of the grinding machine. In order to generate a regulating wheel form which, at the level where it touches the work, will produce a straight contact line, the truing diamond must follow a path which is displaced by the same amount from the plane of the wheel axis as the height of the work contact above the machine centerline.

In the case of small diameter workpieces, for practical purposes, that distance is considered equal to the height setting of the work center above the C/L. However, for larger workpieces the value of D to d decreases, causing the level of the actual contact to differ from the work center height.

These conditions are illustrated diagrammatically in Fig. 2-6.5 showing the case of a 6 to 1 ratio between regulating wheel and work diameters. It can be seen that although the work center is $\frac{1}{2}$ inch above the C/L, the actual level of contact between work and regulating wheel is closer to the plane of the wheel centers (about $\frac{15}{32}$ nd inch). That actual distance between the contact level and the C/L must be transferred to the diamond position, adjusting the "set-over"

Table 2-6.1 Regulating Wheel Truing Angle Setting for Centerless Thrufeed Grinding*

Incline Angle of the Regulating Wheel	Ratio $\frac{D}{d}$ $\left(\dfrac{\text{Regulating-wheel diameter}}{\text{Work diameter}}\right)$									
	3	3.5	4	5	6	7	12	18	24	48
	Regulating Wheel Truing Angle Setting									
1°	50′	50′	55′	55′	55′	55′	55′	1°	1°	1°
2°	1°45′	1°45′	1°50′	1°50′	1°50′	1°55′	1°55′	2°	2°	2°
3°	2°35′	2°40′	2°40′	2°45′	2°50′	2°50′	2°55′	2°55′	3°	3°
4°	3°30′	3°30′	3°35′	3°40′	3°45′	3°45′	3°50′	3°55′	4°	4°
5°	4°20′	4°25′	4°30′	4°35′	4°40′	4°40′	4°50′	4°55′	5°	5°
6°	5°15′	5°15′	5°25′	5°30′	5°35′	5°40′	5°45′	5°55′	5°55′	5°55′
7°	6°10′	6°10′	6°20′	6°25′	6°30′	6°35′	6°45′	6°50′	6°55′	6°55′

*With permission from *Cincinnati Milacron, Inc.*

with the aid of the graduations on the diamond holder.

For calculating the diamond set-over the following formula is used:

$$g = h\,\frac{\sin \alpha}{\sin \theta}$$

Where:

g = the required diamond set-over
h = the height of the work center above the C/L
α = regulating wheel truing angle
θ = incline angle of the regulating wheel axis

Based on this formula the approximate values of Table 2-6.2 have been calculated.

An alternative method for calculating the effect on the set-over of the two variables, namely (a) the height of the work center above the C/L of the wheels; and (b) the ratio between the diameters of the regulating wheel and the work, is often used to avoid the reliance on tables.

When using this method the conditions are considered in a manner illustrated in the diagram of Fig. 2-6.6 and the following formula is used for calculating the set-over:

Table 2-6.2 Set-over of the Diamond Point for Different Work Height Settings in Centerless Thrufeed Grinding*

Height of Work Center Above Wheel C/L, in.	Ratio $\frac{D}{d}$ $\left(\dfrac{\text{Regulating wheel diameter}}{\text{Work diameter}}\right)$									
	3	3.5	4	5	6	7	12	18	24	48
	Diamond Point Set-over, inch									
1/8″	7/64	7/64	7/64	7/64	7/64	1/8	1/8	1/8	1/8	1/8
1/4″	7/32	7/32	7/32	15/64	15/64	15/64	15/64	1/4	1/4	1/4
3/8″	21/64	21/64	11/32	11/32	11/32	23/64	23/64	23/64	3/8	3/8
1/2″	7/16	7/16	29/64	29/64	15/32	15/32	31/64	31/64	1/2	1/2
5/8″	35/64	35/64	9/16	37/64	37/64	19/32	19/32	39/64	5/8	5/8
3/4″	21/32	21/32	43/64	11/16	45/64	45/64	23/32	47/64	47/64	47/64
7/8″	3/4	49/64	25/32	51/64	13/16	53/64	27/32	55/64	55/64	55/64
1″	55/64	7/8	29/32	59/64	15/16	15/16	61/64	63/64	63/64	63/64

*With permission from *Cincinnati Milacron, Inc.*

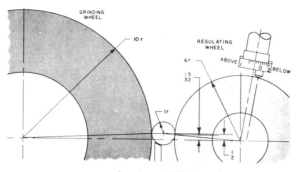

Fig. 2-6.5 Diagram showing the actual level of the point of contact between the work and the regulating wheel (15/32 inch) in relation to the level of the work center (1/2 inch). Both levels are referenced from the common centerline of the grinding wheel and the regulating wheel.

$$B = \frac{H}{1 + \frac{d}{D}} = \frac{H \times D}{D + d}$$

Where:

B = Set-over (shift of the diamond point from the center plane of the regulating wheel)

H = Height of the work center *above* the wheel C/L

D = Diameter of the regulating wheel

d = Diameter of the workpiece

When the work center is distance H below the C/L of the wheels, then the calculated set-over value is applied in the opposite direction, as shown in the bottom diagram of Fig. 2-6.6.

The Tools and Equipment for Truing of Regulating Wheels

Single-point diamonds are the only kind of tools which are used in general practice for the consistently fine truing of the regulating wheel. This operation has to be carried out much less frequently and, as a rule, to a shallower depth than in the case of the grinding wheel. While the breakdown of the grinding wheel is inherent in the grinding process and is related to the volume of stock removed from the work surface, the regulating wheel, having no cutting function, wears only at a slow rate.

Consequently, the truing conditions for the regulating wheel do not call for tools which remove wheel material faster or more economically than single-point diamonds. Nor is there justification for using crush or rotary diamond truing, often preferred for the grinding wheel when grinding intricate part profiles, since the regulating wheel does not have to incorporate all of the details of the final work contour (see Fig. 2-6.7).

However, the regulating wheel truing device must be mounted on a slide base which can be swiveled to the required truing angle to be set by means of a graduated segment (see Figs. 2-4.5 to 2-4.7). A typical example of the regulating wheel truing device is shown in the cross-sectional diagram Fig. 2-6.8. In its essential features this device differs only little from the grinding wheel truing device shown in Figs. 2-6.1 and 2-6.2, except in the following respects:

- The diamond holder is not inclined, because the radial position of the diamond point must be precisely controlled for reference purposes.
- The diamond holder is actually a slide which permits lateral adjustment for the set-over, as indicated by the graduations along the edge of the holder.

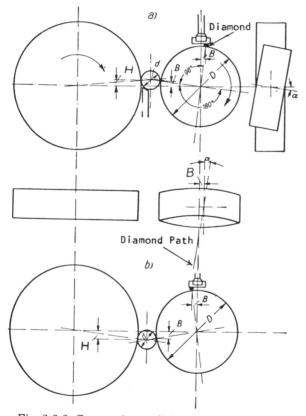

Fig. 2-6.6 Geometric conditions resulting from the tilting of the regulating wheel by the amount of the feed angle. The diagram shows the set-over (B) of the diamond which, in combination with the swivel setting, is needed for producing the correct path of the truing traverse as a function of the distance (H) of the work center from the wheel centerline. (a) Work center above the C/L; (b) Work center below the C/L.

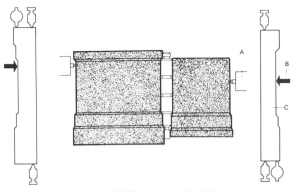

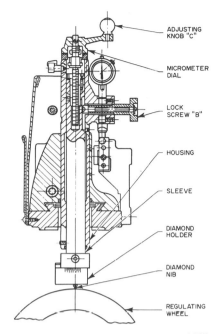

Fig. 2-6.7 Diagram illustrating the trued contour of the regulating wheel (at right) in relation to that of the grinding wheel in centerless form grinding. *A*—Truing diamond; *B*—Follower; and *C*—Profile bar with setting screws and a dial indicator.

- A booster lever is not needed because the truing of the regulating wheel usually does not involve steep profile sections.

A regulating wheel truing device, the design of which represents an uncommon approach toward eliminating the need of a swivel adjustment of the truing slide in conformance with the regulating wheel-head inclination, is shown in Fig. 2-6.9. The slideways of this truing device are mounted on the machine bed and stay in the horizontal position while the regulating wheel head is inclined for producing the required feed angle. Consequently the path of the diamond travel will always remain parallel with

Fig. 2-6.8 Cross-sectional view of a centerless grinder regulating wheel truing unit.

the work axis and, with the adequate height setting, will generate on the surface of the tilted regulating wheel a concave contour corresponding to the contact line of the traversing workpiece. The principles of operation of the machine bed-mounted regulating wheel truing device in comparison to that of the

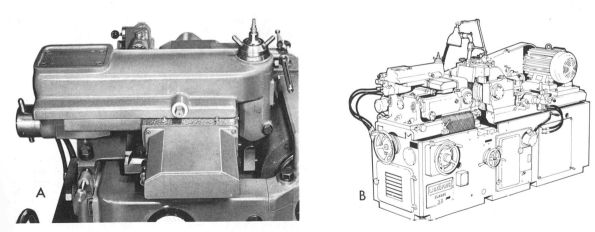

Fig. 2-6.9 Bed-mounted regulating wheel truing device of special design which retains the diamond path in the horizontal plane independently of the tilt of the regulating wheel, thereby dispensing with the swivel setting of the truing slide. A. Close-up view, showing the position of the truing slide with respect to the regulating wheel. B. Line diagram of the centerless grinder, showing in front, the easily accessible location of the bed-mounted regulating wheel truing device.

conventional wheel-head-mounted design, are shown in Fig. 2-6.10.

The truing data applied to regulating wheels differ from those used in the truing of the grinding wheel in actual values and also in that they remain the same for roughing and finishing. The commonly applied feed rate for regulating wheel truing is in the range of 1 to 2 inches (25 to 50 millimeters) per minute.

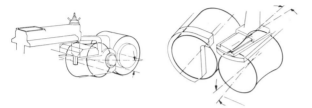

Courtesy of Lidkoeping A.B. — Hirschmann Corp.

Fig. 2-6.10 Diagrams of the operating principles of two systems of regulating wheel truing devices: (*Left*) Bed-mounted design which produces the hyperbolic profile of the tilted regulating wheel without the need of a swivel setting. (*Right*) Conventional design with the device mounted on the tiltable regulating wheel head and swiveled to produce the hyperbolic wheel profile.

Regulating wheel truing for retarded thrufeed traverse of the work. In addition to the basic requirements of the regulating wheel truing discussed above, occasionally, specific conditions have to be produced to satisfy uncommon process requirements. As an example, which also illustrates the flexibility of centerless grinding operations, the thrufeed grinding of relatively narrow parts, with diameters several times larger than the width, is mentioned. In order to assure the stability of the work while being ground, as well as the squareness of the ground periphery to the faces which have been finished in a preceding process, the consecutively advancing parts should be butted against each other. To assure that condition the traverse rate of the work, while passing between the wheels, must be gradually decreased by truing the regulating wheel to a slightly conical shape, with the larger diameter in the front. Consequently, the peripheral speed, and, tied in with it, the axial advance of the parts will be gradually retarded, resulting in a constant push tightly butting the consecutive parts against each other. For achieving the desired truing shape the regulating wheel truing device is equipped with a copying cam.

Automatic Work Feeding and Loading

The term *automatic work feeding* designates, in this context, a function by which individual work-pieces are transferred from a lot, temporarily stored

in a container near the machine, into a location from where they are introduced into the operating area of the grinding machine. Feeding becomes a distinct process function of centerless grinding when it is carried out as an integral element of the operation. In distinction, the transporting of parts by manual action or control is a material handling function, necessary but not integrated into the grinding process.

Considering the generally high production rates in centerless grinding, the automatic feeding of the work has a more frequent application in this, than in most other grinding methods. Automatic work feeding is also needed to utilize the automated operations of many types of centerless grinding machines, thus making it possible to run the entire process essentially unattended. In practice, this may permit having one operator in attendance for several concurrently operating centerless grinding machines, particularly when such equipment is used to produce the same or similar types of parts.

Automatic work feeding is primarily applied in the high volume production of small and medium size parts in either of the basic methods (thrufeed, infeed, and shoe type), although such equipment might exceptionally find uses in the grinding of heavy work-pieces. An example of this latter type of application is centerless bar grinding, which will be discussed in detail in the following section.

The *loading of the work* in centerless grinding designates the transfer of the workpiece into the operating area of the machine. In principle it differs from similar functions carried out on other types of machine tools, due to the self-locating properties of the centerless grinding process. Consequently the loading must result only in placing the work into the approximate operational position. From there on the restraining forces of the centerless grinding process take over the minor corrections in the positioning of the workpiece. By obviating the need for accurate centering and adjustment the loading operation in centerless grinding can be very fast.

Another factor which facilitates the work loading in centerless grinding is the absence of the clamping function that is needed in most other types of grinding machines, such as by the gripping of a chuck or collet, or by the axial pressure of centers applied following their retraction while placing the part into position. These characteristics associated with work loading in centerless grinding aid in the automation of the process and contribute to its high productivity.

Depending on the applied variety of the centerless grinding method, whether thrufeed, infeed, or shoe type, the operational requirements of work feeding and loading will vary. Other factors affecting the selection of the appropriate feeding and loading system are the dimensions, weight, and configuration of the workpieces, the desired production rates, the need for part orientation, and so forth.

Table 2-6.3 presents a general survey of the work feeding and loading systems which are frequently used in different types of centerless grinding processes. Other combinations of process types and work-handling systems may occasionally be needed or deemed advantageous. A more detailed discussion of a few basic work-feeding and work-loading systems follows:

Feeding of Individual Workpieces in Thrufeed Grinding

When individual parts are ground in the thrufeed process the best results, both in quality and productivity, are predicated on a continuous flow of the successive workpieces, with little or no space in between. The latter condition is designated as the butting of adjoining parts.

Considering the rather fast traversing speed, particularly of small diameter parts, in an effeciently operating thrufeed process, it is improbable that a continuous flow of the work could be maintained by manual feeding of the parts to the centerless grinder. Such an uninterrupted manual servicing of the operating machine would in any case defeat one of the major advantages of the thrufeed centerless grinding, that of automatic, with only occasionally controlled, operation.

Automatic devices for the continuous feeding of individual parts are most frequently designed to operate by gravity. A sufficient number of parts are stocked in an essentially vertical tube or chute, which curves into the horizontal just ahead of the grinder's work-rest blade. The weight of the stacked parts will exert the pressure needed to introduce the part which is at the front of the stack into the space between the grinding wheel and the regulating wheel, from here on the action of the regulating wheel, set at an appropriate feed angle, will take over the role of traversing the work.

For the satisfactory operation of the gravity feeding system, the weight of the workpiece stack must be maintained within specific limits; too few parts will not produce the force needed to move the parts through the final elbow section of the feed tube, while too many parts can cause jamming. In order to maintain the proper number of parts in the feed tube ahead of the grinding machine it is practical to install two sensing devices (mechanically contacting limit switches, induction coil, photoelectric sensor, etc.),

signals from which will, by energizing a solenoid, actuate a gate. When the level of the stacked parts reaches the top sensor, it will close the gate, shutting off the entry of additional parts until the stack recedes to the lower level sensor which then will cause the gate to be opened again.

For storing and feeding the parts into the guide tube or chute, some type of hopper or *bowl* is generally used, which may be equipped with a lifter arm, a belt lifter, or a rotating paddle directing the parts into a funnel at the bottom of the bowl, or a vibrating device causing the parts to move up on a peripheral ramp inside the bowl. Vibratory feeder bowls (see Fig. 2-6.11) are getting increasing acceptance because they offer many advantages, including the simpler design of the part orienting elements should such be needed for asymmetrical parts, and their capacity of handling even delicate parts without causing damage.

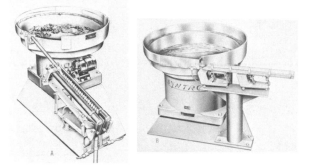

Courtesy of FMC Corp. — Material Handling Equipment Div.

Fig. 2-6.11 Examples of vibratory part feeder bowls with internal spiral track, which are frequently used in the centerless thrufeed grinding of small and medium size workpieces. A. Vibratory feeder bowl combined with power driven orienting roll chute—a device used, e.g., for tapered bearing rollers; B. Vibratory feeder bowl with attached vibrating horizontal chute for transferring shafts into a centerless grinding machine for continuous thrufeed grinding.

Feeding and Loading of Individual Workpieces for Infeed Grinding

In the case of small parts of essentially regular shape, the feeding to the loading station may be identical with the systems used for thrufeed grinding. Parts of irregular shape or those having larger dimensions, usually require specially adapted feeding devices, frequently of the magazine type (see Fig. 2-2.3).

In distinction from the thrufeed process in which a continuous flow of the work into the grinder is maintained, in infeed grinding the parts must be individually transferred into the grinding position either

Table 2-6.3 Systems of Work Feeding and Loading Commonly Used in Centerless Grinding

METHOD	PROCESS	PURPOSE AND SYSTEM	DISCUSSION OF PRINCIPLES AND OPERATION
Thrufeed grinding	Work feeding[1]	Arranging, orienting, and dispensing the randomly stored parts for feeding in a direction along the part's axis.	Bowls of the vibratory type, discharging the parts by causing them to ascend along a spiral ramp, optionally equipped with part-orienting elements. Bowls with paddles discharging by gravity. Hoppers with cleated belt lifters.
		Feed lines operating by gravity, relying on the part's own weight.	A channel, commonly of U cross section, with a slope selected to provide advance without excessive speed.
		Feed lines operating by gravity supplemented by the weight of stacked parts.	When the weight of the part is not sufficient to overcome the frictional drag, stacking in a confined channel, preferably a tube, is used; often complemented with stack height controlling devices.
		Guide rails for hand feeding. Guide rails operating by weight assisted feeding.	Rails, which represent the functional extension of the work-rest blade, and work guides form a trough along which the parts may be pushed manually, or advanced by a pusher being forced against the workpieces by the pull of a cable through pulley, the free cable end being loaded by attached weights.
		Guide rails with continuous power advance of the parts.	Disc-shaped parts, placed or fed into the trough, are advanced along their axis by the force component produced by a pair of power-rotated hyperbolic rolls in contact with the starting section of the butting row of workpieces.
		Bar feeding devices. Guides with supporting rollers, freely turning or power rotated.	Long bars and tubes must enter and leave the grinding area in alignment with the operational position of the work. Guide rails, usually equipped with supporting rolls to reduce friction of the advancing work, must have lengths at least equal to that of the workpiece and are arranged at both ends of the grinder. The translating action of the regulating wheel is not sufficient to advance heavy workpieces, which require a power drive for the inclined supporting rolls.
	Work loading[2]	Pushers for individual parts transferred from the end of the nonaligned feed line into the operating position.	Needed in special cases only, such as for the thrufeed grinding of tapered rollers, a process which uses a drum with a helical groove, substituted for a regulating wheel.

Table 2-6.3 *(Cont.)* **Systems of Work Feeding and Loading Commonly Used in Centerless Grinding**

METHOD	PROCESS	PURPOSE AND SYSTEM	DISCUSSION OF PRINCIPLES AND OPERATION
Infeed grinding	Work feeding	Chute, operating by gravity.	For infeed grinding, the work feed line ends at the transfer station where the parts are picked up manually, or by a loader; only at that point must the part be essentially in alignment with its position during the grinding. Consequently, the part may either slide or roll from the storage area to the transfer station at the end of the sloping feed line.
		Rails at an incline providing a track for rolling workpieces.	Infeed ground parts frequently have a larger length than the diameter, therefore parallel rails must be used to provide a track for the rolling workpieces. For profiled parts or those having sections with different diameters, tracks composed of rails with heights and spacing adjusted to the supported work surface elements, are used.
	Work loading[2]	Axial loading— Parts loaded by transfer along their axis and essentially in the horizontal plane of their position during grinding.	While one of the wheel slides is retracted, the work is pushed into the grinding position, sliding on the work-rest blade and advancing to the end stop. After grinding, the part is usually discharged along the same line of transfer, the end stop often serving as an ejector pin. May be operated manually (lever) or by power (e.g., hydraulic cylinder), the latter usually part of an automatic cycle. Applicable only for parts which are essentially cylindrical in shape.
		Overhead loading— Parts loaded by being lowered essentially vertically to the level of their grinding position.	Requires more elaborate devices than the preceding system, but is adaptable to most work configurations, as well as to multiple part loading. The part may be presented to the loader in the plane of the grinding position, or loaders are used which also incorporate transfer movement, picking up the part at a transfer station. The discharge of the ground parts may be carried out by the loader, plain or transfer type, or small parts may also be permitted to roll down into a discharge chute, by retracting the regulating wheel when the grinding is finished.

(Continued on next page.)

Table 2-6.3 *(Cont.)* **Systems of Work Feeding and Loading Commonly Used in Centerless Grinding**

METHOD	PROCESS	PURPOSE AND SYSTEM	DISCUSSION OF PRINCIPLES AND OPERATION
Infeed grinding *(Cont.)*	Work loading[2] *(Cont.)*	Swing loading— The part is loaded by being moved along an arc, its axis pointing toward the pivot point; the part attains a horizontal position when deposited on the work-rest blade.	Simpler in execution than overhead loading and is adaptable for work of any length within the width of the grinding wheel; however, not suitable for parts with sections having substantially different diameters. Designs are available for manual, and also for power operation.
Shoe type centerless grinding	Work feeding	Vibratory bowls— for random storage, orientation, and controlled discharge.	Disc of ring-shaped parts, for which this grinding method is primarily used, require orientation prior to introduction into the feed channel, and are best handled by vibratory bowls.
		Channel chutes	The essentially disc or annular shape of the parts makes them adaptable to rolling down along sloping chutes. These are often designed with helical intermediate sections, many convolutions high, providing in-process work storage with efficient floor space utilization.
	Work loading.[2]	Loading arms with basic swing movement, combined with auxiliary motions.	The parts must be loaded and unloaded individually to and from the grinding position on the backing plate. The loader has to deposit the work in a position from which radial contact with the shoes is reached promptly when the wind-in is produced by the rotation of the work spindle. Designs of single and double arm loaders are used, with gripping and stripper actions, sometimes with work transfer by a hinged section of the feed chute, all aimed at minimizing the work-handling time.

[1] In feeding for thrufeed grinding the work is usually advanced, at least in the terminal section of the feed line, essentially aligned with the operational traverse position. The feeding action must bring the work into effective contact with the face of the regulating wheel, but generally no separate loading is required.

[2] Loading, manual or power actuated, is always needed in infeed grinding. For automated processes, work feeding is arranged so that a workpiece is always presented for loading in the transfer station.

Courtesy of Lidkoeping A.B.—Hirschmann Corp.

Fig. 2-6.12 Hydraulically operated loading device for the automatic infeed grinding of profiled workpieces which are presented by a magazine feeder to the fingers of the loader carriage for transfer into the grinding area. Subsequently, the ground part is picked up again, transferred to a gaging station, and then into the discharge chute.

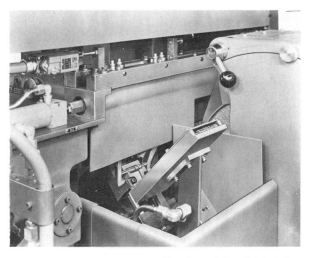

Courtesy of Landis Tool Co.

Fig. 2-6.14 Close-up view of the work transfer position of a ram-type automatic loader which picks up the part from an inclined magazine and transfers it between the wheels for the simultaneous grinding of three diameters.

manually or, more frequently, by an automatically operating mechanism. Such a mechanism will pick up the part at the end of the feed line and at intervals, which are tied in with the retracted position of the wheel slide, place the part on the work-rest blade.

Examples of such automatic work loading devices are shown in Figs. 2-6.12 to 2-6.15.

Of course, the advantages derived from the speed and unattended operation of automatic work loading devices, do not eliminate manually controlled loading, which is preferred or actually needed for small runs and for heavy parts which are not adapted to automatic handling. Manual in this context, implies only occasional work positioning by hand, possibly by

Courtesy of Landis Tool Co.

Fig. 2-6.13 Automatic hydraulically operated ram-type loading device installed on a centerless grinding machine, which is also equipped with loading magazine and discharge chute.

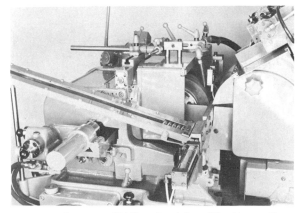

Courtesy of Lidkoeping A.B. — Hirschmann Corp.

Fig. 2-6.15 Push-rod type automatic work loader for cylindrical shafts stored in an inclined magazine. An air or hydraulic cylinder operates the push rod which advances the part to an end stop; at the end of the operation the end stop retracts and the workpiece is ejected on the exit side of the grinder.

using a wooden rod; a somewhat hazardous procedure considering the restrained position of the workpiece in the centerless grinder. The loading is actually carried out by hand-controlled devices such as a lever-operated work-loading slide, or by manually controlled power-operated equipment. The example of an uncommon work loader specially designed for certain types of heavy parts, is the device shown in Fig. 2-6.16.

Fig. 2-6.16 Special manually controlled hydraulic loading device for railroad axles the journals on both ends of which are centerless infeed ground in subsequent operations, requiring the intermediate turning of the heavy workpiece by the special loader.

Work Size Gaging In Centerless Grinding

As in the case of all process-connected gaging operations, the purpose of gaging in centerless grinding is to avoid rejects by controlling the size of the parts during grinding or immediately following it. The term "process-connected" should cover only the gaging operations which are essentially integrated into the grinding process. Consequently, it will not comprise measurements on ground parts which are transferred individually and occasionally, either by hand or by mechanized means, to measuring equipment which is separate from the grinder, where the part size or other parameters, such as roundness are inspected, for monitoring the operation or checking the setup.

In-process gaging in centerless grinding is, in comparison with most other grinding methods, hindered by the close retainment of the workpiece, a process characteristic considered beneficial in other respects. An exception in this aspect would be the shoe type

centerless grinders which, due to the open face of the workpiece and the relative narrowness of the ground surface, usually offer convenient access for the sensing members of in-process gages.

The limitations caused by the restrained position of the work are, in general, counterbalanced by the high degree of consistency of centerless grinding, a characteristic which often permits achieving the size-holding objectives by post-process gaging, while other less consistently performing methods require the application of in-process gaging.

Although limited in its use, in-process gaging is not excluded in centerless grinding. Certain, partly special, gaging methods belonging to the in-process category are applied successfully in some centerless grinding processes, both of the infeed and the thrufeed type.

To permit a systematic evaluation of the objectives and possible applications of different process-related gaging methods in centerless grinding, the various commonly used systems are surveyed in Table 2-6.4. However, this listing does not include the special systems and equipment.

A few of the typical gaging methods and instruments used in centerless grinding, will now be discussed in greater detail. Unless otherwise indicated these gages are of the electronic type with sensing heads located within, or adjacent to, the grinding area. The sensing heads are connected by a cable with amplifier-indicator which is located remotely, in a position convenient for observation, at the same time it is protected from the immediate effects of grinding fluid and grit.

Chordal height measuring gages (Fig. 2-6.17). Variations in the diameter of a cylindrical part, such as are generally produced in centerless grinding, cause changes in the chordal height of a segment which is bounded by a chord of constant length. Although the rate of the chordal height change is not proportional to the reduction of the work diameter, and its magnitude is sensibly smaller than that of the diameter changes it reflects (e.g., for a chordal length of $\frac{1}{2}$ diameter the change in chordal height is about $\frac{1}{7}$th of the actual diameter variation), a sensitive gage, air or electronic, can be calibrated to indicate with adequate resolution when the diameter is reduced to the required finish size of the part.

Such gages ride on the work surface, contacting it at two points sufficiently far apart to embrace an adequate chordal length, and carry in the center the sensing probe. The gages, being self-referencing, require only controlled orientation, but are not mounted rigidly. Chordal height measuring gages are well adapted for the infeed centerless grinding of parts of

Table 2-6.4 Systems of Process Connected Gaging in Centerless Grinding

CATEGORY	PROCESS PHASE	CENTERLESS GRINDING METHOD		
		THRUFEED	INFEED	SHOE TYPE
In-Process— Carried out while the grinding is in progress.	DIRECT— Measurement of the surface being ground	Conditions not suitable for use	Chordal height gage	Single probe— shoe referenced Fork type— floating contact self-referencing
	INDIRECT— Measuring an equipment element which controls the work size	Monitoring the distance of the grinding wheel face from a reference plane which corresponds to the finish size of the work	Indicating the position of the advancing wheel slide	Size control by grinding wheel truing with single point or rotary diamond roll, located in accurate correspondence with the finish work surface.
Post-Process— Carried out immediately after the workpiece emerges from the grinding area.	CONTINUOUS— Gaging of parts while moving in an uninterrupted stream out of the grinding area	Single contact with external referencing, e.g., from the work-rest blade. Fork type with self-contained referencing, restraining the part laterally.	Applicable for parts which are pushed out or roll off the grinding area.	Used exceptionally only for plain annular shapes, as a substitute for in-process gaging.
	INDIVIDUAL— Measurement of the part transferred onto a gaging stage	Application generally limited to non-cylindrical parts (e.g., tapered rollers)	Profiled parts may be transferred after the grinding by the overhead loader on the gaging stage.	Applied when parameters other than the diameter (e.g., concentricity, squareness, etc.) require monitoring.

medium or large diameter, although in some cases chordal height type gages are also finding application in thrufeed grinding.

Post-process gages for inspecting continuously traversing parts in thrufeed grinding. The measurement is carried out immediately after the parts leave the grinding area and present themselves for gaging while in motion. Two different systems are used: (1) *Single probe gage head with external referencing.* The part is led along a fixed supporting surface (e.g., the extended work-rest blade), and contacts the gage probe during its traverse movement. To assure the

proper positioning of the workpiece on the inclined top surface of the blade, the work guide on the regulating wheel side must also extend so as to support the part firmly while it passes along the gage probe. Due to the gradual wear of the supporting elements used in this type of gaging setup, the occasional readjustment of the probe position with the aid of a master, is required.

(2) *Gage head with self-contained referencing.* (Fig. 2-6.18) Two gage probes are in simultaneous contact with diametrically opposite points on the work surface. The diameter measuring gages of this design are

Courtesy of Cincinnati Milacron

Fig. 2-6.17 Work surface riding sensing head of a chordal height gage in operational position. It produces continuous signals for indicating the diameter changes of a large workpiece while the infeed grinding process is in progress.

not sensitive to the wear of the work support member because that does not have a gage reference function. However, the lateral position of the work in a plane normal to that of gaging, should be maintained to assure the measurement of the true part diameter. That lateral positioning is frequently accomplished by means of two parallel carbide rods which act as a V-type support. The sensing head of the gage may operate by one of two different systems: (a) Two live (signal producing) probes with reversed polarities producing, through a differential circuit, signals proportional to the variations in the distance between the two probes which are mounted on individually adjustable arms. (b) A diameter gage with a single, reed-supported measuring head which has two arms. One of these carries a rigid reference contact which is riding on the work surface and thus keeps the gage head in a position which corresponds to the momentarily contacted surface. The second arm, contains a live probe in contact with the opposite side of the work surface, and will produce signals which reflect the distance between the two probes.

The contact force of the live probe is substantially smaller than the force needed to deflect a reed which supports the gage head. Consequently, this second system, while simpler to build, is limited to workpieces which are not sensitive to deflections or to the marring of the surface by the higher contact pressure which is exerted by the rather stiff reed spring acting through the rigid reference contact.

Diameter gages for shoe type centerless grinding machines. The retention of the workpiece in a fixed

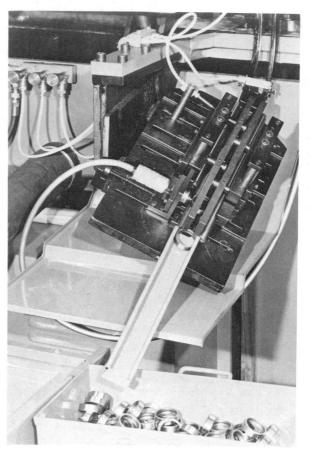

Courtesy of Marposs Gauges Corp.

Fig.2-6.18 Continuously indicating post-process gage used in centerless thrufeed grinding for monitoring the diameter variations of the workpiece immediately following its exit from the grinding area. The workpieces are gliding down in an inclined chute, along a carbide reference rail installed opposite the gage probe, whose signals are led to an amplifier with meter (not shown), displaying the diameter variations of the individual parts. The signals may also trigger automatic functions, such as segregation, size correction by slide adjustment, etc.

plane by the backing plate, and supporting it radially by two shoes sufficiently far apart to embrace a chord of close to two-thirds of the diameter, are conditions which lend themselves particularly well to diameter checking by chordal height gaging. It is feasible, and the method actually has been widely applied, to mount the sensing member of the gage (an air nozzle or an electronic gage probe) on the shoe support block halfway between the shoes, an arrangement that obviates the need for special gage arms which have to be retracted during the loading of the workpiece.

Of course, there are the inherent system limitations in the gaging of chordal height variations, such as the nonlinear nature and the reduced ratio of the measured variations as compared with the actual changes of the work diameter. Furthermore, the single probe gaging by referencing from the shoe surface is affected by the wear of the shoe surfaces and requires occasional adjustment to compensate for these variations. For that reason, when very close work diameter tolerances have to be achieved consistently, two-point diameter gages are preferred, because these carry their own reference point mounted on a common bracket with the sensing probe (Fig. 2-6.19). The rigid contact point will ride on the work surface during the grinding process, thus maintaining a floating reference point for the gage head. This arrangement makes the operation of the gage independent of the positions (spread) and the condition (wear) of the work support shoes.

Courtesy of Marposs Gauges Corp.

Fig. 2-6.19 Fork type in-process gage on a shoe type centerless grinder for continuously measuring the diameter of the workpiece, a taper roller bearing cone (shown held in hand), while it is being ground. The reed supported gage head has two contact fingers, one with the referencing point and the other with the probe of the electronic gage.

The function of process-connected gages in controlling the centerless grinding process. The operation of the process-connected gages by sensing and converting into dimensional values the variations of the work diameter can serve two major purposes:

(1) To provide indications for the guidance of manual adjustments aimed at controlling the size of the ground workpiece
(2) To originate command signals for the automatic control of the machine functions which affect the work size. Frequently, the indicating function of the gage is retained in this second category of gage utilization, mainly as a means for providing continual information on the progress of the process.

The indications are generally supplied by the position of the pointer on the dial of a remotely located indicator. In some gage designs two dials are contained in a common amplifier cabinet, arranged side by side and graduated in different increments. The coarse scale dial will show the progress of the grinding during the roughing stage of the process, while the adjacent higher amplification dial will start indicating only in the final section of the wheel advance, showing the size changes in the work with a resolution commensurate with the tolerance specifications.

In a few types of gages of special design digital displays replace the more conventional dial (see Fig. 2-5.11). As an additional convenience for the operator, the indicating functions of some gages are supplemented with light signals announcing the attainment of significant stations of the grinding process, such as the end of the coarse feed section, where the fine feed should stop and when the sparkout has produced the finish size.

The control initiating functions of the process-connected gages are the key to process automation for the purpose of size control. Such functions can be limited to the occasional adjustment of the slide position or of the infeed movement, a purpose which post-process gages with feedback commands are intended to serve. In the in-process gaging, particularly on shoe type centerless grinding machines, the measuring functions of the gage actually can be the basis for the control of all major phases of the automated grinding process; such as the travel of the rapid approach, the distance of the coarse feed, the position for terminating the fine feed, and the retraction of the wheel when the finish size of the work has been attained.

While several of these control functions can be programmed, in combination with the automatic compensation for wheel truing, the final sizing phase of the process is always gage controlled when tight tolerance specifications make guidance by the actual work size mandatory.

Optional Accessories and Equipment for Centerless Grinding Machines

Accessories and equipment available from the manufacturers of centerless grinding machines on customer's option will be discussed for two major reasons:

(1.) To indicate the high degree of process versatility and machine adaptability of the centerless grinding method in general

(2.) To point out the large number of specific processes which may be accomplished by the centerless grinding method just by equipping the basic machines with appropriate, often readily available, special accessories.

In order to permit a clearer appraisal of the wide and diverse array of special accessories, Table 2-7.1 presents an indicative listing subdivided into groups according to the principal application purposes. The term special used in this context indicates that such accessories are not part of the basic machine equipment, either because they have to be chosen from different alternatives (e.g., grinding wheels), or are needed for particular operations only. Of course, no attempts were made to embrace in that indicative listing all those special accessories and tooling which grinding machine manufacturers supply either as catalog items or develop to meet specific process requirements.

Complementing the listing of the table a more detailed description of a few typical accessories is presented in the following. These descriptions serve as examples; their selection does not reflect a preference based on relative importance or frequency of application.

Centerless Bar Grinding

Bars and tubes, in lengths varying from a few centimeters to several meters, ground along their entire surface to tightly toleranced diameters, combined with precise roundness and good finish, are required for many uses in industrial production. The starting condition of this kind of material may be hot rolled; drawn, or turned (peeled); but when the dimensional requirements are more severe than such processes can assure, centerless grinding is selected as the finishing operation.

Thrufeed centerless grinding is particularly well adapted for such operations, primarily because of the excellent dimensional and geometric control it provides, but also from purely economic aspects it can often successfully compete with other types of secondary sizing processes.

It is customary to distinguish between three groups of work which are finished by centerless bar grinding: (a) short bars and tubes in lengths less than about one meter; (b) light bars and tubes in diameters up to about $1\frac{5}{8}$ inches (40 millimeters) and lengths up to about 13 feet (4 meters); and (c) heavy bars and tubes, up to about 6 inches (150 millimeters) in diameter and lengths to about 25 feet (7.5 meters).

Each of these groups requires the appropriate types of centerless grinding machines, which will be selected on the basis of the work diameter range as the primary consideration, but also taking into account the intended stock removal rates defined by two factors: (a) the material allowance on the work surface, and

(b) the applied thrufeed speed. The material allowance is generally dependent on the thickness of the surface layer which should be removed from the raw or semi-finished state in order to safely obtain the required size and roundness, at the same time eliminating the metallurgically unsound surface which may be affected by carbon depletion, seams or similar deficiencies. A typical value for the reduction of the diameter in centerless grinding is 0.020 inch (0.5 millimeter). The thrufeed rates on high-production centerless bar grinding machines may be in the order of about 8 feet (2.5 meters) per minute for larger bars of e.g., about $1\frac{5}{8}$ inches (40 millimeters) diameter, and 33 feet (10 meters) per minute for bars of about $\frac{5}{8}$-inch (16 millimeters) diameter.

The essential equipment which distinguishes the bar type centerless grinding machines is the support which long bars or tubes require before entering and after leaving the grinding area. Such supports must have sufficient length to maintain the entire workpiece in a position assuring excellent alignment with the section being ground.

For shorter bars of less than about 1 meter length, brackets may be mounted on the machine bed for holding the work support members, while longer bars require support stands or tables, attached to the grinder, but also having their own legs, standing or anchored to the floor.

For short, thin bars the support may be just a set of guide rails, often with exchangeable hard inserts on the contact surfaces which sustain the rotating work while sliding along the rails, one of which must be the functional extension of the work-rest blade.

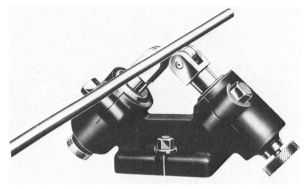

Courtesy of Lidkoeping A.B. - Hirschmann Corp.

Fig. 2-7.1 Roller rest for work support in centerless bar grinding, equipped with idle rollers.

For longer heavier workpieces, however, bar rests with hardened rollers running on antifriction bearings, are used, thus nearly eliminating the braking effect of a sliding support. Figure 2-7.1 shows a typical bar rest with idle rollers, and Fig. 2-7.2 illustrates a centerless grinding machine arranged for bar grinding with long support tables attached to both the ingoing and the outgoing sides of the grinder. Each of these support tables carries an appropriate number of roller rests. For very thin rods whose own weight would not be sufficient to keep the work resting on the supporting rollers which are arranged in pairs, a special roller rest equipped with an arm carrying a third roller is used. This arm can be swung out of the way when introducing the work, in a manner similar to that in the center rests used on the center type cylindrical grinding machines (see Fig. 1-5.6).

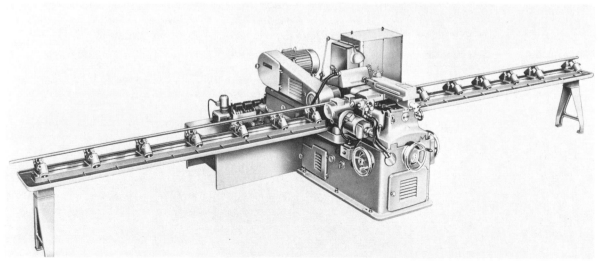

Courtesy of Lidkoeping A.B. - Hirschmann Corp.

Fig. 2-7.2 Bar support tables attached to the ingoing and outgoing sides of a centerless grinding machine arranged for the thrufeed grinding of long bars.

Table 2-7.1 Examples of Optional Accessories for Centerless Grinding Machines

APPLICATION PURPOSE	TYPICAL ACCESSORIES
Grinding and Regulating Wheels. Wheel Mounting Elements	— Grinding and regulating wheels in different shapes, sizes, and compositions. — Spacers to compensate for narrow wheels and to control the lateral location of the mounted wheel. — Interchangeable wheel spindles for the selective use of alternate wheels and for reducing the lost machine time due to wheel change.
Wheel Truing and Dressing	— Automatic grinding wheel truing devices, also with stroke counters, rapid traverse between spaced wheels, usually combined with automatic wheel position compensation. — Profile truing devices, profile cams and followers; side truing attachments. — Truing tools for single-point truing, diamonds, holders, hydraulically operated diamond turners, combination holders for truing in different planes. — Crush truing and rotary diamond roll truing devices, tools.
Work Feeding and Loading	— Hoppers with chutes for the continuous feeding of short parts. — Troughs for in-line feeding with manual, gravity, or power-actuated work advance. — Roller stands with or without individual drive for thrufeed grinding of long bars and tubes. — Push type loading and ejecting devices for infeed grinding. — Overhead loaders and swing-in loaders, manually or power operated. — Walking beam type loading devices for transferring single or multiple parts.
Work Locating and Support	— Work rests with interchangeable blades for both thrufeed and infeed operations. — Work rests combined with loading slide, end stop, and ejector. — Special work-rest blades designed for noncylindrical parts in solid design or with carbide tips. — Solid and adjustable work guides, with interchangeable jaws in various configurations. — Attached auxiliary supports for long parts, also with power drive. — Shoe-holder blocks, shoes, and backing plates for shoe type centerless grinding.
Gaging of Machine Elements and Workpieces	— Auxiliary gages for machine setup adjustment, work-blade height setting, truing device swivel and tilt, table swivel, etc. — Machine slide position indicating gages, e.g., as a guide for sensitive manual work-size compensation. — Post-process gages for thrufeed grinding operations, with indicators and, in special cases, as the originator of size compensation commands. — In-process gages for infeed type grinding operations, with continuous measurement of the work surface being ground, for indication and for supplying signals directing the size control functions.

(Continued on next page.)

Table 2-7.1 (*Cont.*) Examples of Optional Accessories for Centerless Grinding Machines

APPLICATION PURPOSE	TYPICAL ACCESSORIES
Work Size Control	— Size compensation for thrufeed grinding, with pushbutton or gage directed actuation, operating either one way (compensates for the wheel wear), or both ways (also withdraws the wheel slide when needed to balance thermal effects on work size). — Electro-hydraulic automatic control device for infeed grinding, operating by single or continuous cycling, with adjustable travel and feed rates, occasionally also equipped with compensating attachment. — Automatic size control for shoe type centerless grinding, wheel slide advance controlled by direct measuring in-process gage.
Related Accessories	— Balancing stands and balancing arbors. — Electro-dynamic wheel balancers. — Coolant clarifiers with rotating magnetic drums, with moving filter paper, etc. — Jib cranes for machine servicing (e.g., installation of spindles with mounted grinding wheels).

For the centerless thrufeed grinding of long and heavy bars the simple supporting of the work is not sufficient for maintaining the effective centerless grinding operation. Bars weighing about 450 lbs (about 200 kg) or more, cannot be brought to the required high rotational speed instantaneously by the regulating wheel, which must also impart the traverse movement to the work. For such workpieces the centerless bar grinding machines are used with support tables which have self-contained drives for the roller rests. The rollers are turning at a rate which will impart to the work a peripheral speed equal to that of the regulating wheel, in addition to an advance speed approximately equal to that resulting from the tilt setting of the grinder's regulating wheel. For this latter purpose the driven feed rollers have adjustable angular setting.

The feed rollers are driven by an extended shaft consisting of several sections, each equipped with universal joints (see Fig. 2-7.3). The feed rate of the driven support rollers at the outgoing side is usually set slightly higher than the regular thrufeed rate of the grinding, in order to avoid the succeeding bar bumping into the end of the preceding one.

Automatic Power Infeed Control by Air-actuated Accessory Unit

Existing centerless grinding machines built originally for hand controlled infeed grinding, can be converted to automatic feed control by means of a special air-powered device, which is adaptable for retro-fitting. Such attachments can be installed on several popular types of centerless grinding machines and operate by the properly sequenced, timed, and adjusted turning of the feed screw for regulating the wheel slide.

Fig. 2-7.3 Power driven support rollers with angular setting and crowned roller periphery for the purpose of feed action. The power is transmitted through a sectional drive shaft equipped with universal joints ahead of and behind each roller rest to permit setting of the required feed angles.

The power infeed movement comprises the following steps: (a) rapid approach; (b) slow infeed; (c) retraction, all applied sequentially at individually set rates and over distances which can be precisely adjusted. While the rapid movements are carried out by the plunger of an air cylinder, the infeed is controlled hydraulically to assure the uniform and sensitively adjustable rate of that motion.

The operational cycle can be limited to a single one which requires actuating a switch for restarting, or the cycle can be set to repeat itself automatically, when mechanized part loading is provided. Timers are also supplied to halt the motion at different stations over an adjustable dwell time: (a) at the end of the feed stroke for spark-out; (b) at the end of the retraction movement, for work discharge and loading, and (c) optionally, close to the end of the infeed for an extended work rounding period preceding the final sizing.

Such accessory units can be easily installed, and then can be temporarily disconnected to be used alternately with manual infeed, selecting for each operation that particular system which is best suited for obtaining optimum performance from the centerless grinder.

Automatic Size Control in Centerless Thrufeed Grinding

The thrufeed grinding process, because of the rigidity of the setup which requires no slide movements for work loading and discharge, is inherently repetitive, assuring a high degree of consistency in holding the size of the work. While the wear of the grinding wheel is generally slow, particularly in the case of finish grinding of small diameter parts with moderate stock allowance, it is still bound to gradually affect the work size. The changes in the diameter of the ground work, generally occurring as a trend, can be detected by automatically operating post-process gages, whose indications or light signals will alert the machine operator to the need for corrective slide adjustment.

With the aid of appropriate equipment such adjustments can also be carried out automatically without the need for the operator's action. The two widely accepted systems of automatic size control in centerless thrufeed grinding operate in the following manner:

(a) *Stepping motor*. This operates by turning the wheel slide feed screw by a small incremental amount. The size variations detected by the gage will produce proportional signals for actuating the stepping motor by that number of increments which the compensation of the measured work diameter changes require. Such increments used to be in the order of about 0.0001 inch (0.0025 millimeter), but special equipment of more recent origin can operate with increments as small as 0.0000125 inch (0.0003 millimeter)

when work of extreme dimensional accuracy is required and the grinding machine is built to respond to such particularly small input values. Preloaded ball-screw type feed screws are used to assure the precise translation of the rotational input into a linear slide movement of corresponding value, and also for the purpose of reducing friction and backlash. For such extremely small incremental movement, machine tool slides of special design, operating as hydrostatic bearings, are required.

(b) *Magnetostrictive slide adjusting devices*. The attachment acts directly on the wheel slide for its positional adjustment and operates by utilizing the magnetostrictive effect which produces precisely repetitive expansions, followed by contractions of a nickel armature when it is submitted to momentary magnetization. That core expansion which is effective only during the magnetized state, will always represent the same, originally selected amount. The core movements, which are triggered quickly and are of short duration, can be harnessed by hydraulically actuated clamps, to be utilized in either direction along the axis of the core for advancing or retracting the machine slide by a fixed increment, typically 0.000050 inch (0.00125 millimeter), although smaller or larger incremental values can also be selected. For compensating the work size variations detected by an appropriate post-process gage, very accurately controlled slide movements can thus be created, in the required sense and by the needed number of increments.

The operation of these size control devices is usually designed to disregard single workpieces which, due to improper gage staging or other isolated reasons, may cause gage signals which are not characteristic of the preceding and following parts. In order to avoid unwarranted slide adjustments which subsequently will have to be canceled or actually retracted, the automatic size control devices are usually equipped with a counter, adjustable to a selected number of consecutively sensed part sizes which are needed to cause the emission of signals for slide position adjustment. As an example, five consecutively measured parts may be considered to indicate size variations requiring correction.

Limiting the issuing of slide adjustment commands to those size variations which represent a trend does not exclude the elimination of single parts which are, or appear to be, outside the tolerance range. Frequently the size control devices are equipped with a rejection mechanism, which will automatically discard every single out-of-size part, independently of whether it originated a slide adjustment or constituted an isolated size deviation only.

Centerless Grinding Fixtures

The principles of centerless grinding, namely supporting and concurrently rotating the workpiece by a slowly running regulating wheel, and also restraining the work by a rest blade while it is being ground by a grinding wheel, can also be accomplished with the aid of special fixturing on a horizontal spindle type surface grinding machine.

Fixtures, actually supplementary grinding attachments with individual drive motors for the regulating wheel, are commercially available and are adaptable to both thrufeed and infeed type centerless grinding. The attachments may be mounted on the table of a horizontal spindle type surface grinding machine of adequate size, and are equipped with a variable speed drive motor for accommodating different workpiece diameters, a typical range being about $\frac{1}{8}$ to $1\frac{1}{4}$ inches (3 to 30 millimeters). Because these attachments are designed with the regulating wheel beneath the grinding wheel, with the work lying on the top of the regulating wheel periphery, a second blade, similar to the upper guides in regular centerless grinding, is also needed. That second blade has the role of preventing the workpiece from rolling away before being restrained by the combined action of the wheels and of the regular work-rest blade.

Such a fixture may provide a useful service in small shops and toolrooms, where only occasional centerless grinding is needed and the tying up of a surface grinder is an acceptable alternative. Of course, neither the performance, nor the adaptability of such fixture-equipped surface grinders is comparable to that of a regular centerless grinding machine.

The outlined system of having the grinding wheel above the regulating wheel in a vertical arrangement, in distinction to the conventional horizontal plane, is not limited to the described fixtures. There are actually complete centerless grinding machines built with this system and claiming certain advantages, although the application of centerless grinders in that special design has been limited to particular types of workpieces, such as thin rods made of fragile material.

The Operation and Setup of Centerless Grinding Machines

This section will be limited to three important topics related to the operation of centerless grinding machines in general: installation; setup; and common faults in the centerless grinding process.

The discussions will cover aspects which are considered characteristic of the centerless grinding method, in distinction to other types of grinding processes.

Installation of Centerless Grinding Machines

In selecting the *location* of the machine those aspects which may affect the operation of the equipment should, whenever possible, take precedence over most other considerations, such as the convenience of material handling and flow, the availability of plant engineering facilities, etc. A location should be selected where external factors with potentially harmful effects on the operation of the machine, such as vibrations originating from external sources, extreme temperature variations, etc., are safely avoided.

The transportation of most types of centerless grinding machines to the installation location may be carried out by lifting with a crane, using a four-point suspension system. The machines are usually provided with lifting holes (covered when not in use), or temporarily mounted brackets used as sling attachment points.

A foundation is generally of advantage and positively recommended for heavy grinders, weighing about 4 tons or more. Well built plant floors may have sufficient strength to safely support smaller grinders used for light work, without requiring the preparation of special foundations. However, in each case, the actual bolting of the machine to the floor or special foundation must be preceded by the precise leveling of the grinder.

Leveling of the centerless grinding machine is essential for its proper operation. For that purpose the machines are usually supplied with leveling jacks, consisting of a bolt supported in a ball-cup shaped washer for permitting a sensitive adjustment of the machine support pad positions. First, the four screws at the corners of the machine bed are adjusted, and then the screws of any additional pads, which are needed for larger machine beds, are tightened.

The leveling of the machine must be guided by a sensitive machinist's or laboratory type spirit level, placing it on the particular surfaces of the machine which are designated by the manufacturer as the applicable leveling planes. In certain types of centerless grinders, e.g., those with inclined bed (see Fig. 2-2.4), special fixturing, such as sine plates with specific support bars, may be needed to translate the position of critical machine surfaces into the corresponding horizontal plane. The leveling instrument should have a sensitivity of 10 seconds of arc per graduation (about 0.0005 inch/foot or 0.001/200 mm), preferably even better.

It is after the leveling of the machine has been accomplished and rechecked, that the bolting to the foundation is to be carried out. For that purpose combination leveling jacks are available. These have hollow jack screws through which the threaded end of the anchor bolts penetrate to accept the hold-down and locking nuts on the top of the level adjusting screw. That system of combination leveling and anchoring with detachable elements is of particular advantage during the recommended procedure of occasional verification of the machine's level position and adjustment when required.

The Setup of Centerless Grinding Machines

Several more or less regularly required or useful setup operations will now be described to provide assistance in determining the best procedures for a particular job. Some of these setup operations are alternatives which are not applicable concurrently, while others are needed in special cases only.

Finally, while specific instructions detailed on process sheets are desirable for repetitive or continuous processes, for short run jobs centerless grinding machines are often set up very successfully by experienced operators who rely on empirical data, possibly combined with some amount of experimentation. As a matter of fact, a reliance on the observation and examination of the grinding results cannot be entirely avoided in the final adjustment of the work size, a phase of the setup operation which must be guided by the gaging of the actually ground part.

Observation of the operational condition and of the resulting product parameters are particularly necessary as a guide to the adjustment of the swivel table, for correcting the taper errors of the workpiece in infeed grinding or assuring the uniform cut along the face of the grinding wheel in thrufeed grinding.

Table 2-8.1 reviews the major phases of setup procedures applied in centerless grinding in general. The order of that listing is not necessarily the best suited for all cases, nor is it necessary to carry out all these steps in each case. However, several of the preceding chapters discussing the design and execution of centerless grinders contain additional information on the involved machine elements, thereby complementing the brief survey of the table. Some of the setup phases have been discussed in greater detail in other sections of this chapter, while some others will be elaborated upon in the following.

The Incline Setting of the Regulating Wheel

The tilting of the regulating wheel head is always needed for thrufeed grinding as well as for endfeed grinding, and is used in some cases for infeed grinding as well. The incline setting of the regulating wheel head (see Fig. 2-8.1) controls, in combination with the peripheral speed, the axial feed rate of the work in thrufeed grinding. The relationship of these two controlling factors, namely the incline angle and the peripheral speed of the regulating wheel, to the resul-

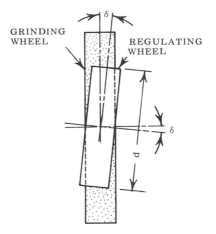

Fig. 2-8.1 Principles of producing the traverse feed of the work by the incline setting of the regulating wheel. If δ = incline angle; d = diameter of the regulating wheel (inch or meter); n = rpm of the regulating wheel, then V = peripheral speed of the regulating wheel (inch or meter) = $\pi\, d\, n$, and F = traverse feed of the work (inch or meter) = $V \sin \delta$.
NOTE: Where F is a theoretical value, assuming no slippage of the work in contact with the regulating wheel.

ting traverse feed rate can be expressed by the following formula:

$$f = V \times \sin \delta$$

Where:

f = traverse feed rate of the work in meters (or inches) per minute

V = the peripheral speed of the regulating wheel, in meters (or inches) per minute*

δ = the incline (tilt) angle of the regulating wheel.

As an example, assume the use of a regulating wheel of 300 millimeter (about 11.8 inches) diameter,

---

*For convenient reference, the simple formulas generally used for calculating the peripheral speed of the wheel are as follows:

Inch system

$$V = \frac{\pi\, DN}{12}$$

Where:
  V = Peripheral speed, fpm
  D = Diameter, inches
  N = Regulating wheel speed, rpm

Metric system

$$V = \frac{\pi\, DN}{1000}$$

Where:
  V = Peripheral speed, meters per minute
  D = Diameter, millimeters
  N = Regulating wheel speed, rpm

Table 2-8.1 Survey of Frequently Required Operations in the Setup of Centerless Grinding Machines

(*NOTE:* The order of the listing does not necessarily represent the actually applied sequence. Some of the operations mentioned are alternatives, or needed for specific conditions only.)

GROUP OF MACHINE ELEMENTS	PURPOSE AND OBJECTIVES	ACTIONS INVOLVED— ASPECTS CONSIDERED
Preparatory actions, often also involving machinist's work.	The selection and installation of the grinding wheel and regulating wheel will precede the actual machine set-up. The preparatory work will also include the attachment and connection of feeding and loading devices should the use of such equipment be necessary for the process.	The installation of the grinding wheel will be preceded by its mounting and out-of-machine balancing. Due to the weight of grinding wheels commonly used in centerless grinders, and particularly when the wheel is installed together with the wheel spindle, special holding devices to be lifted by hoist, are required.
Selecting and setting the rotational speed of the regulating wheel.	General starting values, such as 30 rpm for average operations on medium size machines, are chosen frequently and then corrected to better suit the operation. Work-adapted values are based on specific peripheral speed, translated into rpm value according to the diameter of the regulating wheel.	Most centerless grinders have steplessly variable regulating wheel speeds, which can be adjusted by turning a knob to a position where the resulting speed of the regulating wheel reached the desired level, as indicated by a tachometer. The highest value in the available speed range usually serves the purpose of regulating wheel truing, carried out at much higher than regular operating speed.
Incline setting of the regulating wheel head.	The incline of the regulating wheel, in combination with its peripheral speed, controls the traversing speed of the work. The specified work traverse speed should balance the requirements of high productivity with those related to quality, such as size, finish, and roundness.	The tilt setting of the regulating wheel head to the specified incline angle, is usually guided by graduated segments on the wheel base and on the wheel head. More accurate incline setting can be made by the use of gage blocks, when the grinding machine design comprises the necessary support surfaces at known distances from the pivot of the tilt movement.
Work-rest blade selection, installation and height setting.	The general dimensions (length and thickness), the material (e.g., steel or carbide inserts), and the top angle (30° being a general purpose value) are selected. The height setting of the blade is the primary controlling factor of the level of the work relative to the C/L of the wheels.	The adjustment of the blade height is frequently aided by the use of inserts or shims. Some models of centerless grinders have work rests with sensitive height adjustment by means of a handwheel. The gap between the blade and the grinding wheel also affects the resulting height position of the work and should be precisely controlled.

Table 2-8.1 *(Cont.)* **Survey of Frequently Required Operations in the Setup of Centerless Grinding Machines**

GROUP OF MACHINE ELEMENTS	PURPOSE AND OBJECTIVES	ACTIONS INVOLVED— ASPECTS CONSIDERED
Work guides for thrufeed grinding; installation, and adjustment.	In thrufeed grinding the work guides assure the precise alignment of the ingoing and the discharged work with its position during the grinding. The guides must be set to provide the proper clearances before and after grinding. Improper guide setting affects the geometry of the ground work.	The use of a precise alignment bar, as an aid, is recommended for setting the operating faces of the work guides parallel with the face of the regulating wheel, and adjusted to the required clearance values. Most centerless grinding machines are supplied with work guide holders and inserts, permitting sensitive adjustment for both parallelism and radial positions.
Gages and special attachments; installation and adjustment.	Process connected gages are used as guides for setting and adjustment of the wheel slides or may be required to direct the automatic size control. The accurate gage adjustment, generally with setting masters, is a critical step in the machine setup procedure.	In automatic operations the progressive dimensional changes of the continually gaged workpiece will have to trigger changes in the slide infeed movement (approach, roughing, finishing, sparkout, retraction) and other related functions, e.g., work loading, wheel truing, compensation, etc. The adjustment of the respective switching positions may be a part of the machine setup.
Grinding wheel truing; profile generating.	For work of plain cylindrical shape only a true-running, straight wheel face, having the appropriate free-cutting characteristics is needed. For profiled parts, the setup also includes the installation of the profile cams or other form-truing devices (crush rolls, rotary diamond rolls), followed by generating the required wheel profile.	When automatic grinding wheel truing is used, the machine setup will also comprise the setting of the stroke length, the traverse speed, the diamond advance at reversals, the number of strokes per truing operation, the frequency of truing, the wheel position compensation, the periodic diamond rotation, the turning on and off of the coolant line, etc., actions generally needed, but controlled manually in operator-attended processes.
Regulating wheel truing, preceded by setting the angle and location of the diamond path in thrufeed grinding.	For infeed grinding the actions are similar to those listed for the grinding wheel. For thrufeed grinding a hyperbolic contour is needed for assuring a straight line of contact between the traversing cylindrical work and the inclined regulating wheel, at the level of the wheel-work engagement, which is usually above the wheel C/L.	For producing a regulating wheel face having the correctly curved profile along the contact with the traversing work, the regulating wheel truing unit has to be set to a swivel angle corresponding to the incline angle and the diamond point set over by an amount related to the work height above the wheel C/L. Correction factors are used to arrive at the required truing setup values, which vary according to the ratio of the wheel/work diameters.

(Continued on next page.)

Table 2-8.1 (*Cont.*) **Survey of Frequently Required Operations in the Setup of Centerless Grinding Machines**

GROUP OF MACHINE ELEMENTS	PURPOSE AND OBJECTIVES	ACTIONS INVOLVED— ASPECTS CONSIDERED
Work size controlling elements; slide position for thrufeed; stop positions for infeed. Optionally automatic size-control devices.	In thrufeed grinding the conditions established in the setup determine the work size which will be maintained by occasional compensations for wheel wear. For infeed grinding the positions for the slide movement stops also have to be set. For automatic operations the setup will also comprise the setting of the intermediate positions.	The final phase of the size adjustment steps is generally guided by the accurate measurement of the produced work. Subsequently, adjustments to compensate for wheel diameter changes due to truing or wear, are part of the process control, but in the case of automatic machine operation the adjustment of such compensating movements may also be comprised in the machine setup procedure.
Swivel table of the regulating wheel head (when used); setting and adjustment.	The swivel adjustment serves the final, optional correction of the basically cylindrical work shape in infeed grinding. It is also a means of bringing about the uniform distribution of the cut over the total wheel face in thrufeed grinding. An alternative means for similar objectives may be the swivel of the grinding wheel truing unit.	The adjustments are commonly carried out by experimentation, sometimes requiring partial withdrawal in the case of overcompensation. Better control can be attained by using a sensitive dial indicator, attached to the bed and contacting an anvil surface on the swivel table. Similar indicator arrangements are also used for the alternate method, consisting of swiveling the truing unit of the grinding wheel.

rotating at 30 rpm, producing a peripheral speed of about 28.2 meters (about 1110 inches) per minute. When selecting an incline angle $\delta = 3°$ (often recommended as a good starting average), the calculated value of the resulting traverse feed rate will be $f = 1.48$ meters (58.3 inches) per minute. The actual thrufeed traverse rate is, as a rule, very close to the calculated value, the delaying effect of the small amount of slippage between the periphery of the regulating wheel and the work usually does not exceed 1 to 2 per cent.

Of course, the $3°$ incline angle setting selected for that example is not a rule, and other incline angles over a range to about 8 degrees are generally feasible and actually applied, in order to produce the optimum thrufeed rate for the operation. In selecting the applicable thrufeed rate the following conditions are considered:

(a) Increasing the thrufeed rate:
  (1.) Raises the productivity (number of parts ground per minute)
  (2.) Reduces the number of revolutions the workpiece is making while passing along the face of the grinding wheel

(b) The larger the work diameter, the less revolutions it will make during the grinding pass at a given thrufeed rate.
(c) Increasing the number of revolutions the part is making during the grinding pass
  (1.) Reduces the ratio of stock removal per work revolution to the total material allowance
  (2.) Contributes to the expected rounding effect of the process which, theoretically, is a function of the total number of work revolutions during the centerless grinding operation.

The incline angle produced by tilting the regulating wheel head generally refers to a condition in which the wheel axis is raised at the feed end; this is often referred to as the positive incline because it causes the work to advance along the faces of the wheels. Much less frequently a negative incline may be needed, produced by lowering the front end of the regulating wheel, and that will result in a reversed feed action. The latter may be required when, e.g., for fixturing reasons the work feeding from the rear of the machine is chosen. The angular adjustment scales of the regulating wheel incline setting, as well as that of the swivel setting for the truing unit, are graduated both ways (usually over a wider range in the "positive" sense).

Work Guides for Thrufeed Grinding; Purpose and Setting

In thrufeed grinding the work must enter and leave the operating area of the machine along the same straight-line path in which it travels during the grinding process. For assuring that position of the workpiece, the rest blade usually extends at both ends beyond the wheels, while the restraining role of the two wheels is taken over by the work guides which are usually equipped with adjustable jaws.

The principles of the proper work guide setting are shown in the diagram of Fig. 2-8.2 illustrating, in relation to the work, the position of these elements which should provide the following operational conditions:

● When entering the grinding area the work shall be guided along a plane aligned with the operating face of the regulating wheel, except for one-half of the stock allowance, by which amount the work guide is set back.

● When leaving the grinding area the work shall be supported along a plane aligned with the face of the regulating wheel, except for a very small clearance of about 0.0004 to 0.001 inch (0.01 to 0.025 millimeter).

● On the grinding wheel side the work guides shall be parallel with the face of the wheel but set back by about $\frac{1}{64}$ to $\frac{1}{32}$ inch (0.4 to 0.8 millimeter). The role of the work guides on the grinding wheel side is more that of a guardrail than of an actual guide, since the position of the workpiece is controlled by the rest blade and the guides on the regulating wheel side.

The improper alignment of the work guides can affect the straightness of the thrufeed ground workpiece. When the work guides are deflected toward the regulating wheel the thrufeed ground work will tend to assume a barrel shape; deflection in the opposite direction will cause a concavity along the work surface, making a nominally cylindrical workpiece faintly resemble a slender hourglass.

The Truing of the Grinding Wheel

For each new setup, the truing of the grinding wheel is recommended, even when the basic cylindrical shape of the wheel is retained. Grinding wheel truing as part of the setup is, of course, indispensable

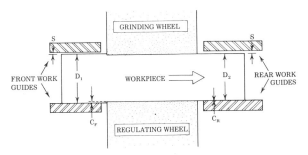

Fig. 2-8.2 Principles of work guide setting for centerless thrufeed grinding. Where: D_1 = Work diameter before grinding; D_2 = Work diameter after grinding; S = Setback of work guides on the side of the grinding wheel about 0.15 to 0.30 inch (0.4 to 0.8 millimeter); C_F = Clearance (Front) of the work guide on the side of the regulating wheel = $(D_1-D_2)/2$; and C_R = Clearance (Rear) of the work guide on the side of the regulating wheel about 0.0004 to 0.001 inch (0.01 to 0.025 millimeter).

when a new wheel profile is required which differs from the plain cylindrical, or a leading relief is needed for the entering workpiece, such as that recommended for the thrufeed grinding of disk-shaped parts. In the case of profile trued grinding wheels the setup will also comprise the installation and adjustment of the profile bar, sometimes referred to as the profile cam, of the truing unit.

The Truing of the Regulating Wheel

Important aspects of this procedure are discussed in Chapter 2-6 of this section. The truing of the regulating wheel, preceded by the proper setting of the truing unit, is particularly critical when the incline angle of the regulating wheel has been changed for assuring a proper grinding condition for a specific thrufeed grinding operation. Because changes of the regulating wheel profile can be quite extensive in the case of substantially different setup conditions, that factor alone may warrant a production and sequence scheduling which reduces the frequency and the extent of regulating wheel profile changes.

Setting the Reference Positions of the Slide Movements

Centerless grinding is essentially a repetitive kind of operation, greatly reducing the need for manual size control. In thrufeed grinding only occasional adjustments of the operating machine are needed, and even in manually controlled infeed grinding the retraction and advance of the wheel slide are against

Table 2-8.2 Process Deficiencies in Centerless Grinding—Occasionally Experienced Conditions and Probable Causes

DEFECTIVE PERFORMANCE		POTENTIAL CAUSES REQUIRING CHECKING AND CORRECTION
CATEGORY	AFFECTED CONDITION	
Geometric shape of the workpiece	Out-of-round work.	Poor initial out-of-roundness, and/or insufficient number of work revolutions. Improper setup; workpiece held too low or blade-top angle too small. Stock allowance not enough. Regulating wheel runout. Varying wheel penetration due to deflections (e.g., of the work-rest blade).
	Taper of the basically cylindrical work.	Unstable position of the part during grinding. Excessive external feeding force. Nonuniform rise of temperature in the part due to unbalanced stock removal. Improper adjustment of the work guides.
	Barrel shaped work.	Work guides deflected toward the regulating wheel.
	Hourglass (concave) shaped work.	Work guides deflected toward the grinding wheel.
Size-holding difficulties	Inconsistent size holding.	Poor condition of machine elements (see last category). Unreliable gage operation or compensation when automatic size-control is used.
	Rapid size drift.	Too-soft grinding wheel. Excessive or nonuniform stock allowance.
	Unreliable size correction.	Wear or dirt in the feed mechanism or in the slide ways.
	Insufficient grinding action causing generally oversize work.	Inadequate sparkout in infeed grinding; excessive stock removal per pass, in thrufeed grinding.
Poor surface condition of the work	Chatter marks on the work surface.	Work supported too high above wheel C/L. Intermittent cut due to wheel runout. Vibrations within the system or transmitted. Work blade too thin, too long, or improperly clamped.

Table 2-8.2 *(Cont.)* **Process Deficiencies in Centerless Grinding—Occasionally Experienced Conditions and Probable Causes**

DEFECTIVE PERFORMANCE		POTENTIAL CAUSES REQUIRING CHECKING AND CORRECTION
CATEGORY	AFFECTED CONDITION	
Poor surface condition of the work *(cont.)*	Surface finish inadequate or inconsistent.	Wheel truing done with too-fast speed or with uneven advance of the truing slide. Diamond chipped, dull, or loose. Too-high rotational speed of the regulating wheel. Coolant of poor lubricity or dirty. Improper blade material causing pickup.
	Feed lines of work surface in thru-feed grinding.	Grinding wheel not adequately relieved on outgoing side. Work guides improperly aligned. Nonuniform cut along the wheel face.
	Fish tails on work surface.	Cutting fluid unsettled, not sufficiently clarified.
Wheel specifications, condition, and truing	Burn marks on the work surface.	Wheel too hard. Wheel trued with too-slow diamond traverse. Coolant quantity or application inadequate. Work speed too slow.
	Loading of the wheel surface.	Bond too hard, grit too fine. Coolant has insufficient lubricity or is improperly chosen for the work material.
	Sizing troubles.	Rapid breakdown of the too-soft wheel.
	Regulating wheel runout ("camming").	Out-of-roundness of spindle, worn bearings, loose wheel mounting.
Machine operation or condition	Erratic size holding.	Worn ways, improperly adjusted gibs, play in the feed screw/nut couple.
	Jumpy traverse of truing slide.	Ways inadequately lubricated, gibs improperly adjusted in the truing unit. Air trapped in the hydraulic line.
	Work chatter.	May be caused also by nonuniformly tightened leveling screws, or vibrations due to unbalanced motors, spindles, etc.

positive stops which are set to produce the required work size with occasional adjustments only.

Consequently, the setup of the centerless grinding machine must include the assurance of the proper worksize by the positioning and then locking of the wheel slide for thrufeed grinding, or by setting the endstops of the slide movements for infeed grinding. In the case of automatic operation, particularly when it includes size control, several additional adjustments will be needed.

The Swivel Adjustment of the Regulating Wheel Slide

Several popular types of centerless grinding machines whose work rest is mounted on the regulating wheel slide, are designed with a table between the machine bed and the slide of the regulating wheel head. That table, often referred to as the swivel plate, has a fixed pivot plug around which it can be swiveled by a small amount in either direction, and adjusted with the aid of two heavy set screws (see Fig. 2-3.6). The swivel adjustment of the regulating wheel is of advantage:

(1.) In infeed grinding as a means of compensating for minor parallelism errors which may tend to produce a part with a slight taper instead of the intended cylindrical shape

(2.) In thrufeed grinding where a gradual stock removal has to take place during the passage of the work between the wheels. Even minor deviation in the setup can cause a heavier cut in the front or in the rear, depending on the direction of departure from the required conditions. Such small, yet undesirable, setup deviations may be easily corrected by the swivel adjustment of the regulating wheel head. The amount of the swivel adjustment can be measured by a properly mounted dial indicator, or by observing the distribution of the grinding sparks for checking the accomplished correction of the grinding contact.

On centerless grinding machines not equipped with a swivel table the functionally comparable corrections of the setup may be achieved by the swivel movement of the grinding wheel truing unit, as guided by a sensitive dial indicator.

Common Faults and Potential Causes in Centerless Grinding

Imperfections in the operation or in the performance of machine tools are not uncommon, particularly when the quality of the work has to satisfy high standards, such as are frequently set for centerless grinding operations.

Recognizing the presence or, preferably, the early stages of imperfect operation is an essential requirement for assuring a production with a minimum of defective parts. However, in order to maintain a substantially uninterrupted production, the *potential causes* of defective performance must also be diagnosed and removed or corrected.

The diagnosis of the causes of defective operation in centerless grinding may be considered a task more dependent on particular process know-how and experience than in the case of most other types of grinding processes. The main reason for that dependence on specialized information is the rather unconventional method of metalworking which centerless grinding in general involves, particularly the role of balanced forces as a substitute for mechanically fixed work location, holding, and moving.

While it is impossible to condense into a short chapter section the vast experience which may be needed for the optimum operation of centerless grinding processes, a brief survey of the common faults and of their potential causes can be of definitive value to the less experienced, and also serve as a reviewing aid for the experienced technicians responsible for operating centerless grinding machines.

Such a survey is presented in Table 2-8.2, using a method of grouping based on five major categories of defective process performance. The categories embrace various families of workpiece imperfections and also deficient operation of the equipment. Within each category frequently occurring defect types are mentioned, with potential causes listed for each.

The purpose of establishing categories for both the product and the equipment defects is to facilitate the identification of fault conditions. However, the examination of deficiencies from two aspects makes a limited number of redundancies unavoidable.

The Wheels and Operational Data of Centerless Grinding

The close functional interrelation of the two individual subjects listed in the title justifies the discussion within a common section.

Grinding Wheel and Regulating Wheel Selection for Centerless Grinding

Most of the grinding wheel recommendations published by wheel and machine tool manufacturers are intended to supply general information only. Such publications are the basis of Table 2-9.1. In actual centerless grinding practice often substantially different grinding wheel specifications will be found to provide the best service, representing the most suitable compromise between conflicting operational purposes and wheel properties.

As examples, a few conditions are mentioned in random order, which will affect the specification of the grinding wheel, often to the extent of substantial variations from the general recommendations.

(a) *Very small parts*, particularly in thrufeed grinding, due to their limited area of grinding contact and the high ratio of edges to the work length, have a crushing effect on the grinding wheel, which calls for a very hard bond.

(b) *Excessive variations in the size and conditions of the raw work* (as an extreme example: cold headed parts with flash), may call for grinding wheels with elastic bond, such as resinoid or rubber, in distinction to the commonly used vitrified bond.

(c) *Crush dressing*, when applied, increases the free-cutting properties of the grinding wheel, which may permit the use of harder bond, finer grain or denser structure than recommended for similar applications when truing by the conventional method with a diamond.

(d) *Profile grinding*, particularly of workpieces with intricate shape and/or fine details, will require wheels with finer grit and harder bond than commonly used for the same work material. In such cases, developing and retaining the proper wheel profile takes precedence over productivity which would result from a higher rate of stock removal.

(e) *Tight work size tolerances* will require wheels with a harder bond for reducing the rate of the wheel breakdown in the interest of more accurate size control. Often this use of a harder bond for the grinding wheel must be counterbalanced by smaller stock allowance for the operation; as an example, only about ten times the equivalent of the final work size tolerance range may be left to be removed in certain thrufeed finish-grinding operations.

(f) There are centerless infeed grinding operations which, in addition to the cylindrical surface of the workpiece, commonly referred to as the *OD*, also include the *grinding of a shoulder*. Similar conditions arise in the so-called conjugate grinding, mentioned in Chapter 2-5 of this section, where surfaces which are mutually perpendicular, are ground in a single operation. For such centerless grinding processes combination wheels, also designated as "layer," or "sandwich" wheels, are generally used. These wheels consist of sections having different specifications, adjusted to the conditions controlled by the area of grinding contact, which results from the geometric form and location (e.g., substantially parallel with or normal to the work axis) of the work surface section. Softer

Table 2-9.1 Grinding Wheel Recommendations for Centerless Grinding

WORK MATERIAL	DIAMETER OF THE WORK GROUND			
	Less Than 3/4″	3/4″ to 2″	2″ to 4″	4″ to 5″
Aluminum and Aluminum Alloy (hard) (soft)	A60-M6-VL C46-J6-VP	A60-L6-VL C46-I6-VL	A60-J6-VL C46-H6-VP	A60-H6-VL C46-G6-VP
Brass and Bronze (soft)	A60-K6-VL	A60-J6-VL	A60-I6-VL	A60-H6-VL
Bronze (hard)	A60-M6-VL	A60-L6-VL	A60-J6-VL	A60-H6-VL
Cast Iron	A60-M6-VL	A60-L6-VL	A60-I6-VL	A60-H6-VL
Glass (Tubing)	C180-M5-VP	C180-K5-VP	C180-15-VP	C180-H5-VP
Plastics	C46-K6-VP	C46-J6-VP	C46-I6-VP	C46-I6-VP
Porcelain	C180-L5-VP	C180-K5-VP	C180-H5-VP	C180-H5-VP
STEEL Hard Soft High-Speed	A80-L6-VL A60-M6-VL A60-L2-VL	A80-K6-VL A60-M6-VL A60-K6-VL	A80-I6-VL A60-L6-VL A60-I6-VL	A80-H6-VL A60-J6-VL A60-H6-VL
Stainless—300 series (free machining)	A60-K6-VL	A60-I6-VL	A60-H6-VL	A60-H6-VL
Stainless—400 series (hardened)	A60-L6-VL	A60-K6-VL	A60-I6-VL	A60-H6-VL
Tubing (steel) Thin wall Thick wall	A60-L6-VL A46-M6-VL	A60-K6-VL A46-M6-VL	A60-J6-VL A46-L6-VL	A60-I6-VL A46-K6-VL

REGULATING WHEEL SPECIFICATIONS:

All regulating wheels 8″ dia. and less, ½ to 2½ ins. wide: A80–R3–R

All other sizes of regulating wheels: A80–R2–R

NOTE: Table based on recommendations by Cincinnati Milacron. The abrasives selected are either semifriable aluminum oxide (A), or black silicon carbide (C). The last letter of the symbol indicates the specific type of vitrified (V) bond selected by the originator of these specifications.

bond and coarser grains are used for the shoulder section, where the area of contact is large and some "wiping" action also occurs, than for the OD of the part.

(g) *For obtaining a fine finish* the use of fine grain wheels seems to be the logical choice. However, other factors, too, affect the resulting work surface finish, such as: the ratio between the peripheral speeds of the work and of the grinding wheel; the amount of stock removal, and the number of work revolutions during the grinding pass; the method and rates of wheel truing; the composition and cleanliness of the coolant; the rigidity of the work, of its support and of the entire machine; and various other factors. As a matter of fact, extremely fine-grain wheels may, due to adverse effects of other contributing factors, produce a poorer finish than obtained, e.g., with an 80-grit wheel, when all the other pertinent grinding data are properly chosen and maintained.

(h) *The amount of stock removal per pass* or, more specifically, per revolution of the workpiece during the grinding, is an essential factor in determining the suitable grain size. For a high specific rate of stock removal the grain size of the wheel should be definitely coarser than for a comparable operation applied with a low stock removal rate.

The regulating wheels of the centerless grinding operation do not participate in the process of stock removal, their role being limited to supporting, to controlling the rotation and, in thrufeed grinding, to feeding; that is, axially advancing the work being ground. From that role of the grinding wheel stem the alternate designations of "control wheel" (the British usage) or "feed wheel."

With very few exceptions, rubber bonded wheels with aluminum oxide grain in grit size 80 are used for centerless regulating wheels.

The Handling and Mounting of Centerless Grinding Wheels

Grinding wheels used in centerless grinding machines are usually of large diameter and, in most cases, have substantial width, as required by a grinding process offering the advantage of high productivity. Another general characteristic of the centerless grinding wheels is the relatively large bore, reflecting the practice that when wear reduces the original diameter by about 20 per cent, or even less, and consequently the peripheral speed correspondingly diminishes, the wheel is considered worn and will be exchanged. For accepting wheels with large bore size, either properly dimensioned wheel mounts are used as intermediate members or, in the case of spindles supported on both ends, the body of the spindle has the required large diameter.

Grinding wheels held on mounts have effective bore lengths which vary by a limited amount only. When wider wheels are needed, they are made with recesses, either on one side only or on both sides, depending on the difference between the effective bore length as determined by the mount and the actual width of the grinding wheel. Spindles with straddling bearing support are free from these limitations and wheel bores having lengths equal to the width of the wheel can be fully supported.

Because of the sizes of grinding wheels which are used in centerless grinding machines in general, and more particularly in the heavier models, special wheel and spindle handling devices are usually supplied by

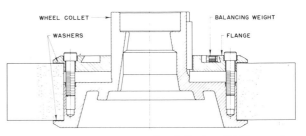

Fig. 2-9.1 Wheel mount of a centerless grinding machine with cantilever type wheel spindle support. The wheel is secured by a flange against the shoulder of the wheel collet, which fits over the nose of the machine spindle. The diagram shows one of the balancing weights which can be moved in the annular groove and locked by screw in the appropriate balancing position.

the grinding machine manufacturers. Most of these devices operate in conjunction with a hoist, mounted either on a jib crane, with which larger centerless grinding machines may be equipped, or on an available overhead crane.

Depending on the design of the wheel spindle, the methods of wheel mounting differ. Centerless grinding machines with cantilever wheel spindle support, which permit the installation of the wheel from the free end of the spindle (see Fig. 2-3.3) have the wheels installed on a mount (Fig. 2-9.1) whose collet is then slipped over the nose of the machine spindle.

Centerless grinding machines with wheel spindles supported on both ends (see Fig. 2-3.4) require the wheel to be removed from the machine together with the entire spindle assembly in order to carry out the wheel change outside the machine. For this process the grinding wheel is supported, preferably in a wide-

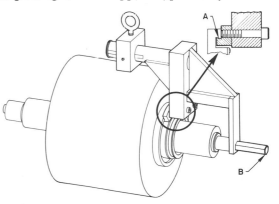

Fig. 2-9.2 Wheel-changing device for a centerless grinding machine requiring the removal of the spindle assembly. for changing the wheel. The wedge *A* fits into the balancing weight groove of the flange and the bolt *B* is tightened securely to the threaded end of the spindle.

angle wooden V-block, and the spindle assembly is suspended on a hoist by means of a special bracket; an example of such a wheel-changing device is shown in Fig. 2-9.2.

The bore of the grinding wheel should be about 0.006 to 0.010 inch (0.15 to 0.25 millimeter) larger than the diameter of the mount or of the spindle over which it must slip without binding but also without excessive clearance. When the wheel is balanced in the grinding-wheel manufacturing plant that clearance in the mounting fit is taken into consideration and the wheels are marked showing the side which should be UP or TOP during the mounting process.

The Balancing of Centerless Grinding Wheels

For centerless grinding machines of older vintage, operating with wheels much smaller than the high productivity machines of recent origin, the conventional wheel balancing is still being used. In that process the wheel, installed on its mount, is first trued in the grinding machine, then removed, mounted on a balancing arbor replacing the machine spindle, and placed on a balancing stand. That latter has either two parallel rails with narrow edges, or two overlapping and freely rotating disks on both sides, forming a V-block type cradle in which the ends of the balancing arbor rest. The freely turning grinding wheel will take a position in which its heavy portion is in the lowest spot. By installing the four balancing weights (see Fig. 2-9.1) in appropriate positions, sometimes with further adjustment for altering the amount of counterbalance, it is possible to accomplish a good balancing of the grinding wheel as indicated by the absence of any noticeable "heavy side."

Of course, that process of wheel balancing is time consuming, and having the wheel balanced when first installed into the grinding machine does not guarantee an equally well-balanced condition over its entire service life.

Automatic Grinding Wheel Balancing

Balancing the grinding wheel while it is mounted on its spindle in the grinding machine, by applying an automatic process, offers several significant advantages over wheel balancing on a separate stand: (a) the wheel is balanced dynamically, in distinction to static balancing on a stand; (b) the operation is much faster and requires much less effort on the part of the operator than conventional balancing on a stand; (c) it permits the rebalancing of the wheel over several periods of its useful life, thus eliminating the po-

tentially harmful effects of structural variations within the wheel.

One of the automatic grinding wheel balancing systems, which is widely used, operates by the following principles: The regularly rigid support of the wheel spindle is temporarily disengaged to produce a flexibly retention of the rotating spindle. At the free end of the spindle three hardened steel balls are retained in an annular raceway in which they can freely roll when not clamped. In the case of wheel imbalance the flexible support will permit the true center of the spindle, together with the integral raceway, to shift to a new center of rotation, thus raising centrifugal forces which will cause the balls to move toward the light side of the wheel. These conditions are shown in Fig. 2-9.3.

In this position the balls will be locked in their raceway by actuating a spring loaded plunger and subsequently the entire mechanism is rigidly clamped in the wheel head. In the balanced condition (see Fig. 2-9.3/B) the center of the wheel rotation will coincide with the true center of the raceway.

The whole process takes only about a minute and is controlled by a lever in the front of the machine. The attainment of the balanced state of the wheel assembly can be observed on a dial, when its needle, which vibrates in the unbalanced state, comes to rest.

The application of this system of wheel balancing is limited to grinding machines with cantilever type wheel spindle support.

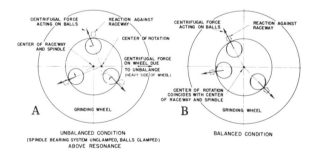

Courtesy of Cincinnati Milacron, Inc.

Fig. 2-9.3 Principles of operation of the ball type wheel-spindle balancing device. A. Unbalanced condition (spindle bearing system unclamped, balls clamped) above resonance; B. Balanced condition.

Grinding Wheel Balancing by Vibration Analysis

The balancing of the grinding wheel is less consistently applied in general shop practice when the centerless grinding machine used has a straddling type wheel spindle support whose inherent rigidity reduces the harmful effects of minor wheel unbalance. At the same time the technology of grinding wheel manufacturing has advanced, resulting in wheels of more uniform structure, which are supplied dressed to precise size and generally in a well-balanced condition.

Besides these positive factors which reduce the need for wheel balancing in the user's plant, several of the commonly used methods are not applicable at all, such as the balancing by freely rolling balls; or are too cumbersome because of the weight and the bulk of the heavy spindle assembly. Finally, with regard to the use of balancing stands, the results of static balancing may not be sufficiently reliable, because the actual behavior of a large wheel under the dynamic conditions of regular operation may differ from the balance in a static state.

Although not needed for general work, a fine balancing of heavy centerless grinding wheels may be desirable when the work to be carried out must satisfy high requirements with respect to size holding, roundness, and surface finish, all these conditions being sensitive to vibrations caused by even a small amount of imbalance.

A method of grinding wheel balancing by vibration analysis, which is applicable to grinding wheels mounted on the spindle in the machine tool and provides guidance for creating dynamically balanced conditions, operates on the following principles.

A portable vibration analysis instrument is used whose seismic pickup is held against or attached to an element of the operating grinding machine, such as the housing of the wheel spindle bearing, which is transmitting the vibrations caused by an out-of-balance grinding wheel. A stroboscopic lamp is fired by the pickup at the same frequency as the vibrations being analyzed. Identifying numbers, corresponding to hour numbers in the respective positions on the clock dial, are generally applied on the side of the wheel, e.g., by using a grease pencil. These marks will assist in determining the peripheral location of the heavy portions of the wheel which are observable under the stroboscopic light as those of the greatest runout. The effects of the heavy spot in the grinding wheel can then be neutralized by the appropriate shifting of the balancing weights.

The advantage of the described system is its capability of detecting the presence of transmitted vibrations originating from exterior sources, which might be not less harmful to the quality of the produced work than vibrations caused by an improperly balanced grinding wheel. The process of vibration analysis may be repeated at intervals, even several times during the service life of a grinding wheel.

Finally, a general rule which is applicable to all methods of grinding wheel balancing, as well as to grinding wheel usage in general: Avoid the presence of retained grinding fluid in the wheel, because that may cause severe imbalance before it is eliminated during the regular operation of the wheel. Such temporary imbalance can indicate the need for shifting the balancing weights, causing an actual imbalance as soon as the wheel has thrown out the retained coolant. Therefore, it is recommended that the grinding wheel is kept running for a few minutes after the coolant supply has been shut off, before stopping the wheel drive motor.

Grinding Data for Centerless Grinding

Defining grinding data for the centerless grinding method is a much less straightforward proposition than in the case of most other grinding methods. First, there are two major process varieties which are primarily related to each other, only by the common application of the method's basic characteristics and by the use of the same machine tools, although in a substantially different setup. As far as the third variety, the shoe type centerless grinding is concerned, the grinding machines used are of fundamentally different design, but the mechanism of the applied metalworking process is essentially an infeed grinding.

In centerless grinding, except for the shoe type variety, the rotational speed of the work is not a directly controlled magnitude, as it is in the conventional types of machine tools which operate with rotating work. In the centerless grinding method the rotational speed of the work has to be derived from the regulating wheel speed, applying the ratio of their respective diameters.

The traverse speed of the work, when applicable in centerless grinding, has to be calculated. In conventional traverse grinding the work traverse rate is equal to that of the machine table, whereas in centerless thrufeed grinding of any specific work diameter, it is the composite result of three independent variables, namely the diameter, the rotational speed, and the incline angle of the regulating wheel.

In view of these unique factors which in centerless grinding control the actual operating conditions, the usual system of tabulated, readily applicable values will not be used. Instead, for each major method, the process factors which primarily control the grinding data will be discussed and their use illustrated by a few examples.

While the values used in the examples were carefully selected to be realistic, they are not necessarily the best suited for particular processes in which factors not considered in these examples might have a major role.

The principal purpose of the following discussion of process data development is to show the interrelations between major constituent elements and to demonstrate one method for evaluating centerless grinding operations and selecting the applicable grinding data.

Depending on the amount of stock allowance or on the specifications for the finished part, centerless grinding operations frequently require more than a single pass; sometimes as many as three or four passes are used. Although multiple passes are occasionally also called for in infeed grinding, their application is more frequent in the thrufeed grinding process. The examples listed refer to single passes, which may have to be complemented by additional ones. In a single example for thrufeed grinding both operations of a two-pass process are listed, for illustrating that operational variety.

The productivity results expressed for infeed grinding by the number of parts ground per hour are, on the other hand, improved by another process variety, namely by grinding more than a single part in one operation. Particularly on powerful centerless grinding machines operating with wide wheels, it is possible to simultaneously infeed grind several parts whose combined length does not exceed the wheel width. Even parts with intricate profile may be ground concurrently when the grinder is equipped with the appropriate work loading and unloading device.

Infeed Grinding Data

The infeed advance of the grinding wheel related to each revolution of the workpiece will produce the depth of cut, the optimum value of which will vary from a minimum of about 0.000040 inch (1 micrometer) to a maximum in the order of ten to twenty times greater, 0.0004 to 0.0008 inch (0.01 to 0.02 mm), or even more, such as for sturdy parts rough ground on powerful machines.

Factors, often acting cumulatively, which require a reduction of the depth of cut are: tight diameter tolerances, fine finish, strictly controlled geometry of form and profile, great hardness of the work material, fragile or distortion-sensitive parts, insufficient rigidity or limited power of the grinding machine, etc.

The rotational speed of the work is derived from its desired peripheral speed, whose best value will vary depending on the dimensions and configuration of the part (vibrations must be avoided), the required degree of size and finish control, and various other factors. In general practice the suitable work speed is within a range of 1 to 3 per cent of the peripheral speed of the grinding wheel. The selected peripheral speed of the work has then to be translated into work revolutions per minute for arriving at the pertinent infeed rate.

Following the procedure used for developing the centerless infeed grinding examples of Table 2-9.2, the consecutive steps are commented on in the following:

1. The values used in the examples for the work peripheral speed have been determined by the controlling factors which have just been discussed. The work speed, first expressed as a percentage of the grinding wheel peripheral speed, is then transferred into a meters/minute value for the convenience of the related calculation steps. One of these defines the corresponding rpm of the regulating wheel which will produce the same peripheral speed that is required for the workpiece.

2. The work peripheral speed is also the basis for expressing the corresponding rotational speed of the workpiece in rpm, a value needed for the next consecutive step.

3. The infeed rate is first expressed as a specific value, selected for the optimum depth of cut. Subsequently, by a simple multiplication, that specific value will then be transferred into the more common term of infeed rate per minute.

4. Knowing the amount of stock removal (using one half of the allowance which is expressed as an addition to the diameter), by applying a simple division, the theoretical value of the basic grinding time can be expressed.

5. The time of the sparkout during which no infeed advance takes place, has been set in these examples uniformly at an average value of 10 per cent. The actual value will vary depending on factors such as the rigidity of the workpiece and of the setup, the cutting properties of the grinding wheel, the attainable work surface finish, etc.

6. The time provided for work handling has been selected arbitrarily, primarily to indicate that this element, too, must be included in estimating the operational time for infeed grinding.

7. The term "at 100% efficiency" means that the stated values do not contain any allowance for machine setup, adjustments, wheel truing, the operator's personal time, etc.

Thrufeed Grinding Data

Thrufeed grinding differs in several essential aspects from centerless infeed grinding and its process

Table 2-9.2 Development of Grinding Data for Infeed Centerless Grinding

STEP NO	WORK FACTORS	EXAMPLES		
		1	2	3
	Work material.	Plain carbon steel	Tool steel, hardened	Tool steel, high-alloyed, hardened
	Workpiece diameter, millimeters.	100	50	10
	Stock removed on the side (one half of the allowance on the diameter), millimeters.	0.10	0.04	0.03
	Geometric characteristics of work.	Plain cylinder	Intricate profile	Slender taper
	Finish size (diameter) tolerances ± mm.	0.01	0.003	0.002
	GRINDING DATA			
1	Peripheral speed of work (and of the regulating wheel).			
	— expressed as per cent of wheel speed	2.5	1	0.8
	— expressed in meters/minute (based on 30 m/sec wheel speed).	45	18	14.4
2	Workpiece revolutions per minute (Based on surface speed as per Step 1).	143	114	475
3	Selected amount of infeed per work revolution, millimeters on the side.	0.004	0.002	0.001
	Infeed rate per minute, millimeters.	0.572	0.228	0.475
4	Theoretical time of stock removal, in minutes. (Stock on the side/infeed per min.)	0.1840	0.1760	0.063
5	Added for sparkout (e.g., 10 per cent).	0.0184	0.0176	0.007
6	Added for work changing (slide retraction, work handling, slide approach) (*NOTE:* the value used is an example; actual time can widely differ), minutes.	0.150	0.100	0.050
	Total theoretical cycle time, minutes.	0.3524	0.352	0.120
7	Production rate, at "100 per cent efficiency," pieces/hour.	170	170	500

Table 2-9.3 Development of Grinding Data for Thrufeed Centerless Grinding

Step No.	Note	Work Factors	Examples			
			1	2	3	4
	General information	Workpiece type	Steel rod	Ball bearing race		Bearing roller
		Operation	Roughing pass	1st pass	2nd pass	Finishing pass
		Work diameter in millimeters	50	80	80	15
		Size tolerance of the ground diameter ± millimeters	0.050	0.025	0.010	0.002
		GRINDING DATA				
1	Select	Stock removal by the pass, mm — from the diam. (stock allowance).	0.50	0.30	0.06	0.05
2	Select	Peripheral speed of the work — in per cent of grinding wheel speed.	5	2.5	2	4
		— expressed in meters/minute (based on 30 m/sec grinding wheel surface speed.)	90	45	36	72
		— expressed in work rpm.	573	179	143	1530
3	Select from equipment range	Regulating wheel diam./width, mm.	400/300	350/250	350/250	300/200
		Express the rpm of the regulating wheel for producing the surface speed as per Step 2. rpm.	72	41	33	76
4	Select	Number of required part revolutions during the pass.	36	36	48	120
5	Calculate (see Note A)	Work traverse feed rate in m/min (Should produce the number of revolutions as per Step 4).	4.8	1.25	0.75	2.55
6	Calculate (see Note B)	Incline angle setting of the regulating wheel. (To produce the traverse feed rate as per Step 5).	3°4′	1°36′	1°11′	2°2′

Formulas for Calculating the Stated Figures Based on the Selected Values of the Adjustable Variables

Note A

Traverse Feed Rate f of the Work in Meters per Minute $f = \dfrac{S \times W}{T}$

Where: S = Work revolutions per minute (see Step 2)
W = Width of the regulating wheel in meters (see Step 3)
T = Total number of part revolutions during the pass (see Step 4).

Note B

Incline Angle δ of the Regulating Wheel   arc sin $\delta = \dfrac{A}{V}$

Where: A = Advance (feed) rate of the traversing work in m/min (see Step 5)
V = Peripheral speed of the rotating workpiece in m/min (see Step 2).

evaluation requires a particular approach, such as has been used for setting up the data shown in the examples of Table 2-9.3. Again, only a limited number of examples have been chosen; the following, more generalized discussions are numbered in accordance with those steps in the table to which they are related.

1. For defining the number of passes and the amount of stock removal per pass, factors such as the following are considered:

(a) The stock to be removed from the surface of the part (substantial stock allowance may require two or more passes in subsequent operations on the same machine or, when additional machines are available, in a continuous process)

(b) The capacity of the grinding machine (general dimensioning, driving power, etc.), the width of the grinding wheel, which determines the length of the work contact area, is also a factor of the stock removing capacity

(c) The specifications for the finished work dimensions and conditions (size tolerances, finish, etc.) may call for a finishing pass even when the total amount of stock could be removed in a single pass. Modern centerless grinders, accepting very wide wheels, will in some cases permit the roughing and finishing passes to be combined by mounting grinding wheels with different specifications, side by side. That approach, however, is not always applicable because other factors besides the composition of the grinding wheels may have to be altered to accomplish the finish grinding objectives.

2. The peripheral speed of the work, expressed first in a ratio to the grinding-wheel speed will be decided by considerations similar to those of the infeed grinding, however the selected values will be, in general, somewhat higher and extend over a wider range. Percentage values vary from about 1.5 to about 5 per cent, and exceptionally, even higher. Subsequently in the process, the corresponding work peripheral speed in meters per minute (or feet/minute) values should also be stated.

3. The dimensions of the regulating wheel are listed as reference values on the basis of which the peripheral speed of the regulating wheel, which is considered equal to that of the work, can be expressed in revolutions per minute. The speed of the regulating wheel is then adjusted, by observing the indications of the tachometer on the regulating wheel head, to agree with the calculated rpm value.

4. The number of part rotations per pass should express the number of revolutions made by any single point on the work surface while in the grinding area. The selected value controls both the specific depth of wheel penetration (depth of cut) as well as the expected degree of roundness correction. Values in the range of 30 to 120, sometimes even higher for very fine finish, are usually selected.

5. The thrufeed rate of the work is an essential value because it expresses the rate of performance of the operation. When a systematic process development is applied, this will actually be a resulting value whose factors are the rotational speed of the part and the number of its revolutions during the passage between the wheels. The range of thrufeed rates may vary from about 1 to 30 meters (about 3 to 100 feet) per minute.

6. The regulating wheel incline angle is the function of the selected regulating wheel peripheral speed and of the thrufeed rate (both related to a common time unit of one minute). The value of the incline angle is listed because it represents an important machine setup parameter for centerless thrufeed grinding.

INTERNAL GRINDING

General Characteristics and Applications

The grinding of internal round surfaces represents a major category of grinding that has several distinguishing characteristics which set it apart from the rest of the grinding methods and warrant its discussion as a distinctly defined process.

These characteristics are essentially related to three particular sets of restrictive operational conditions under which the process has to be carried out, namely:

1. *The restrained area* in which the grinding of internal surfaces must be accomplished, that area being entirely enclosed by the work surface.

2. *The limited access* of the work area which can only be reached from the end of the hole.

3. *The transfer of the wheel into the grinding position* along a path which is perpendicular to the direction of the infeed advance.

In addition, thore aro several functional character istics which have a controlling effect on the capabilities and operational requirements of internal grinding, such as:

(a) The large area of contact between the grinding wheel and the work creates conditions to which the composition of the grinding wheel has to be adjusted.

(b) The cantilever support of the grinding wheel at the free end of a spindle, whose diameter is restricted by that of the hole being ground.

(c) The generally small diameter of the grinding wheel which requires high rotational speeds for operating the wheel with the necessary surface speed.

Considering these limiting characteristics, inherent to internal surfaces and determining the conditions under which the grinding process must be carried out, the question may be raised: Why is internal grinding still the preferred, and is often the only adequate method for finishing internal surfaces? For providing the proper answer the question should be investigated from the following two aspects:

1. Why are internal round surfaces with high geometric and dimensional accuracy needed?

2. Why is internal grinding the proper method for finishing these surfaces?

By considering the following general conditions, a convincing explanation will be found for the use of internal grinding and its importance as a basic metalworking process.

1. Round surfaces, whether cylindrical, tapered or profiled, are economical to machine because they are produced by a continuous tool path, permitting the uninterrupted engagement of the tool with the work surface.

2. Round surfaces having accurate shape and dimensions are needed primarily for the following types of applications:

(a) Assembly surfaces requiring accurate fit

(b) Surfaces in precise compliance with a mating member to produce tight, well-sealing contacting surfaces

(c) Sliding guides for rotational or axial displacement with uniform and precisely controlled clearance between the corresponding surfaces

(d) Tracks for rolling elements, particularly in antifriction bearings, with near perfect profile, roundness, and diameter.

Generally, one of these applicational conditions is present in workpieces for which internal grinding is applied.

3. While other methods of metalworking may also be capable of producing precise internal surfaces, there are several conditions which, individually or in combination, make internal grinding the preferable or actually the only technologically satisfactory method for such operations. Following are listed a few of the conditions favoring internal grinding. This list also serves as a means of indicating the type of operations which are usually carried out by this method:

(a) Hard materials, such as hardened steel parts, cannot be worked, or are only worked very inefficiently by any method other than grinding.

(b) Grinding permits the consistent attainment of the highest degree of accuracy with regard to location, form and dimensions, which is generally required for machined parts.

(c) Surface finishes can be produced at the desired level within a rather wide range, and with upper limits close to the highest technically attainable degree.

(d) Small allowances left for finishing by the preceding rough machining or forming process are acceptable for grinding.

(e) The small cutting force which results in grinding with light cuts makes this method adaptable for delicate parts which are sensitive to distortions from the pressure of the work holder or of the tool.

(f) Profiles, even complex ones, can be produced efficiently by grinding, often requiring only an appropriate wheel truing profile bar, and with no need for expensive expendable tools.

(g) Automation of the process, particularly when it involves tight tolerances, can be accomplished more easily and with higher efficiency by grinding than by most other metalworking methods. Certain systems of internal grinding permit the application of very fast-acting, work-loading, locating and holding devices; the continuous monitoring of the work size; and the in-process reconditioning of the tool by automatic wheel truing and size compensation.

Thus, the need for accurate, round internal surfaces and the favorable process characteristics of internal grinding, have resulted in a wide variety of regularly applied types of internal grinding operations, both in small- and large-scale production. Table 3-1.1, which presents and discusses typical internal grinding operations, can be helpful in visualizing and evaluating the many different work surface configurations which can be successfully, and often very efficiently, produced by properly designed internal grinding operation and equipment.

Work Capacities of Internal Grinding Machines

The grinding of internal surfaces, similarly to other metalworking methods, has its areas of preferential application which, however, due to the earlier outlined inherent process characteristics, is more limited in scope than are most other methods. Actually, the boundaries of these limitations are not rigid, they may even be widely extended, usually with the aid of special equipment.

Nevertheless, an appraisal of the controlling process conditions will occasionally have to precede the selection of internal grinding as a readily adoptable and best suited method. In such cases, a general review of the work capacities of internal grinding may prove a useful guide.

The factors which affect the suitability of the internal grinding method are multiple, consequently, the major controlling aspects will be discussed separately.

Workpiece Dimensions and Configurations

Basically, and in the generally used systems of application, internal grinding operates with a rotating workpiece. Consequently, the maximum work diameter that can be rotated by the machine without interference, a dimension commonly termed *the swing*, is the limiting factor of any external work size. The only exceptions in this respect are the planetary type internal grinders which operate with a nonrotating workpiece.

Typical values for maximum swing over the table are:

1. 16 inches (about 400 mm) in a popular model of universal type internal grinding machine

2. 4 to 10 inches (about 100 to 250 mm) in different models of centerless type production internal grinders

3. up to 36 inches (about 900 mm), occasionally even greater, in large-capacity internal grinding machines.

The swing over the table is merely indicative of, and not equivalent to, the maximum acceptable work diameter. This latter is usually smaller because of the space required by the work-holding device. Further reductions of the work size, as compared with the swing, may be caused by work-loading fixtures such as are needed in machines equipped for automatic operation.

Dimensions of the Worked Surface

The capacity data applicable to different models of internal grinding machines are usually based on the grinding of plain cylindrical holes. The two characteristic dimensions are the diameter and the length of the hole, values which, at the upper limits of the capacity, are interdependent. Internal spindles with

Table 3-1.1. Examples of Work Surface Configurations Finished on Internal Grinding Machines

GENERAL SHAPE	DESIGNATION	DIAGRAM	DISCUSSION
CYLINDRICAL	PLAIN CYLINDRICAL HOLE		The basic form of surfaces ground on internal grinding machines are plain cylindrical holes, commonly bores of essentially ring-shaped parts, concentric with the outside surface. Such parts are located on the OD and ground with reciprocating wheel movement. Traverse grinding for other than through holes requires adequate relief at the closed end, otherwise plunge grinding is needed. Plain cylindrical holes are well adapted for automatic operations with size control to tight tolerances.
	CYLINDRICAL HOLES IN RECESSED LOCATIONS		Holes behind a shoulder require an extended approach movement of the wheel, generally a part of the automatically controlled sequence. Commonly, plunge grinding, perhaps with very short stroke oscillation, is used, particularly when the hole being ground is bounded on both ends by shoulders. In special cases the face grinding of one or both shoulders, too, may be carried out on machines equipped with an appropriate wheel truing device and longitudinal feed control.
	PLANETARY CYLINDRICAL HOLE GRINDING		Holes in workpieces which cannot be rotated around the axis of the hole are finished by planetary internal grinding, with the wheel rotating around both its own axis and that of the hole. The radius of the second rotation is adjusted during the grinding for producing the infeed movement. The reciprocating movement of the wheel is also used when traverse grinding is needed. The center of the planetary rotation which controls the location of the hole can be set with great accuracy in jig grinders, also operating on the planetary principle.

(Continued on next page)

Table 3-1.1. *(Cont.)* Examples of Work Surface Configurations Finished on Internal Grinding Machines

GENERAL SHAPE	DESIGNATION	DIAGRAM	DISCUSSION
CYLINDRICAL	EXTERNAL SURFACES OF CYLINDRICAL FORM.		The grinding of external surfaces on an internal grinding machine may be applied for the purpose of convenience, or required for technical reasons, such as the recessed location of the hub being ground, accessible only by a small wheel at the end of a long spindle. Although not an effective method of OD grinding, this capability of internal grinding machines is under specific circumstances, a valuable equipment characteristic.
TAPER	PLAIN TAPER		Tapered internal surfaces ground to accurate angle and size are needed in many mechanical elements; important but not unique examples are the pathways of tapered roller bearing outer races. In addition to the universal internal grinders many models of production type internal grinding machines are equipped with workheads which can be swiveled a controlled amount to produce a tapered work surface of specific angle. Some models have sine bar type setting elements for the very accurate control of the swivel angle by means of gage blocks.
	BACK TAPER ALONE OR IN COMBINATION WITH FRONT TAPER		A back taper, that is, a tapered hole surface located on the end of the part away from the wheel head, has to be ground when turning the part around is not feasible due to work holding or locating limitations. Another example of back taper grinding is workpieces with tapers on both ends, to be ground in a single operation for assured coaxiality, such as in the case of double-row tapered roller bearing outer rings.

Table 3-1.1. *(Cont.)* **Examples of Work Surface Configurations Finished on Internal Grinding Machines**

GENERAL SHAPE	DESIGNATION	DIAGRAM	DISCUSSION
CURVED OR COMBINED PROFILES	RADIUS (CIRCULAR ARC) PROFILES		"Radius" in common shop usage designates a circular arc cross-sectional contour whose characteristic dimension is the radius of the pertinent circle. This is the form easiest to generate, outside the straight, in truing the wheel. Radius profiles may be produced by plunge grinding with an appropriately dressed wheel or on oscillating grinders whose pivot axis coincides with the center of the circle to which the arcuate profile to be produced pertains. Typical work examples are the ball races of ball bearing outer rings.
	IRREGULAR OR COMPOSITE PROFILES		Irregular profiles, composed of several different elements may be required for internal surfaces to be ground with consistent accuracy. Internal plunge grinding is used, with the wheel properly shaped by: (a) single diamond truing with profile bar and follower; (b) rotary form diamond dressing; or (c) crush dressing (only on special types on internal grinding machines).
INTERRUPTED OR ECCENTRIC HOLES	INTERRUPTED INTERNAL SURFACES		Internal grinding is one of the few methods by which the finish machining of holes interrupted by either axial or radial grooves, can be carried out accurately and efficiently. The production grinding of such internal surfaces in an automatic process is now feasible due to sensitive gaging instruments with probes designed to be unaffected by interruption of the work surface. Such instruments indicate the size of the envelope representing the maximum material condition.

(Continued on next page)

Table 3-1.1. (*Cont.*) **Examples of Work Surface Configurations Finished on Internal Grinding Machines**

GENERAL SHAPE	DESIGNATION	DIAGRAM	DISCUSSION
INTERRUPTED OR ECCENTRIC HOLES	ECCENTRICALLY LOCATED HOLES		When the axis of the hole to be ground does not coincide with the center of the workpiece, the hole can still be ground on a regular internal grinding machine with swing capacity permitting the rotation of the workpiece around the hole axis. For holding the work in single-piece production a face plate may be used, applying tramming with an indicator for lining up the hole with the workhead spindle. In continuous production, special fixtures with fixed locating elements are needed.
FLAT SURFACES	EXTERNALLY LOCATED FLAT SURFACE		Flat surfaces representing the face of a part (also wide angle convex cones) can be ground efficiently on internal grinding machines when technical reasons make them preferable to regular surface grinders (e.g., a protruding hub or shoulder limits access, also when special squareness must be assured). Grinding wheels with relieved or recessed face are used and a special face dresser device is needed. The grinder must also have sensitive axial feed and a positive end stop for that movement.
FLAT SURFACES	INTERNALLY LOCATED FLAT SURFACES		Flat surfaces inside a bore and at right angles to the bore axis, such as internal shoulders, can only be ground on internal grinding machines equipped with the appropriate spindle and face-dressed grinding wheel. Such surface may be at the end of a hole section open to the front, or in a recessed location adjacent to the hole surface which has a shoulder in the front (e.g., a U-type roller bearing outer ring).

Table 3-1.1. (*Cont.*) **Examples of Work Surface Configurations Finished on Internal Grinding Machines**

GENERAL SHAPE	DESIGNATION	DIAGRAM	DISCUSSION
FLAT SURFACES	HOLE WITH ADJACENT INTERNAL OR EXTERNAL FACE		By grinding both surfaces in the same operation, using sequentially the periphery and the face of the same wheel, efficient production and a high degree of squareness accuracy can be achieved. The cylindrical hole section may be ground with short reciprocating strokes when the part design provides sufficient end clearance, while a slow axial infeed is used in the grinding of the ring shaped face section.
	ROTARY SURFACE GRINDING		Grinding a ring-shaped surface which is concentric with the axis of the part, by using the periphery of the grinding wheel, is an efficient method for assuring excellent geometric relationship with the axis of the rotating part. The universal type internal grinding machines, with swivel capacity of the wheelhead to 90°, are well adapted for such operations which can produce surfaces at exactly right angles to the axis, or with less than 90 degrees for producing a wide angle cone which can be either convex or concave.
MULTIPLE OPERATIONS	DOUBLE END INTERNAL GRINDING		Workpieces with coaxial but individual internal surfaces at the two ends, can be ground in a single operation on special double end internal grinding machines which assure both excellent geometric relationship and efficient operation. Another possible application of such equipment is the indexing of the worktable around a vertical pivot axis, to present the same part end subsequently to two separate wheel spindles for carrying out two mutually complementary grinding operations on the same internal surface.

(*Continued on next page*)

Table 3-1.1. (*Cont.*) **Examples of Work Surface Configurations Finished on Internal Grinding Machines**

GENERAL SHAPE	DESIGNATION	DIAGRAM	DISCUSSION
MULTIPLE OPERATIONS	TWO-WHEEL INTERNAL GRINDING MACHINES		Two grinding spindles arranged side-by-side, carrying different wheels and having independent retraction movements, in combination with a work-head slide with optionally operated alternate indexing positions, permitting grinding sequentially, or simultaneously, several internal and external surfaces (OD or face), with a single operation, thus assuring excellent geometric interrelations.

adequate rigidity are needed for grinding deep holes. These spindles operate with wheels of correspondingly large diameters which, in turn, determine the minimum limit of the hole diameter into which such wheels can be introduced. The following informative values must be appraised without losing sight of the fact that the practical diameter of the worked hole is, of course, always smaller than the outside diameter of the workpiece, whose maximum limits have been previously discussed.

Machine Type	Maximum Limits for Typical Models	
	Hole diameter	Hole length
Universal internal grinder	18 inches (450 mm)	12 inches (300 mm)
Centerless type production grinder	4 inches (100 mm)	6 inches (150 mm)
Large capacity internal grinding machine	24 inches (600 mm)	72 inches (1800 mm)

Work Material and Conditions of Work Preparation

In principle, internal grinding, just as other grinding methods, is applicable to a very wide range of different work materials, such as steel in both its soft and hardened condition; nonferrous metals including cemented carbides; and hard nonmetallic materials, such as ceramics, plastics, etc. However, the reduced productivity of internal grinding due to the dimensional limitations of the applicable grinding wheels and other inherent characteristics of the process, generally confines the actual use of internal grinding to

those materials, primarily hardened steel, which cannot be worked economically by any other than abrasive processes.

The limitations of the process productivity also influence the required condition of work preparation. It is generally not considered efficient processing to use a limited productivity method for removing work material which could have been worked off by a more productive method. For example, the bore of a workpiece made of steel, which has been machined prior to heat treatment, should not have excessive stock left on its surface for internal grinding. The amount of stock to be ground off, termed "grinding allowance" should, as a rule, not exceed the combined value which results from providing for the pertinent technical variables such as: (a) size tolerances of the preceding process; (b) the maximum distortions of the hole during heat treatment; (c) the depth of the decarburized layer which must be removed by grinding. The commonly recommended grinding allowance, as a function of the hole dimensions, will be discussed in Chapter 3-8. Those factors should be considered when determining the condition of the workpiece which is required for the efficient application of internal grinding.

Some other factors which can harmfully affect the efficiency of internal grinding, and should possibly be avoided are:

1. *Substantial variations in grinding allowance from piece to piece*

2. *Distorted, badly out-of-round holes* caused by improperly controlled preceding operations, such as machining, forming or heat treatment (e.g., quenching thin-walled rings without the use of appropriate fixtures)

3. *Poor locating surfaces* which affect the proper presentation of the hole to the internal grinding wheel—in relation to its position and its preset paths of linear movements.

The Capabilities of Internal Grinding Processes

Internal grinding, with its wide variety of processes and equipment, does not permit the establishment of uniform values for the "limits of capability," a term designating the consistent accuracy level of the produced work. A few of the parameters which are often required to be held to specific tolerances will be reviewed as an approximate guide in appraising the potentials of this general method from the perspective of particular accuracy requirements of internal surfaces.

Roundness of the Ground Internal Surface

An essential purpose of internal grinding operations in general, is to produce a functionally excellent surface with regular geometric form. Consequently, the nominally round surfaces, such as are usually produced by internal grinding, are required to have only a minimum amount of out-of-roundness, often spelled out as a toleranced value.

Out-of-roundness in internal grinding is mainly due to the following causes, often acting in combination and prone to produce cumulative errors:

1. The runout of the workhead spindle
2. The runout and/or vibrations of the grinding wheel spindle
3. Deflections of the quill or spindle, caused primarily by excessive grinding forces, resulting from, e.g., inadequate grinding wheel, too-high feed rate, etc.

The actually accomplished roundness of the ground surface may also be affected by inadequate work preparation, improper locating or clamping surfaces on the work, the type and condition of the work-holding devices, etc. Consequently, it is difficult to establish performance values of the equipment, expressed as the achievable roundness accuracy of the worked surface. For that reason guaranteed performance values are generally specified in terms of maximum runout for the workhead spindle and the grinding-wheel spindle.

Internal surfaces can be produced with a maximum out-of-roundness in the order of 0.000 010 inch (0.00025 mm) in gage making, and for similar exceptionally critical surfaces which are ground on special types of super-precision internal grinding machines, possibly equipped with hydrostatic spindles. In general manufacturing practice, however, out-of-roundness in the order of 0.000 050/0.000 100 inch (0.00125/00250 mm) for holes of about 1 to 2 inches (25 to 50 mm) diameter, is consistently attainable with modern grinding machines in well-controlled processes.

In centerless type internal grinding, the conditions of the external locating surface (on which the work is supported on shoes or rolls) is a major factor in determining the resulting roundness of the ground hole. As a general rule (whose final effects can only be slightly mitigated), the roundness of the ground internal surface will be a replica of the corresponding condition of the external supporting surface. *Cylindricity*, designating the parallelism of the surface elements in combination with the basic roundness of the hole, is a concept related to roundness but applies that geometric condition to a three-dimensional figure instead of to a single plane. Deviations from cylindricity of an otherwise round hole can result in a figure having a larger central section, called "barrel shape," or expanded end sections, known as "hourglass" shape. The profiles of such internal surfaces are concave or convex, respectively, instead of straight as is the element of a perfect cylinder.

Deviations from the nominal cylindrical form of an internal surface are due primarily to imperfections in the alignment of functionally interrelated grinding machine members and of their movements. This matter is discussed in greater detail in Chapter 3-6.

Modern internal grinding machines can produce holes with walls which are parallel to about $0° 0'$, $20''$, or even better, when the setup and the operational variables are adjusted to such rigorous requirements. Some models of internal grinding machines designed to carry out rough and finish grinding in a single operation, in two consecutive steps, have built-in devices to compensate for the expected amount of spindle deflection in the roughing phase of the operation. These systems operate by a swivel movement of the wheelhead, adjusted according to the operational conditions and maintained during the roughing phase to offset the quill spring. Thereafter, the wheelhead returns automatically to its aligned position for the wheel dressing, which precedes the finish grinding.

Wall thickness variations is a term expressing the lack of concentricity between the essentially cylindrical outside surface and the hole, such as in a ring-shaped part. In the regular, fixed-axis type internal grinding, the obtainable concentricity is mainly dependent on the mounting of the work controlled by the regularity of the clamping surfaces and the accuracy of the work-holding device. The optimum limit of concentricity when mounting the part by a previously finished outside surface, is about 0.000 040 inch (1 micrometer), a value which may be improved in exceptional cases when grinding both the external and the internal surfaces in a single operation.

In centerless-type internal grinding, particularly in its shoe-type variety, the uniform wall-thickness results from the basic system of the process, and under optimum conditions, can be consistently maintained in the order of about 0.000 040 inch (1 micrometer). However, these are accuracy values required for special purposes only, and the concentricity tolerances in general-type internal grinding are usually multiples of the above-cited extreme values.

Size (Diameter) Control

Size control in internal grinding refers to the capacity of the process and equipment to consistently produce holes whose actual size is within the specified tolerance limits. The principal factors which determine the accomplishments of that objective in continuous production, are the following:

(a) *Size setting*, the accuracy of which is dependent on the sensitivity of the machine slide adjustment. Incremental adjustments in the order of 0.0001 inch or 0.0025 mm, is a typical value in modern internal grinding machines, although special models provide the means for even more sensitive size-setting adjustments.

(b) *Size holding*, that is, repeating the set size consistently during the continuous operation of the machine, is mainly a matter of gaging in combination with the reacting capacity of the grinder to the commands issued by the gage. In-process gaging in internal grinding will be discussed in detail in Chapter 3-6.

As a matter of general information, current internal grinding machine capabilities for holes in the size range of about 25 to 50-mm diameter (1 to 2 inches) have a repeat accuracy in the order of 0.006 mm (0.00025 inch) — exceptionally, even better accuracy — may be achieved under optimum operational conditions.

Squareness is the commonly used term for the mutual perpendicularity of the worked surface and of a reference surface. In many types of workpieces the orientation of the hole axis at right angles to one of the faces, or to both—when they are precisely parallel —is a major functional requirement. Typical, but not exclusive examples are the rings of antifriction bearings which, in a widely used system of processing, have both the outside and the internal surface ground in subsequent operations, in relation to a face which

is regarded as a common reference plane. The centerless type internal grinding with backing plates, commonly of the magnetic type, is generally applied for such operations. The produced squareness, that is, bore orientation relative to a reference face, may be accurate within a fraction of a minute of arc (e.g., $90° \pm 0° 0'30''$).

In operations which hold the workpiece by clamping on its outside surface, such as in a collet, the attainable squareness is controlled by several factors, for example, the accuracy of the holding device, the condition of the clamping surface on the work, and the rotational accuracy of the workhead spindle in the machine. That latter condition is evaluated as the *axial runout*, often termed as "end camming." In internal grinding machines of the super-precision type, the value of end camming may be as low as (0.00002 inch/1 inch). Such extreme accuracies are, of course specified only rarely and are neither required nor available on internal grinders used for general industrial products.

The Finish of Surfaces Produced by Internal Grinding

The smoothness of ground internal surfaces is measured almost exclusively by means of tracer-type, surface-finish measuring instruments, because the application of alternative methods is hampered by the location and limited access of holes or similar internal surfaces. Surface condition measurement by using a stylus which traces the specimen surface and an amplifier which electronically averages the amounts of surface deviations from a mean reference plane is, of course, the method generally used for determining the standard surface finish values which are designated as Arithmetical Average, AA (in the USA), or Center Line Average, CLA, or currently R_a (in European countries).

In generating the surface finish (micro-geometry) the process variables have the major role, while the general geometric conditions of the ground surface, sometimes termed "macro-geometry," are predominantly the result of equipment design and performance. Such process variables include the composition of the grinding wheel, the coolant, the ratio of work speed to wheel speed, the rate of stock removal and the time allowed for sparkout; this last factor being of particular importance in internal grinding due to the extended wheel spindles, often provided as a design necessity, with reduced stiffness.

Since any surface finish value specified as the required result of an internal grinding operation is strongly dependent on appropriate process factors, the following indicative values should be of interest:

(a) The surface finish of the bore of a gage ring ground on a super-precision internal grinder: 2 microinches AA $(0.05\mu m R_a)$, exceptionally, even better.

(b) The surface finish of the bores of ring shaped industrial parts, with inside diameters in the range of 1 to 2 inches (about 25 to 50 mm), ground on modern shoe type centerless internal grinding machines in high volume production: 10 to 12 microinches AA $(0.25\text{-}0.3\mu m\ R_a)$.

It is possible to further improve on these values by modifying certain process conditions, such as by reducing the stock allowance, the work speed and the infeed rate, also by extending the sparkout time. However, for economic reasons such changes in the process data, which may substantially lower the productivity, are seldom applied. In cases where a better finish of the internal surface of industrial products is needed, the application of a subsequent finishing operation, commonly honing, is generally preferred for economic, and frequently also for technological reasons.

Actually, an excellent smoothness is not always necessary on internal surfaces which are finished by internal grinding, for reasons of form and size control. Surface finishes of ground internal surfaces in the range of 16 to 32 microinches AA $(0.4\text{-}0.8\mu m R_a)$, are quite common practice and are entirely satisfactory for many applications.

The Adaptability of Internal Grinding Machines

Operating with cantilever supported grinding-wheel spindles, a distinctive characteristic of internal grinding machines, while unfavorable with regard to substantial stock removal, offers great adaptability in the grinding of limited access work surfaces. Such work is often accomplished by applying different wheel approaches, sometimes consecutively in a single operation.

The work examples shown in Table 3-1.1, although by far not a complete listing, are indicative of the variety of surface configurations and locations which are adapted for grinding on internal grinders of either the standard, the universal, or one of the special types.

The grinding can be carried out:

(a) With the periphery (straight or profiled), or the face of the grinding wheel, or with a transitional surface between these two basic surfaces

(b) On internal or external surfaces of the work, generally parallel with, but also perpendicular to the axis of work rotation, or in an intermediate orientation for producing angular surfaces.

This flexibility of adaptation extends the use of the essentially internal type grinding machines to grinding surfaces whose location or configuration makes them inaccessible for processing by grinding machines designed expressly for operations on external work surfaces.

Principal Elements of Internal Grinding Machines

The primary operating elements of the basic types of internal grinding machines are the workhead (with its drive) and the wheelhead, essentially an internal grinding spindle, which is integral with or connected by a belt to its individual drive motor. These two principal members must be installed in the machine in specific relation to each other, and the grinding process requires their controlled relative movements. Because internal grinding does not provide direct access of the wheel to the worked surface, the number of movements which the principal members must carry out in internal grinding generally exceeds those needed for other grinding methods. The major operating elements of a modern automatic internal grinding machine are illustrated in Fig. 3-2.1. These movements, generally linear and mutually interrelated, are essential and characteristic elements of the internal grinding process.

Traverse and Feed Movements in Internal Grinding Machines

In addition to the basic rotational movements of the grinding wheel and of the workpiece the internal grinding process also requires several linear movements for controlling the relative positions of the wheel and the work at various stages in the process.

The traverse movement will bring the grinding wheel into the work area inside the workpiece and retract the wheel at the end of the operation. Frequently, wheel retraction during operation is also needed for wheel dressing, manual gaging, etc.

The reciprocating movement, sometimes also termed "oscillation," is coincident with the direction of the traverse, and is needed for the traversing type of internal grinding, but is not used in plunge grinding.

The transverse movement has the double role of (a) a wheel approach to the surface of the work which, in some cases, may be in a recessed location involving substantial approach travel and (b) a feed movement during the stock removal phase of the operation, sometimes at different rates for roughing and finishing, which is maintained up to a point corresponding to the finish work size.

The purpose of these movements is to produce changes in the relative positions of the grinding wheel and the workpiece. Consequently, the functional effectiveness of the movements is independent of which particular machine member is actually moving with respect to the bed of the machine.

In most of the basic types of internal grinding machines the traverse and the feed movements are carried out by the machine table on which the grinding-wheel head is mounted. While the machine slide itself serves for the traverse movements, in a transverse direction, the cross slide of the table is used for the approach and retraction, as well as for the actual wheel infeed. The infeed progresses in specific increments at the end of the strokes in traversing-type internal grinding, or continuously, at a selected rate, on machines designed for automatic plunge grinding. Of course, manual operation of these movements is still used in single or low-volume operations on internal grinding machines designed for such applications.

The workhead, too, may be mounted on a cross slide whose travel, either manual or by power, is considered an auxiliary movement when serving work positioning or workhead retraction for part loading.

However, variations from the described basic arrangement are found in various models of production-type internal grinding machines. Both the traverse movement, comprising approach and reciprocation,

WORKHEAD

WORKHEAD MOTOR

TABLE OSCILLATOR

TABLE

WORKHEAD OIL MIST
LUBRICATOR

LOADING CHUTE

WHEELHEAD OIL MIST
LUBRICATOR

WHEELHEAD MOTOR

WHEELHEAD

WHEELHEAD CROSS SLIDE

Courtesy of Cincinnati Milacron, Inc., Heald Machine Div.

Fig. 3-2.1 Front view of a production type internal grinding machine designed for automatic operation. Major operating elements are indicated.

as well as the cross feed, or either of these movements, may be incorporated into the workhead slide. Such a distribution of the movements, which will result in the same relative displacement between work and wheel, is considered preferable for certain models of internal grinders, for reasons connected with other aspects of the machine's functions.

Compensation for wheel wear is a supplemental infeed movement, often incorporated into the automatic grinding cycle. Certain types of internal grinding machines have a separate compensation slide which is positioned with a pre-loaded ball screw driven by a stepping motor. Programmed compensation, in increments as small as 0.000 050 inch (about 0.001 mm), may be applied.

Controlled Force is the designation used for a feed system (see Fig. 3-2.2) which substitutes a constant feed force of preset amount, for the commonly used constant feed rate which is generally produced by a plain hydraulic system or by mechanical elements. In the Controlled Force System the advance of the cross slide is effected by the pressure, transmitted by the piston of a hydraulic cylinder, which is exerted against an adjustable damping action. The applicable feed force is usually established experimentally by seeking the optimum balance between high productivity, by increased rate of stock removal, and by the

quality requirements of the work, including geometric accuracy, size holding, good finish, etc.

This feed system represents an adaptive control which adjusts itself to variations in the cutting ability of the grinding wheel, influenced by sharpness and peripheral speed; and the conditions of the work surface, such as a high degree of waviness or variations in the stock allowance. Adjusting the feed rate automatically to the amount of stock to be re-

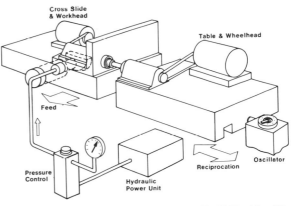

Cross Slide
& Workhead

Table & Wheelhead

Feed

Pressure
Control

Hydraulic
Power Unit

Reciprocation

Oscillator

Courtesy of Cincinnati Milacron, Inc., Heald Machine Div.

Fig. 3-2.2 Diagram showing the operational principles of a "controlled force" type of adaptive feed system applied on internal grinding machines.

moved is particularly valuable in the case of workpieces which have been produced by methods causing a wide range of prefinish size variations, such as cold forming without subsequent machining.

The Controlled Force System has proved also of advantage in operations using sequentially, two wheels on the same quill, the first with a coarser grain for roughing and the second with a finer grain for finishing the same work surface, each wheel advancing with the appropriate feed force. Very substantial grinding wheel savings are claimed to be achieved by this method on particular workpieces, such as on the tracks of ball-bearing outer rings.

In operations required to maintain a very tight finish-size tolerance, the controlled force feed is sometimes applied only to the first phase of the process in which most of the stock is being removed, and then the rate control takes over for attaining the final size as signalled by an in-process gage.

The successful application of the controlled force principle is predicated on a nearly frictionless cross slide movement, otherwise the frictional forces would severely interfere with the effectiveness of a constant feed force. Internal grinding machines built for the application of the controlled force feed principles have hydrostatic slideways which satisfy the requirement of a very low frictional drag. (See Figs. 3-2.6 and 3-2.7.)

The Reciprocating Movements of the Grinding Machine Table

For the grinding of internal surfaces with straight line elements, whether cylindrical or tapered, the traverse grinding process is usually applied. The reciprocating movement, sometimes referred to as oscillation, is generally carried out by the wheel head table, although some models of production type internal grinders have the reciprocation assigned to the work head slide. Plunge grinding without reciprocating movement is generally limited to internal shapes, such as the raceways of ball bearing outer rings, the contour of which excludes the use of traverse movement. In plunge grinding the wheel is advanced to the proper axial position inside the bore (a movement termed "indexing"), but no additional traverse movement takes place until the grinding process is terminated.

The two systems most frequently used for imparting the reciprocating movement to the machine table operate as follows:

1. A hydraulic cylinder whose stroke length is adjusted by the setting of reversing dogs, also called "table dogs" (see Fig. 3-2.3), which act on the direc-

tional change lever of the table movement, thus controlling the directional valves of the drive system. In this system the length of the stroke can be varied over a wide range starting from a minimum, such as 0.125 inch (about 3 millimeters).

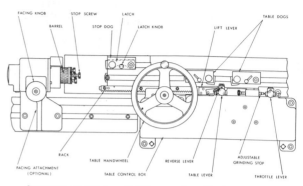

Courtesy of Cincinnati Milacron, Inc., Heald Machine Div.

Fig. 3-2.3 Table movement control elements of a universal internal grinding machine, using adjustable table dogs for controlling the length and the location of the reciprocating stroke.

2. A rotating cam with follower which is kept in contact with the cam by spring pressure or hydraulic force that can impart very rapid reciprocating movement; a typical medium speed range being 50 to 250 strokes per minute. The applicable stroke length may be close to zero, when needed, but its practicable maximum length is limited, e.g., to 0.250 inch (6 millimeters).

The retraction of the wheel slide can be carried out with the same hydraulic cylinder which actuates the reciprocating movement during the grinding. When cam actuation is used, an additional hydraulic system is required for table retraction, which is operated for work loading and unloading, as well as for wheel dressing. Some designs obtain the retraction and the force bearing on the follower by using the same cylinder, yet with a different direction of action for each type of motion.

The grinding stroke must be adjusted, when setting up the machine, for both the location and the length of the reciprocating movement. The length of the stroke may have to supplement the width of the wheel when it is narrower than the length of the hole. As a general rule, in the extreme positions of the stroke, the wheel should extend beyond the ends of the bore by about ¼ inch (approximately 6 millimeters) for wheels of about ¾ inch (about 20 millimeters) width, or narrower. In the case of wider wheels, the length extending over the faces of the bore can be increased, but it should never exceed one-half of the wheel width.

When the wheel is wider than the length of the hole, the traverse should leave the wheel at the ends of the stroke in contact with about three-quarters of the bore length. The location setting of the stroke must assure that the wheel extends at both ends of the hole by an equal amount.

The speed of the reciprocating movement depends on the type of system. Typical values, stated as examples, are:

(a) For hydraulic table movements reversed by dogs, the steplessly adjustable speed may have a maximum of about 25 feet/minute (about 7.5 m/minute)

(b) Cam controlled reciprocating movements used on high-production types of internal grinders, when operated with short stroke lengths may produce a maximum of about 450 strokes per minute.

The Guideways of Internal Grinding Machines

The relatively short operation cycles, the multiple movements, and the frequent use of high-speed reciprocation over short distances are a few of the characteristics of internal grinding, which explains why the machine guideways are generally designed for particularly demanding operational conditions. For that reason the plain sliding guideways, usually comprising an inverted V-shape and a flat track, finished by scraping or grinding, are those used in most types of basic grinding machines, including internal grinders, but are being replaced by more advanced guideway designs in modern production type internal-grinding machines.

The operating accuracy, the consistent performance, and the adequate maintainability of such advanced production equipment being strongly dependent on the appropriate type of guideways, a few of the designs used by reputable manufacturers will be briefly reviewed. The order of listing is random and it is not intended to indicate either qualitative superiority or scope of adoption.

Hydrostatic tongues of the slides are guided in a special channel between the two supporting guideway tracks and serve the purpose of very precisely controlled lateral restraint, thus assuring the constant direction and straight-line path of the slide movement.

Ball bearing supported slide bars. The slide assembly comprises two parallel bars, one of which has two; and the other, one bearing point. The bearings are hardened steel sleeves which surround the bars; the sleeves and the bars are separated by precision balls mounted in ball retainers. (See Fig. 3-2.4.)

Slide bar and control plate. In this design the slide has a single bar in the front, which is supported in

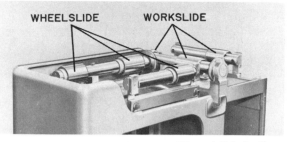

Courtesy of Bryant Grinder Corp.

Fig. 3-2.4 Ball-bearing guided slide bars providing a three-point support for the slides of an internal grinding machine.

three widely spaced bearing boxes, attached to the machine bed. The rear side of the slide is supported on a set of bearing rollers which travel on a hardened and ground plate. The latter, termed the "control plate," can be adjusted for correcting the alignment of the wheel head which rests on the slide, with the work head. (See Fig. 3-3.10)

Hardened and ground bed ways as tracks for ball bearing mountings. These are attached to the bottom of the traversing machine member to provide support and lateral guidance. The balls of the bearing mountings are pre-loaded for reducing friction and backlash. (See Fig. 3-3.11.)

Roller bearing tracks. Used as guideways, these are also adaptable for heavy machine slides of substantial size and provide dependably straight travel with very low friction. Internal grinding machines using roller bearing type guideways are designed with special sealing elements for the effective protection of the rollways.

Ball-supported cylindrical sleeves. The grinding spindle is mounted in the end of a large sleeve (See Fig. 3-2.5) which is surrounded by an outer sleeve. The two sleeves are held apart by a substantial number of bearing balls, mounted in a retainer and pre-loaded between inner and outer sleeve. The balls provide a firm support for the inner sleeve which is free to travel by moving axially inside the outer sleeve. The outer sleeve has as integral elements, and located close to its end which holds the grinding spindle, two trunnions which project at right angles. These trunnions are supported, in the rigid wheel slide housing, by flanges with pre-loaded balls, thus providing an accurate guide for the cross-feed motion. The purpose of this design, which places the members for both the longitudinal and transverse movements in a common plane with the action of the generated radial grinding force, is to assure a high degree of alignment accuracy which is retained over an extended service life.

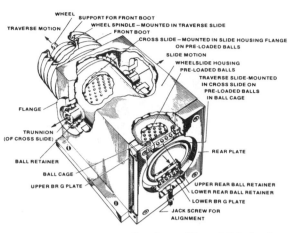

Courtesy of Bryant Grinder Corp.

Fig. 3-2.5 Ball-supported cylindrical sleeves provide consistently aligned guides for both the traverse and the infeed movements of an automatic internal grinding machine.

Hydrostatic machine ways have pressurized pockets of oil supporting the mating machine elements, preventing metal-to-metal contact. Figure 3-2.6 shows the design principles of a hydrostatic table with a V-shaped and a flat guideway, both provided with hydrostatic pockets. The cross slide, too, has hydrostatic support which, in the case of a particular model (see Fig. 3-2.7), consists of a round bar and a flat way. The round bar serves the triple function of a guideway, a hold-down member, and a force cylinder for the hydraulic cross feed.

Hydrostatic guideways for the table movements, both the reciprocation in traverse grinding and the indexing in plunge grinding, have excellent vibration damping properties and practically eliminate wear.

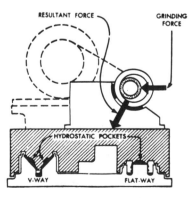

Courtesy of Cincinnati Milacron, Inc., Heald Machine Div.

Fig. 3-2.6 Hydrostatic table ways comprising a Vee and a flat element. The guideways are pressurized with filtered oil introduced at high pressure and throttled by a capillary coil at each pocket.

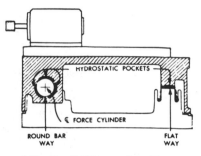

Courtesy of Cincinnati Milacron, Inc., Heald Machine Div.

Fig. 3-2.7 Hydrostatic wheel-head cross slide of a "controlled force" type internal grinding machine. The diagram shows the support on a flat way and on a round bar way, the latter functioning also as the feed and backoff piston of the infeed system .

The Work Head

The work head of internal grinding machines serves two basic purposes: (a) holding the workpiece, and (b) rotating it at a particular, consistently maintained speed.

The work head with its drive may be mounted directly on the machine bed, on an intermediate swivel base or on a cross slide; in some models the latter may also incorporate a swivel support.

A wide variety of work-holding devices is available for holding the work in position; these will be discussed in more detail in Chapter 3-6.

For rotating the workpiece the work head may offer a limited number of steps, particularly in machines of older design or in those intended for specific uses. Other models have a wide range of speeds, which are infinitely variable, to provide the optimum peripheral work speed for any particular operation within the capacity of the machine. Typical speed ranges for internal grinder work heads are:

(1) From 0 to 500 rpm, steplessly variable and reversible, in modern universal type internal grinding machines

(2) Specific speeds within a range of about 200 to 1500 rpm, in production type, medium-size internal grinders. The actually applied speed must be selected from the available values to best suit the operational requirements

(3) A similar range of 200 to 1500 rpm, but infinitely variable, is available on some models of production type internal grinding machines designed to offer a higher flexibility of adaptation to varying operational conditions.

A cross slide of the work head, when part of the design, may be provided for: (a) workpiece positioning; (b) retraction for work loading; (c) cross feed when incorporated into the work-head slide instead of the wheel-head slide.

The swivel adjustment of the work head, a feature not present in all types of internal grinding machines, has for its primary purpose the grinding of taper bores. A typical range of adjustment may be 30 degrees, permitting tapers to be ground to an included angle of 60 degrees, predicated on tooling which does not cause interference.

Universal type internal grinding machines which are designed to assure great adaptability to widely different types of operations, typically have work heads with 90 degrees swivel adjustment. When turned to the extreme position, the work-head axis is perpendicular to the wheel spindle and permits the grinding of a flat work surface with the periphery of the grinding wheel, in an operation known as rotary surface grinding. When the wheel head is turned by somewhat less than 90 degrees, wide angle taper surfaces, either convex or concave, depending on the location of the wheel contact with respect to the work axis, may be ground by the same rotary method.

The swivel base of the work heads, designed for angular adjustment, usually has a graduated ring or segment, the scale of which provides approximate information on the number of degrees by which the head has been swiveled. More precise adjustment can be provided with the aid of a sine-bar type setting which, when used in conjunction with a sensitive dial indicator (see Fig. 3-2.8), permits angular adjustment accuracies to fractions of a minute of arc.

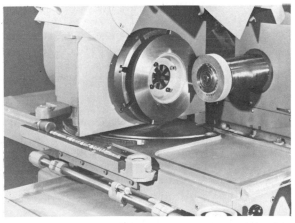

Courtesy of Cincinnati Milacron, Inc., Heald Machine Div.

Fig. 3-2.8 Swivel adjustment of the work head of a universal type internal grinding machine. The photo shows the work head swiveled by 90 degrees for rotary surface grinding. The accuracy of the angular setting is controlled by a sine bar system using end-rods and a sensitive indicator.

General Characteristics of Internal Grinding Spindles

The grinding wheel spindles of internal grinding machines, sometimes termed "wheel heads," are designed and used as readily interchangeable machine elements. This practice is warranted in view of the special conditions which, in general, characterize internal grinding operations. The following may be considered the primary reasons for requiring the interchangeability of internal grinding spindles:

1. The selection of the internal grinding spindle to be used for certain operations is generally guided by the principle of using that type and size of spindle which, within the dimensional limitations determined by the work, will provide the highest degree of stability. The exchange of the grinding spindles is generally desirable when work with substantial variations in dimensions, configurations, or other pertinent characteristics is to be carried out on the same grinding machine.

2. The high rotational speed with which internal spindles, particularly those designed for small grinding wheel diameters, are operated, makes these machine elements prone to wear, at least to a degree requiring regular maintenance or overhaul. This is done away from the machine, frequently in special spindle repair areas.

Internal grinding machines are designed to permit the ready exchange of the grinding spindles, although the connecting dimensions of various makes and models of internal grinders can vary substantially. For that reason the replacement or exchange spindles have to be ordered in accordance with the specifications of the particular grinding machine on which the spindles will be installed.

The grinding wheel spindles of internal grinding machines are commonly supported in ball or roller bearings which are usually of the high accuracy grades, permit pre-loading, and are designed to operate under both radial and thrust load. The bearings are usually mounted in pairs, at least at both ends of the spindle, and are pre-loaded, such as by the action of a spring whose force may be adjusted to suit the operating conditions (See Fig. 3-2.9.)

Internal grinding spindles of more recent design, and intended for low and medium speeds, are generally permanently lubricated during assembly in the manufacturer's plant. Special lubricants and seals have been developed to protect the bearings during the life of the spindle by avoiding deterioration of the lubricant and preventing both leaks from the bearing and penetration of dust from the outside. Spindles

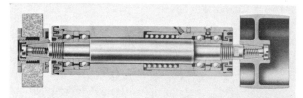

Courtesy of Cincinnati Milacron, Inc., Heald Machine Div.

Fig. 3-2.9 Cross-sectional view of a heavy duty internal grinding spindle with flange-mounted wheel and double-row ball bearings preloaded by spring pressure.

used for higher speeds are designed for oil lubrication, and for very high speeds, oil mist lubrication is used.

The Basic Types of Internal Grinding Spindles

The majority of internal grinding spindles in common use belong, with regard to their general design, to one of the following basic types:

(a) Removable quill type (interchangeable extension arbor type)

(b) Solid projection type (integral projection or solid extension type)

(c) Deep hole type (enclosed spindle or integral projection, deep hole type).

(Designations in parentheses are those used alternately by different manufacturers.)

The removable quill type internal grinding spindles are preferred for general-purpose work such as in job shops or toolroom applications where quick adaptation to the work on hand by interchanging the quill is of definite advantage.

The quills fitting the main spindle are generally available in a wide variety of sizes and can be readily exchanged, thus providing spindle dimensions well suited to the workpiece. As shown in the diagram of a typical quill type spindle (Fig. 3-2.10) taper connections are frequently used; other spindle models are designed to accept quills with cylindrical shanks. The relative advantages of these two alternative designs are discussed in a subsequent section of this chapter.

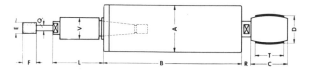

Courtesy of Fortuna-Werke, Stoffel Grinding Systems

Fig. 3.2.10 Diagram of a quill type internal grinding spindle.

The drawback of this type of internal grinding spindle is that it requires the joining of two individual elements, causing a reduction of rigidity. To counterbalance this condition—at least to some extent—quills of the largest diameter and shortest length which still agree with the work dimensions and do not exceed the maximum recommended quill size of the particular spindle, should be selected.

Solid projection type internal grinding spindles represent the variety of spindle most frequently used in the general production grinding of internal surfaces, where the size of the work assigned to a specific machine does not change at all, or only infrequently. The integral extension (in some cases a stub nose only) of this type of spindle, provides a higher degree of rigidity than that of spindles with mounted quill. The example shown in the diagram of Fig. 3-2.11 has a very short extension as it is designed for bores larger than the spindle diameter.

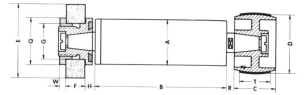

Courtesy of Fortuna-Werke, Stoffel Grinding Systems

Fig. 3-2.11 Diagram of a solid projection type internal-grinding spindle with the wheel mounted close to the body of the spindle.

As a general rule, spindles with extensions should not have a length-to-diameter ratio in excess of 5 to 1, in order to avoid deflections which become noticeable by the imperfections of the ground surface due to the "whipping" action of the running spindle. For particular cases where the bore has a greater length than five times the diameter, the amount of spindle deflection can be reduced by making the extensions of specially rigid metal, such as exotic alloys or cemented carbides. As a practical approximation it can be estimated that the deflection of the spindle extension caused by the radial grinding force, will increase by the third power of the unsupported length, and by the fourth power of the ratio of the original diameter to the reduced diameter.

Deep hole type internal grinding spindles are recommended when the diameter of the bore is larger than about 2-3/8 in. (about 60 mm), particularly when the depth of the bore exceeds about 6 in. (150 mm). This type of spindle (see Fig. 3-2.12) has an additional set of bearings near the grinding wheel and thus provides a stiffness superior to

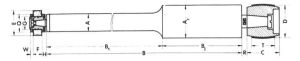

Courtesy of Fortuna-Werke, Stoffel Grinding Systems

Fig. 3-2.12 Diagram of a deep hole type internal-grinding spindle.

types with unsupported extension parts. For grinding very deep holes of relatively large diameter the deep hole type spindles can also be supplied with eccentric design of the extending section, in order to assure great rigidity without interfering with the effective usage of the grinding wheel.

The Drives of Internal Grinding Spindles

Belt drive through flat, or multiple V-belts is used for internal grinding spindles equipped with pulleys, which are generally interchangeable for adjusting the operating spindle speed to the wheel diameter or other process-related conditions.

Motorized spindles are frequently used as space savers and to reduce upkeep. Such spindles with regular motors are available with a maximum speed of 3000 or 3600 rpm, depending on the AC current (50 or 60 cycles per second). For higher speeds, however, the spindles must be equipped with high-frequency motors.

Direct drive for internal grinding spindles by means of high-frequency motors provides advantages under particular operating conditions or when special process objectives must be met. Typical examples of preferred applications of high-frequency grinding spindles are:

(a) To accommodate the spindle with its drive motor in a limited space

(b) For very high rotational speeds (such as in excess of 40,000 rpm), particularly when relatively constant speed under varying loads is needed

(c) For providing the means of infinite speed variations over an extended range, in order to assure optimum operating conditions.

High-frequency motors used for spindle drive, whether directly (see Fig. 3-2.13) or through a belt, require an appropriate power source which can be provided by several alternative means, including frequency converters, alternator sets with fixed or adjustable speed drive, and solid-state frequency regulators. High-frequency grinding spindles are now available for speeds up to about 120,000 rpm.

For extremely high speeds, such as used for miniature size bores, *air turbine* spindle drives are used

which, while sensitive to load variations, are available for speeds of several hundred thousand revolutions per minute.

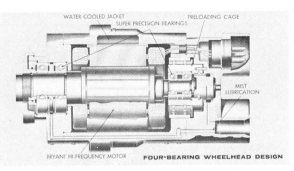

Courtesy of Bryant Grinder Corp.

Fig. 3-2.13 Cross-sectional diagram of a high-frequency wheelhead for rotational speed of 60,000 to 100,000 rpm.

The Selection of Internal Grinding Spindles

One of the essential differences in the equipment of external and internal grinding is the selection of grinding wheel spindles which, in the case of internal grinding, must be adapted to the major dimensional characteristics of the workpiece. For that reason, while in external grinding machines the grinding wheel spindle is a permanently retained element of the grinding machine, the internal grinders are built to accept interchangeable wheel spindles.

First there are the general requirements of effective internal grinding which call for spindles of different designs and dimensions. The properly operating internal grinding spindle must:

(1) Accept grinding wheels of optimum diameter for the work, often considered to be about three-quarters of the work hole diameter, except for very large bores where other factors control the choice of the wheel diameter

(2) Hold the wheel on an extension which is long enough to permit the wheel to penetrate to the farthest end of the bore

(3) Provide that rotational speed, usually expressed in rpm, which produces an efficient surface speed for the operation of the grinding wheel

(4) Assure the proper stiffness of the spindle by appropriate bearing support, a condition which, in certain applications, may require spindles of special design.

Other aspects which may influence the selection of the internal grinding spindle are as follows:

(a) Flexible adaptation to widely different work dimensions, by means of exchanging quills—and perhaps pulleys—yet retaining the same basic spindle.

(b) A relatively high rate of stock removal, calling for particularly sturdy spindles designed for powerful drive.

(c) Face grinding with the wheel, either as the only type of operation, or combined with peripheral grinding, in each case requiring additional thrust support.

(d) Specially high accuracy and/or surface finish requirements of the work, which are to be met by the use of internal grinding spindles expressly designed for such operations, and equipped with selected types and grades of antifriction bearings, or, possibly, hydrostatic bearings.

(e) The grinding of extremely hard materials, such as cemented carbides or exotic alloys, which are difficult to grind so that spindles with high stiffness are needed to avoid major deflection of the operating spindle.

(f) The grinding of complex internal profiles with crush-trued wheels, a truing process requiring particularly sturdy and well-supported grinding spindles.

Finally, the selected internal spindle must fit the grinding machine for which it is being used, with regard to both the connecting and the overall dimensions.

Several important aspects which must receive attention during the selection of internal grinding spindles are reviewed in Table 3-2.1.

The above-mentioned and various other variables which must be satisfied by the internal grinding spindles for assuring simple installation and proper operation, have resulted in hundreds of different types and sizes. Such spindles are made by the manufacturers of internal grinding machines for their own equipment and also by specialized producers of grinding spindles, the latter supplying spindles for most better known makes and models of internal grinding machines.

Aspects Controlling the Selection of Quill Shank Types

Internal grinding spindles of the interchangeable quill type are made to accept quills with either taper or straight (cylindrical) shanks. The essential differences between these two types can be recognized from the diagrams and photos of Figs. 3-2.14 and 3-2.15. The taper shanks usually have the 0.600 in. per foot Jarno taper (1:20 ratio), although other tapers are also used.

The choice must be controlled by the conditions of the operation; the following comparison of ad-

vantages and limitations may be helpful in evaluating those aspects considered important for the job:

Characteristics of Taper and Straight Quill Shanks

Aspect	Taper Shank	Straight Shank
Stiffness	The diameter of the taper shank quill is less than that of the spindle by twice the wall thickness at the end of the taper hole in the spindle.	Quills with straight shanks can have the same diameter as the spindle, thereby assuring superior stiffness.
Interchangeability	When varying jobs require the frequent exchanging of the quills, a taper shank will assure a more precise relocation.	Small centering errors of the newly installed quill can be neutralized by truing the wheel to sufficient depth following the change of quill.
Consistent accuracy	Limited local wear or scoring of the connecting surfaces does not subtract from the locating accuracy of the taper connection, as long as protruding surface elements are avoided.	Wear can affect the locating accuracy of cylindrical connections; therefore, extreme care in handling, and cleanliness, are recommended.
Detachment of the connection	The heating up of the spindle during its operation can cause a tightening of the quill in its seat, making removal difficult.	Temperature variations do not affect the straight shank quills to a degree which interferes with its removal for exchange.

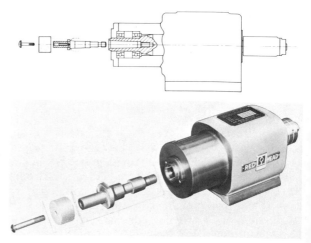

Courtesy of Cincinnati Milacron, Inc., Heald Machine Div.

Fig. 3-2.14　Interchangeable quill type grinding spindle, accepting quills with tapered shank. Shown in diagram and actual photo.

Table 3-2.1 Some Principles of Selecting Internal Grinding Machine Spindles

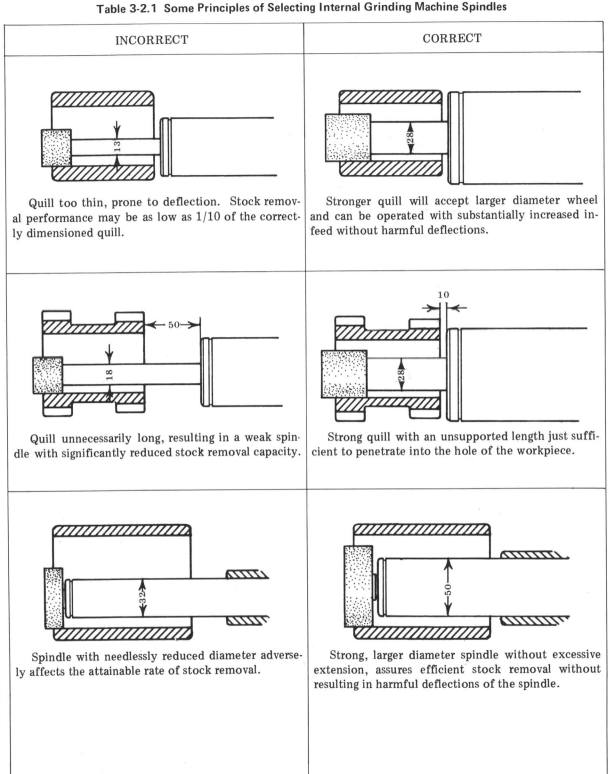

INCORRECT	CORRECT
Quill too thin, prone to deflection. Stock removal performance may be as low as 1/10 of the correctly dimensioned quill.	Stronger quill will accept larger diameter wheel and can be operated with substantially increased infeed without harmful deflections.
Quill unnecessarily long, resulting in a weak spindle with significantly reduced stock removal capacity.	Strong quill with an unsupported length just sufficient to penetrate into the hole of the workpiece.
Spindle with needlessly reduced diameter adversely affects the attainable rate of stock removal.	Strong, larger diameter spindle without excessive extension, assures efficient stock removal without resulting in harmful deflections of the spindle.

(Continued on next page.)

Table 3-2.1 *(Cont.)* **Some Principles of Selecting Internal Grinding Machine Spindles**

INCORRECT	CORRECT
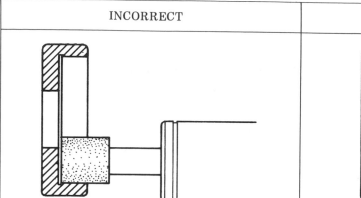	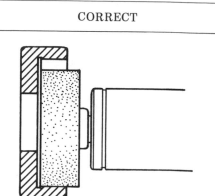
The diameter of the grinding wheel is too small, its width is unnecessarily large. Extension over the bearing support is excessive. Such conditions may reduce the performance to 1/4 that of the correct spindle.	Using a large diameter grinding wheel with width only slightly exceeding the depth of the hole and selecting a spindle which mounts the wheel near the front bearing, results in an efficient wheel and spindle combination.
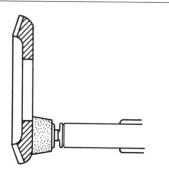	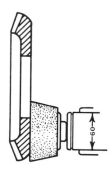
Spindle is too thin and extends too far over the bearing support. The diameter of the wheel is too small, producing poor grinding quality. The performance is about 1/4 that of the correct setup.	Strongly dimensioned spindle, supported close to the wheel, with large diameter grinding wheel, permits efficient operation.
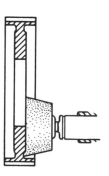	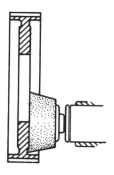
The spindle is too weak in relation to the wheel diameter. Performance is about 1/3 that of the correctly selected spindle.	Large diameter spindle, supported close to the grinding wheel results in high performance and good grinding quality.

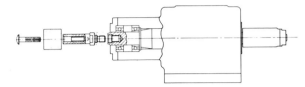

Courtesy of Cincinnati Milacron, Inc., Heald Machine Div.

Fig. 3-2.15   Interchangeable quill type grinding spindle, accepting quills with cylindrical shank. Shown in diagram and actual photo.

Characteristics of Taper and Straight Quill Shanks

Mounting the wheels on internal grinding spindles can be accomplished by one of three commonly used methods, whose choice is generally determined by the size of the grinding wheel. For grinding wheels of 2½ inches (about 65 mm), or larger, diameter, spindles which mount the wheels between flanges are usually selected. Smaller wheels are mounted with the aid of a large-head screw at the end of the spindle or quill, which has a tapped hole in its face for that purpose. Piloted wheel screws (see Fig. 3-2.14) which have a precisely fitting cylindrical section guided by the corresponding ground pilot hole in the spindle, are preferred where applicable. The threaded section of these mounting elements is slightly undersize in order not to interfere with the aligning function of the pilot member.

Wheels with very small diameter are used in the mounted wheel design with integral shanks, which are held in spindle collets.

The Basic Types of Internal Grinding Machines

The basic types of internal grinding machines hold and rotate the workpiece; the axis of rotation, corresponding to the axis of the work spindle, is in the horizontal plane.

The majority of internal grinding operations are carried out on the various models of the basic types of internal grinders. A review of the currently used and available regular machine models with regard to their significant characteristics, including workpiece capacities, is presented in Table 3-3.1 which is complemented with the following comments:

(a) Categories have been established for the purpose of this review, in order to assist the appraisal and comparison. These categories, however, do not represent a standard or industry-wide accepted practice.

(b) The parameters discussed are those which primarily characterize an internal grinding machine, especially with regard to the selection of the appropriate model for handling a particular workpiece, or carrying out a specific operation.

(c) The magnitudes listed represent the pertinent ranges that apply to a group of typical machines within each category.

To complement the limited data of the table, a representative model for each of the machine categories will now be discussed in greater detail.

Internal Grinding Machines for Miniature Parts

Very small workpieces, often referred to as miniature parts, may require internal grinding as one of the operations of their production sequence. Hole diameters in the approximate range of about 3/8 to 1/2 inch (9 to 13 mm) represent the smallest sizes which may be effectively ground on general-purpose internal grinding machines. The upper boundary for the effective operation of internal grinders specially designed to handle very small workpieces for the accurate grinding of their internal surfaces, primarily the bores, lies within this same range.

The model selected as a representative example of internal grinding machines for miniature parts (see Fig. 3-3.1) is recommended for grinding holes in the diameter range of 0.020 to 0.600 inch (0.5 to 15 mm). Workpieces in that dimensional range, when requiring grinding, often also have very tightly toleranced external surfaces. It is convenient and, exceptionally, it may actually be necessary for assuring rigorous geometric interrelations of surfaces, such as: concentricity, parallelism, and squareness—to grind the critical external surfaces of the part on the same

Courtesy of Tripet Precision Machine Work —
Hirschmann Corp.

Fig. 3-3.1 Work area of an internal grinding machine designed for small and miniature parts.

Table 3-3.1 Survey of Basic Types of Internal Grinding Machines
(Based on a limited number of characteristic models.)

DESIGNATION OF THE CATEGORY	GENERAL-PURPOSE AND EQUIPMENT CHARACTERISTICS	RANGE OF CAPACITIES OF DIFFERENT MODELS WITHIN THE CATEGORY				ILLUSTRATION OF TYPICAL EXAMPLE
		MAX. WORK DIAM.	MAX. HOLE DIAM.	MAX. HOLE LENGTH	MACHINE WEIGHT	
Miniature part internal grinder	Specially designed and equipped internal grinders for very small and miniature parts, with hole diameters starting at about 1 mm. Some models are built as universal machines with versatile adaptations. (Miniature part internal grinders are also built as centerless grinding machines, specially developed for very small ball bearing rings.)	4 in. (100 mm)	0.600 in. (15 mm)	1 in. (25 m/m)	1200 lbs. (550 kgs.)	Fig. 3-3.1
Small work type internal grinder	Designed for grinding primarily limited size lots of small parts within the capacity range, with emphasis on easy setup. Operated either entirely manually, or in semiautomatic cycle with hand loading of the work.	6-10 ins. (150-250 mm)	2-4 ins. (50-100 mm)	2-4 ins. (50-100 mm)	4000-6000 lbs. (1800-2700 kgs)	Fig. 3-3.3
Medium size internal grinder	Medium size general purpose internal grinding machines are available either in the plain, or the more versatile toolroom type, with different bed lengths, some are also designed to grind long bores.	16-24 ins. (400-600 mm)	10-14 ins. (250-350 mm)	10-14 ins. (250-350 mm)	8000-9000 lbs. (3600-4000 kgs)	Fig. 3-3.4
Universal internal grinding machine	Offer the capacity and versatility needed in general toolroom work, fixture and machine building. At the same time these machines must satisfy severe accuracy requirements and are used occasionally for grinding external round and flat surfaces.	24-30 ins. (600-750 mm)	18-24 ins. (450-600 mm)	12-18 ins. (300-450 mm)	6000-9000 lbs (2700-4000 kgs)	Fig. 3-3.5

(Continued on next page.)

Table 3-3.1 (*Cont.*) Survey of Basic Types of Internal Grinding Machines
(Based on a limited number of characteristic models.)

DESIGNATION OF THE CATEGORY	GENERAL-PURPOSE AND EQUIPMENT CHARACTERISTICS	RANGE OF CAPACITIES OF DIFFERENT MODELS WITHIN THE CATEGORY				ILLUSTRA-TION OF TYPICAL EXAMPLE
		MAX. WORK DIAM.	MAX. HOLE DIAM.	MAX. HOLE LENGTH	MACHINE WEIGHT	
Production type internal grinding machine	Fully automatic operation is one of the essential criteria of machines classified as production types; that requirement also acts as a limiting factor regarding maximum workpiece size and configuration. Many of the models in this category operate as centerless internal grinders with roll or shoe support.	4-12 ins. (100-300 mm)	2-3 ins. (50-75 mm)	2-8 ins. (50-200 mm)	10,000-18,000 lbs (4500-8000 kgs)	Fig. 3-3.8
Large size internal grinding machine	Large size internal grinding machines have the common characteristic of accepting heavy workpieces and are capable of grinding holes comparable to the OD dimension of that work. Wide range of differences exist, however, between various models with respect to the maximum work length, resulting from the overall design length of the machines.	32-36 ins. (800-900 mm)	20-24 ins. (500-600 mm)	20-50 ins. (500-1250 mm)	20,000-40,000 lbs (9,000-18,000 kgs)	Fig. 3-3.9
Extra large size internal grinding machine	Machines referred to in this category are often custom designed for accommodating specific groups of very large and heavy workpieces. Such components usually have very deep bores requiring extended bridge type machine construction and extra long table travels.	80 ins. (2000 mm)	24 ins. (600 mm)	60-70 ins. (1500-1750 mm)	120,000 lbs (55,000 kgs)	Fig. 3-3.12

(Continued on next page.)

Table 3-3.1 (*Cont.*) **Survey of Basic Types of Internal Grinding Machines**
(Based on a limited number of characteristic models.)

DESIGNATION OF THE CATEGORY	GENERAL-PURPOSE AND EQUIPMENT CHARACTERISTICS	RANGE OF CAPACITIES OF DIFFERENT MODELS WITHIN THE CATEGORY				ILLUSTRATION OF TYPICAL EXAMPLE
		MAX. WORK DIAM.	MAX. HOLE DIAM.	MAX. HOLE LENGTH	MACHINE WEIGHT	
Planetary type internal grinding machine	Planetary internal grinders are used for parts which are too large to be rotated and/or the hole(s) to be ground are far out-of-center. For the acceptable workpiece size the mounting surface of the work-table, the center height of the wheel spindle and the cross slide adjustment are the main controlling factors.	Not a characteristic dimension.	3-14 ins (75-350 mm)	15-30 ins. (375-750 mm)	10,000-30,000 lbs (4500-13,500 kgs.)	Fig. 3-4.1

machine which finishes the internal surface. For that reason, and also to provide a more versatile machine, the very small type internal grinding machines, such as the illustrated model, are frequently designed as universal type internal grinders.

Such versatility of small internal grinding machines is, of course, not always needed, and the many additional features serving extended adaptability may even be undesirable. Typical examples of limited-purpose internal grinders for very small workpieces are the very small centerless type internal grinders for miniature bearing rings, some features of which are discussed in the following chapter.

The universal characteristics of the illustrated model of a small internal grinder result from many design features and accessories, a few examples of which follow:

(a) Interchangeable internal grinding spindles with air turbine drive and with speeds adjustable over a total range from 60,000 to 120,000 rpm

(b) External grinding spindle with belt drive and electric drive motor, as a replacement for the internal spindle, to operate over a speed range from 4900 to 7890 rpm (see Fig. 3-3.2)

(c) Angularly adjustable work head with swivel ranges of 0 to 50° toward the front and 0 to 90° toward the rear, for grinding tapered internal and external surfaces, as well as for rotary surface grinding

(d) Optional accessories in a large variety, particularly many different types of work-holding devices

including centers and driver plates for external grinding between dead centers. An uncommon accessory is the centering microscope mounted on a support which can be swung into alignment position with no need of disturbing the operational setup.

While a major field of application for such universal type miniature internal grinders is in gage and tool making, the illustrated model is also suitable for production work within its capacity range. In repetitive type operations the automatic feed cycling, with different rates for roughing and finishing and adjustable cycle times, as well as the compensating type wheel dressing system using a hydraulically retracted and positioned diamond spindle, are useful characteristics.

Small Work Type Internal Grinding Machine

This designation applies to a category of internal grinding machines (see Fig. 3-3.3) which, due to the need for manual work loading and clamping, is intended for the production of lots comprising a single or a limited number of pieces. However, such machines, particularly when designed for semiautomatic operations, are also used for continuous work with a degree of productivity which, in many applications is found entirely satisfactory. The easy setup of these machines is of advantage for changeovers to different part sizes within the same general dimensional range. The great variety of applicable work-holding devices

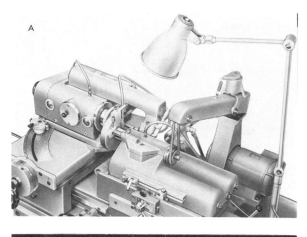

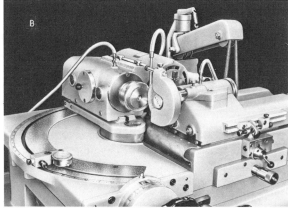

Fig. 3-3.2 Special setups of the miniature part internal-grinding machine shown in Fig. 3-3.1. A. Using special quill for external grinding, B. Work head swiveled at 90 degrees for rotary surface grinding.

substantially widens the scope of use of this category of internal grinding machines.

The illustrated model has a 30-degree swivel adjustment for the work head, permitting the accurate grinding of tapered holes to an included angle of 60 degrees. The availability of diverse attachments further extends the field of economic operations.

Medium-Size Internal Grinding Machines

The illustrated machine (Fig. 3-3.4) is representative of a basic category which comprises models of various capacities, particularly with regard to the diameter of the workpiece and the length of the hole which can be ground as a function of the available table movement. Added application flexibility is provided in models designated as toolroom type, by mounting the work head on a cross slide which, in

Fig. 3-3.3 Plain internal grinding machine for automatic operation in grinding small workpieces.

the illustrated machine, has a 6-inch (150-mm) travel to the front and about 11¾ inches (300 mm) to the rear.

Angular setting of the work head is generally available in all models and the ease of setup is an important characteristic with regard to the general-purpose work for which these models are primarily intended. Other versions of the machine, an example of which is shown in Fig. 3-3.4, are built as plain internal grinders for medium production lots, or as automatic grinders with size-control equipment for extensive production runs.

Typical dimensions of the toolroom models for the swing over the table are 22 inches (560 mm) and

Fig. 3-3.4 Medium size internal grinding machine in a design intended for a wide variety of work.

30 inches (760 mm), to be specified by the user to suit the planned work. The low-pressure, locked-feed type hydraulic system, supplied by a vane type constant delivery pump, provides smooth table movement which remains constant over an extended production period, retaining its preset rate which can be selected from a range of 0 to 18 ft (0 to 5.5 m) per minute. The range of the work-head speed has a ratio of 1 to 3 for the general-purpose type, and 1 to 6 for the toolroom model.

Universal Type Internal Grinding Machines

The term "universal" here expresses as for other types of machine tools using this designation, a particularly high degree of versatility and ease of adaptation to widely different work characteristics. A typical and widely used representative of this category of internal grinding machines is shown in Fig. 3-3.5.

Courtesy of Cincinnati Milacron, Inc., Heald Machine Div.

Fig. 3-3.5 Universal internal grinding machine. A widely used model, designed for flexible applications in toolroom and similar operations.

With regard to universal type internal grinding machines, versatility refers to the ability of the equipment to carry out, with reasonable efficiency and without the need for special accessories, additional operations other than internal grinding proper. Such operations are, for example, the grinding of external rotating surfaces (see Fig. 3-3.6), cylindrical, taper or profiles, and of flat surfaces normal to the axis of rotation of the part, applying a process known as rotary surface grinding (see Fig. 3-2.8). The capability of carrying out such operations on a machine designed primarily for internal grinding may serve any or several of the following purposes:

(a) Grinding the external and internal surfaces of a part in a single setup in order to produce interrelated surfaces with excellent concentricity

(b) Substituting for other types of grinding machines, e.g., rotary surface grinders, which are either not available or not capable of meeting very tight accuracy specifications

(c) Grinding of surfaces in positions or locations which are not accessible to other types of grinding machines in their standard design.

Courtesy of Cincinnati Milacron, Inc., Heald Machine Div.

Fig. 3-3.6 Grinding an external work surface on a universal type internal grinding machine.

Adaptability, in this application of the term, refers to the capacity of the machine to accept a wide range of different work sizes and configurations, on which the required grinding operations can be carried out effectively and with adequate accuracy. These properties are particularly important for machine tools such as the universal type internal grinding machines, whose primary applications are in the toolroom or in small-lot, often jobbing types of production.

To meet the outlined requirements universal type internal grinding machines possess distinctive design characteristics and are usually built to provide the generally high accuracy requirements of the work. A few of such characteristics, also found in the model shown in Fig. 3-3.5, are the following:

(1) The work head is mounted on a positioning slide with a 17½-inch (445-mm) travel.

(2) The drive of the work head is reversible and its rotational speed is infinitely variable from 0 to 500 rpm. The speed below 20 rpm is commonly used for

indicating the work in the mounting phase of the operation to assure runout-free rotation, while the higher speeds are for grinding.

(3) Angular adjustment of the work head over 90 degrees (to a position at right angles to the wheel spindle) which can be set very accurately with the aid of a sine bar attachment.

(4) Long table travel, commonly 20 inches (508 mm) or, in an extended version, by about 50 per cent longer, with a maximum grinding stroke of 16 inches (406 mm) to accommodate very long workpieces. At the same time the minimum stroke of .125 inch (3.2 mm), permits the efficient traverse grinding of narrow workpieces.

(5) Optional accessories, usually available as stock items, in a wide variety, still further increase the adaptability of universal type internal grinding machines. Such accessories include: facing attachment; steady rest (see Fig. 3-3.7); different types of work holders; wheel-truing units for angles, radii, and face grinding; adapter for mounting different wheel heads; etc.

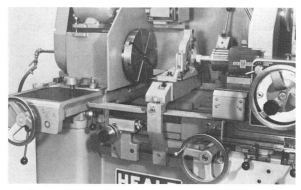

Courtesy of Cincinnati Milacron, Inc., Heald Machine Div.

Fig. 3-3.7 Steady rest with its top jaw retained in a hinged element for supporting long workpieces, mounted on the universal internal grinding machine shown in Fig. 3-3.5.

Although the variety of work which can be performed has been pointed out as the dominant characteristic of universal internal grinding machines, the fact that this kind of machine tool is mainly used for tool and fixture work, as well as for gage making, implies a comparable *performance accuracy*. To meet that requirement the machine shown in Fig. 3-3.5 has hydrostatic table ways and the work head is built for a rotational accuracy of 0.000 030 inch (0.75 micrometer), or even half of that value for special purposes. An ultra-precision version of the same basic machine, built with a hydrostatic work head, has a roundness capability of better than 0.000 010 inch (0.000 25 mm), the cross slide movement is control-

led in 0.000 005-inch (0.000 125-mm) increments and the angular setting for taper is designed to 2 seconds of arc accuracy.

While toolmaking operations are commonly carried out by manual control, the basic machine permits semiautomatic operation by means of preset table reciprocation and cross-feed rate, which is operated at one or both ends of the table stroke. An automatic version of the same model, designed for multi-lot production grinding, performs automatic work sizing; has digital feed setting; timed interval dressing with adjustable traverse rate; and various other elements of automatic operational cycles.

Production Type Internal Grinding Machines

The major requirement of production type equipment generally used in high-volume manufacture, is the fully automatic operation that permits the machine to produce while essentially unattended. Fully automatic operation of internal grinders comprises the entire cycling of the machine movements, at set rates, in proper sequence and over specific distances, except for the final sizing infeed which may be controlled by an automatic in-process gage, monitoring the progress of the grinding. Also included in the automatic controls is the wheel dressing, at preset frequency; as well as the slide position compensation for canceling the effect of the reduced wheel diameter.

Courtesy of Cincinnati Milacron, Inc., Heald Machine Div.

Fig. 3-3.8 Work area of a production type internal grinding machine (shown with protective guards and covers removed), equipped with chutes for work feeding and discharge, also with loading arm for automatic work transfer into the continuously rotating chuck.

Work loading and unloading is an important element of the fully automatic operation, consequently such processes can be used only for workpieces which are adaptable to handling by mechanized feeding, loading, and clamping devices.

The fully automatic internal grinding machine, whose work area is shown in Fig. 3-3.8, is equipped with a gravity chute for feeding the workpieces into the loading position, from which a loading arm transfers the parts into the work-holding chuck designed, in this application, to be unloaded and loaded by means of a special device while the work spindle of the machine is running. Other automatic loading devices operate by bringing the chuck to a stop in a specific position and the workpiece is then loaded either in the front or through the hollow work-head spindle; the latter a very efficient method which, however, is limited to small parts of essentially cylindrical shape.

The outside diameter of the work affects the applicable loading system as indicated by the pertinent capacity values of the illustrated model (Fig. 3-3.5), which accepts work to 4½ inches (114 mm) in diameter for front (end or side) loading, and to 1¼ inches (about 33 mm) in diameter for loading through the work-head spindle.

Internal grinding machines of this type, although very efficient in operation, have applications limited to lots of sufficient volume which warrant the time-consuming setup and the special tooling that may also be required.

Large-Size Internal Grinding Machine

The general specifications of the illustrated model in Fig. 3-3.9, are quoted to indicate, as an example, the size of the largest workpiece which that grinder can accommodate:

Total chuck swing (without guard)	30 inches (about 760 mm)
Total wheel slide movement	20 inches (about 510 mm)
Grinding stroke (maximum depth of hole)	16 inches (about 400 mm)

For the convenient loading and unloading of heavy workpieces, the work head with the work spindle and the chuck operating mechanism is supported on the work slide and can be traveled forward from the grinding position by 16 inches (about 400 mm), using hydraulic indexing for that movement. For grinding slender tapers the work head can be swiveled to a maximum included angle of 15 degrees.

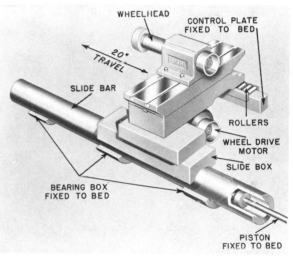

Courtesy of Bryant Grinder Corp.

Fig. 3-3.10 Diagram showing the design principles of the wheel slide used for the internal grinding machine in Fig. 3-3.9.

The long strokes for both the wheel slide and the work slide, combined with the heavy weight of the assemblies which these elements carry, require appropriate construction to assure low friction, smooth travel, and accurate alignment of these movements.

The wheel slide (see Fig. 3-3.10) has its front attached to and supported by a hardened and ground steel slide bar with which it travels in both directions. The rear of the wheel slide is supported on a series of rollers, traveling on a hardened and ground steel plate which can be adjusted to maintain a perfect alignment with the work head. The slide bar travels on a

Courtesy of Bryant Grinder Corp.

Fig. 3-3.9 Large size internal grinding machine for semiautomatic operation, with 30-inch (760-mm) swing and hydraulically moved work slide for facilitating the work loading.

set of three pressure-lubricated bearing boxes which are attached to the machine bed.

The work slide (see Fig. 3-3.11), operating as a cross slide, and serving both the work positioning and feed advance, is supported by two rows of pre-loaded balls, on its side, facing the grinding wheel. These balls have a hardened steel bar for a raceway. The rear end of the cross slide is supported by a self-aligning hardened steel sleeve which is also carried on pre-loaded balls rolling on a hardened steel bar. This bar is attached to the bed by supporting reeds which permit adjustment for aligning the spindle of the work head with the wheel slide.

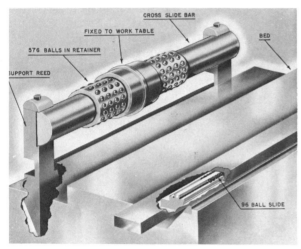

Courtesy of Bryant Grinder Corp.

Fig. 3-3.11 Design of the work slide support of the grinder in Fig. 3-3.9, showing the essential support elements.

Extra-Large Internal Grinding Machines;

The functional conditions of very large workpieces may require the bores (often extremely long ones) to be finished by internal grinding in order to satisfy rigorous tolerances with regard to diameter, round-

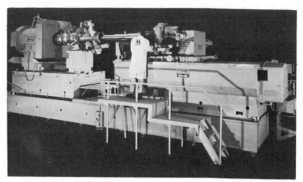

Courtesy of Cincinnati Milacron, Inc., Heald Machine Div.

Fig. 3-3.12 Extra large internal grinding machine with extended bridge permitting longitudinal adjustment over a substantial distance.

ness, straightness, surface finish, etc. Special internal grinding machines such as the model illustrated in Fig. 3-3.12 are needed for example, for accommodating and grinding such typically large parts as airplane landing gears.

While the need for such extra-heavy internal grinding machines, in terms of the total number of this equipment and the companies using them, is limited, a few of the characteristic data are listed, particularly for the purpose of illustrating to what limits the basic principles of internal grinding are applicable in the design and operation of such special grinding machines.

The extended bridge type internal grinding machine shown in Fig. 3-3.12 has a swing diameter of 80 inches (over 2 meters), which is reduced to 48-inches diameter (about 120 cm) when using a chuck guard. The range of work-head adjustment is 144 inches (about 3.65 m) and with a 73-inch (1.85-m) table travel, hole lengths to 60 inches (1.52 m) can be ground. The power of this extra-heavy internal grinding machine is indicated by the rating of its wheel-head motor, which is regularly 10 hp, but optionally, 15 or 30 hp.

Centerless Internal Grinding

The term "centerless," when applied to internal grinding, designates a method in which the axis of rotation of the workpiece is not fixed by rigid clamping in a holding device, but results from fixed position support points in contact with the periphery of the work. As long as the peripheral section of the work in contact with the support elements is perfectly round, the rotational axis of the work will remain constant.

This condition differs from that found in the centerless grinding of external surfaces where the surface being ground serves also to locate the part, consequently the stock removed from that surface causes a proportionate shift in the position of the work axis of rotation.

However, when the outside surface of the part on which it is supported during the centerless internal grinding is not round, that condition will be transferred, causing a similar amount of out-of-roundness in the ground internal surface. The only improvement in roundness which can be accomplished with respect to that of the supporting outside surface, applies to closely spaced lobing; such roundness errors can be diminished by support elements which are wide enough to bridge the gap separating the adjacent lobes. It follows that an external work surface of appropriate roundness, usually produced by shoe-type centerless grinding in a preceding operation, is an essential requirement for successful centerless internal grinding.

There are two basic methods for supporting the work in centerless internal grinding:

(1) On rotating rolls, referred to as roll-supported centerless internal grinding

(2) On non-rotating shoes, referred to as shoe-supported centerless internal grinding.

Roll-Supported Centerless Internal Grinding

A large-diameter, power-driven regulating roll, and a freely rotating smaller support roll form a V-type support for the workpiece, which is held firmly in that rest by a third roll, namely, the pressure roll (see Figs. 3-4.1 and 3-4.2). The work is rotated by the regulating roll whose speed can be varied to provide the optimum surface speed to the work. That speed must be maintained in a sufficiently positive manner, in order to prevent the grinding wheel from imparting an unwanted driving speed to the workpiece.

Roll-supported centerless internal grinding is preferred for long or thin-walled parts, even when they have interrupted bores or annular grooves. The work-holding rolls can be skewed to locate the work either against a rotating backing plate in the rear, or against a stop ring on the front, thus controlling the position of the ground section in relation to one of the

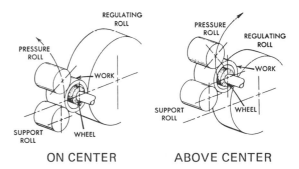

ON CENTER   ABOVE CENTER

Courtesy of Cincinnati Milacron, Inc., Heald Machine Div.

Fig. 3-4.1 Operating principles of roll-supported centerless internal grinding, shown in two alternative setups: On-center: for firmer work support; Above-center: for improved rounding of the ground hole.

Courtesy of Cincinnati Milacron, Inc., Heald Machine Div.

Fig. 3-4.2 Workhead of a centerless internal grinder with roll support and front loading of the work.

workpiece faces. Machines designed for fully automatic operation are usually equipped with retracting devices, often operated by mounting them on air-actuated pivot arms, which can move the rolls out—for releasing, and in—for clamping the workpiece.

Shoe-Supported Centerless Internal Grinding

The principles of operation, as well as the details of the setup, are very similar to those of the shoe type centerless grinding of external surfaces. The major difference results from the direction of the radial grinding force which, in internal centerless grinding, is away from the center of the work. Consequently, as shown in the diagram of Fig. 3-4.3, one of the shoes is placed in, or very close to, the center plane of the workpiece, the plane containing the axis of the wheel spindle. Thereby, the radial grinding force acts through the wall of the workpiece against one of the shoes, which thus prevents that force from affecting the precise location of the workpiece. The second shoe is mounted under the part, but displaced away from its vertical centerline. The two shoes thus act as a V-block support with a center plane essentially coincident with the direction of the radial force exerted by the off-center drive plate.

The workpiece is supported on its back face by the drive plate or backing plate, equipped with magnetic pads which are energized during the grinding phase of the operation. The drive plate and the magnetized pads which hold the part are usually designed to leave the hole area unobstructed, thus avoiding interference with the required overtravel of the reciprocating

wheel. The workpiece is located by the shoes with slightly offset center with respect to the axis of the drive plate (see Fig. 3-4.6), causing a continuous slippage of the work on the pads of the rotating plate. In operation this arrangement exerts a radial force, holding the workpiece firmly against the two shoes.

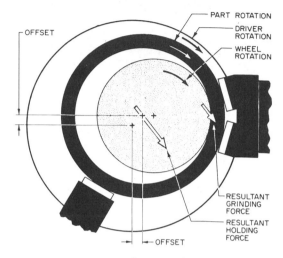

Courtesy of Bryant Grinder Corp.

Fig. 3-4.3 Principles of work holding in shoe-supported centerless internal grinding.

An important characteristic of the setup for internal centerless grinding is the location of the offset work center in relation to the drive plate center. The center of the workpiece is in the first quadrant of the drive plate (see Fig. 3-4.3), while in the shoe-type external centerless grinding the work center is located in the second quadrant of the drive plate.

Shoe-supported centerless internal grinding is a method of unequaled efficiency for the internal grinding of essentially annular shaped parts having a diameter generally larger (often by several times) than the width of the part. Of course, the availability of a precisely finished circular external surface, which is concentric with the bore, is an important requirement. However, that supporting surface need not be cylindrical. The supporting surface may be tapered (such as for tapered roller bearing cones), toroidal, of convex or concave profile (e.g., ball bearing inner rings supported in the raceway), or of some other circular shape, as long as the requirements of roundness and concentricity are met, and shoe shapes complying with the contour of the supported work surface section are used.

Shoe-supported centerless internal grinding offers many advantages, particularly for the high-volume internal grinding of workpieces whose configuration makes them adaptable to the application of this method. Typical, but by far not exclusive examples, are the rings of antifriction bearings (see Fig. 3-4.4) comprising both the bores of the inner rings and the raceways of the outer rings. The advantages of shoe-supported centerless internal grinding are:

(1) Location of the work in the properly centered operational position is an automatic process, accomplished within one turn of the drive plate

(2) Clamping is dispensed with while a free path of access and withdrawal for the part is provided without disturbing any element of the work support so that the loading and unloading can be automated and carried out very quickly, in some cases, within a fraction of a second

(3) During grinding the face of the workpiece is held by magnetic force against the drive plate (which is precisely perpendicular to the axis of the work spindle rotation), so that the accurate squareness of the ground internal surface in relation to the face of the workpiece is assured. In the case of form grinding, the parallel location of the ground feature to the locating face is similarly assured.

(4) By avoiding the need for clamping, the work potential inaccuracies of the work-holding device and distortions of the part caused by clamping forces, are eliminated

(5) The electromagnetic force of the drive plate can be activated and regulated in precise timing with the grinding cycle, providing instantaneous chucking and release, also adjusting the magnetic pull to the grinding force it has to resist—heavier in the roughing and lighter in the finishing phase of the cycle.

Courtesy of Bryant Grinder Corp.

Fig. 3-4.4 Work area of an internal grinding machine equipped for the shoe-supported centerless grinding of roller bearing inner-ring bores.

Design Characteristics of Centerless Internal Grinders

Centerless internal grinding of very small annular shaped parts. Typical examples are miniature antifriction bearing rings with very small outside diameters, in the extreme case slightly more than one millimeter (actually 0.045 inch) which, when ground by the centerless internal method require special work-holding systems. In a widely used model of such special internal grinders the workpiece is supported between two carbide disks and a shoe. One of the disks is directly below the center of the workpiece, while the other, mounted above the part, has its center off the vertical centerline of the work by an amount about equal to the diameter of the part. This offset causes the part to be pushed against the laterally located shoe. The rotating disks drive the workpiece and are mounted with a slight skewing for holding the face of the part firmly against a nonrotating backing plate, thereby producing a bore which is precisely square to the face.

The shoes of centerless internal grinding machines must continuously withstand the friction caused by the sliding contact of the external work surface rotating at a peripheral speed which is necessarily higher, by a substantial amount in thick walled parts, than the operational speed of the work surface being ground. In order to extend service life and to retain the setup accuracy of the shoes, these are usually made with cemented carbide contact surfaces.

The shape of the shoes must correspond to the contacted work surface which may be cylindrical, tapered, or profiled. The correspondence of the shape also applies to the width and orientation of the shoes, whose shape is often flat and must be tangent to the supported work surface. To assure the best compliance between the contacting surfaces of the work and of the shoes, some shoe-holder designs provide self-alignment by having a two-paw shoe supported on a pivot pin, or ball pivot, as shown in Fig. 3-4.5.

The adjustment of the shoes to assure the exact offset of the workpiece, with respect to the center of the drive plate, is a critical setup operation. By mounting the shoes on an interchangeable shoe holder (see Fig. 3-4.6) which is seated on the mounting plate of the machine in a precisely located position, the setting of the shoes can be carried out conveniently and with accuracy, away from the machine, with the aid of appropriate setting masters.

The loading of workpieces in centerless type internal grinding machines designed for automatic operation is usually carried out by transferring the individual parts from a loading chute at the end of a feed line. Parts of annular shape which are ground

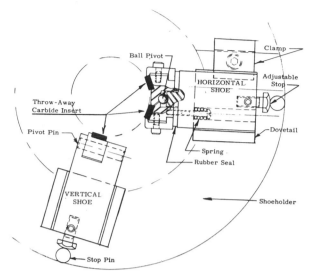

Courtesy of Bryant Grinder Corp.

Fig. 3-4.5 Diagram of a swivel-type shoe holder designed for improved alignment, with easily exchangeable carbide inserts.

on this type of internal grinder are often located by their rear face against the inside surface at the back of the loading chute. They are held in that position by adjusting the front plate to the chute to suit the thickness of the work. When workpieces have unsymmetrical configuration, such as in the tapered roller bearing inner races ("cones"), elements for detecting reversed parts are usually added to the loading chute

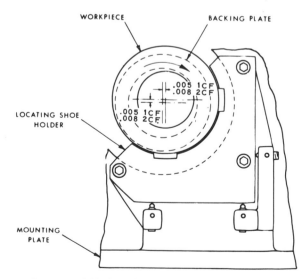

Courtesy of Cincinnati Milacron, Inc., Heald Machine Div.

Fig. 3-4.6 Simplified diagram of an interchangeable shoe-holder which permits the adjustment of the shoes away from the operating area of the grinding machine.

to prevent the work from being loaded facing in the wrong direction.

On traversing work-head type grinders, in order to avoid the need for flexible feed chutes which would follow the retraction of the work head during loading and wheel truing, the work feed lines are sometimes designed with a gap between the end of the chute and the work head. This latter carries a funnel on its upper part for catching and leading into the loading chute the parts released by the feed line. The release action is operated by an air cylinder and is applied to a single piece at a time, its operation being interlocked with the movements of the work-head slide.

Face clamping by mechanical action. There are workpieces with a general configuration which makes the location of the part against a backing plate desirable. However, the retention of the workpiece by magnetic force is impractical and two alternate methods may be used in such special cases:

(a) The roll-clamping method with two special clamp rolls placed one below, and the other above, the bore of the workpiece. The rolls are mounted on vertically adjustable brackets which can be set to suit the workpiece diameter. The brackets are held on pivoted arms whose clamping action is controlled by the piston of a hydraulic cylinder.

(b) The pressure-ring clamping uses a hard ring actuated by a hydraulic or an air cylinder to hold the part against the backing plate during the grinding. (See Fig. 3-4.7.)

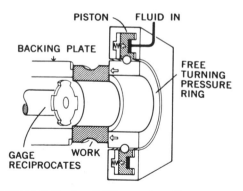

Courtesy of Cincinnati Milacron, Inc., Heald Machine Div.

Fig. 3-4.7 Pressure clamp in freely rotating design, holding the part against the rotating backing plate when actuated by hydraulic pressure. Springs are used for releasing the holding action.

Chucking Type Work Holding

The term "centerless type internal grinding machine" designates, in common usage, a category of

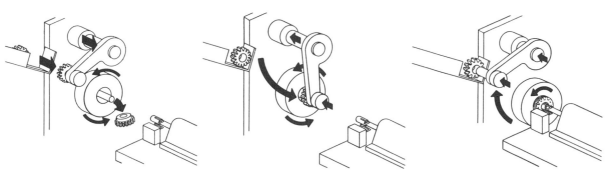

Courtesy of Cincinnati Milacron, Inc., Heald Machine Div.

Fig. 3-4.8 "On-the-fly" loading of workpieces for internal grinding. Diagrams showing the sequence of characteristic positions: (left to right) Eject finished piece; End-load work; Bore grind with reciprocating wheel.

grinders designed for grinding the bores of relatively narrow workpieces of essentially annular shape. Generally these machines operate by holding the work in a position which is related to, but not mechanically coupled with, the axis of work rotation. The lack of rigid linkage with a fixed axis has led to the adoption of the "centerless" designation.

Most of the characteristics of centerless internal grinding machines, developed primarily for high-volume production are also desirable for the internal grinding of a large group of parts which, however, due to their general configuration, overall dimensions, or other characteristics are not adaptable to being supported on shoes or rolls. Examples of workpieces not suitable for centerless holding are small parts whose outside surfaces are not adequate for supporting the rotating work, long parts with narrow bore sections to be finished by internal grinding, or parts with protruding or receding elements on the periphery, etc.

For grinding the bores of such parts some models of production type internal grinding machines which are generally supplied for centerless operations, can be made available with chucking devices. Certain special designs of these chucking type work holders permit "on-the-fly" loading while the chuck continues to rotate, thereby reducing the loading time, often to one second or less. Loading devices for the automatic loading and withdrawing of the part may be either of the front arm type (see Fig. 3-4.8) or the through-the-spindle type; the latter type permits a very fast loading and is particularly advantageous for long and slender parts. The operation of these loading devices is fully interlocked with other movements of the automatic grinding cycle and may be used for workpieces with unusual locating surfaces, including, for example, front face, rear inside face, an internal groove, or an external feature. There are also various special loading devices, such as those operating by

side loading, inserting the work into the side of an end-clamping chuck.

Gap eliminators, also termed "stock sensors," frequently used on centerless internal grinding machines, are devices which permit the rapid approach of the grinding wheel to the actual point of contact with the work surface. The use of such devices eliminates the need for starting the slower infeed advance at a preset point which has to be established according to the workpiece with the maximum stock allowance.

In one widely used system of gap eliminators, the valve solenoid of the device is energized before the cross slide starts its advance movement, thereby bypassing the feedbox throttles and permitting the slide to travel at a fast rate. When the wheel contacts the workpiece a load change occurs which will be detected by a power sensing unit, causing the valve solenoid to become de-energized. The feed control is now taken over by the feedbox throttles and the grinding is carried out with the preset feed rates.

In order to assure the proper functioning of the power sensing unit the differential of the electric current required between load and no-load states must be sufficiently significant, such as can be expected in grinding bores of about one-inch (25-millimeter) diameter, or larger. Other types of gap eliminator devices operate by using accelerometers as sensing devices. These will detect the vibrations in the wheel spindle caused by the wheel making contact with the work surface.

Two-Wheel Quill Internal Grinding

For workpieces which are presented for internal grinding with relatively large stock allowance, while requiring very tight size, form, and surface-finish tolerances, the operation can become more efficient by subdividing it into a roughing and a finishing phase with each phase carried out by different grinding wheels. Other types of workpieces which are preferably ground in two steps are those with internal contours, only one section of which must be ground to very tight limits while the rest of the contour has less rigorous requirements.

Applying two individual internal grinding operations, however, involves additional equipment and work handling, also, the separate mounting of the work can cause concentricity errors between the surfaces ground in subsequent operations.

By mounting several wheels, one behind the other on a common quill (see Fig. 3-4.9), and controlling the cycle of the special internal grinding machine for bringing the wheels consecutively into operating position (possibly adjusting the feed rates to suit the different process requirements), substantial savings in cycle time, grinding-wheel usage, etc., can be achieved. A further advantage of producing the parts in a single setup is the assurance of excellent concentricity. The wheels have different specifications, using coarser grain, and probably also softer bond, for roughing. The finishing wheel is mounted closer to the spindle bearing, thus reducing the deflections resulting from the cantilever construction of all internal grinding spindles.

Courtesy of Cincinnati Milacron, Inc., Heald Machine Div.

Fig. 3-4.9 In-line arrangement on a common quill of several grinding wheels used in consecutive phases of the raceway and seal surface grinding of a ball-bearing outer ring. First grind the entire bore surface with the three wheels, which are dressed with rotary diamond rolls and are mounted at the end of the quill. Subsequently, the wheel head advances axially; the single wheel, mounted farther back on the quill, will finish grind the ball track. The circular profile of the finisher wheel is trued with a single-point diamond installed in the swiveling arm of a retractable dresser.

Special-Purpose Internal Grinding Machines and Processes

Apart from the basic types of internal grinding machines, including the centerless varieties, a rather wide choice of special-purpose machines has been developed for carrying out internal grinding operations for which equipment considered as a basic type is either inefficient or otherwise unsuitable.

Examples of conditions which made necessary the development and application of various types of special-purpose internal grinding machines are the following: The overall size and configuration of the workpiece; the characteristics, dimensions, and locations of the internal surfaces to be ground; the special profiles or the inter-relations of the internal surfaces with other elements of the part's surface which are also finished by grinding; and multiple holes on the same workpiece.

The more important varieties of internal grinding machines serving particular purposes will be reviewed in the following. To make this survey more comprehensive, certain types of special internal grinding machines, which are discussed in greater detail in other parts of this book in conjunction with the particular purpose they serve, will also be mentioned briefly. As long as the work surface being ground is an internal one, such special machines are subjected to the same basic conditions as are all internal grinding machines.

Planetary Internal Grinding Machines

Most of the commonly used internal grinding machines operate by rotating the work around the grinding spindle which, after having brought the wheel into contact with the work surface, retains the posi-tion of its axis during the grinding, except for the in-feed movement when carried out by the cross slide of the wheel spindle.

There are, however, certain workpieces which, because of their weight, bulk, non-symmetrical general configuration, the location of the bore being ground, or for other reasons are not adaptable for being rotated during the process.

The holes in these workpieces, when grinding is a required or preferable process, must be ground while the workpiece is stationary, and the relative movement between the grinding wheel and the round hole's surface is achieved by an orbiting movement of the grinding wheel spindle around the theoretical axis of the hole being ground. Internal grinding machines designed for that type of operation are known as "planetary grinders," and less frequently, "orbital grinders." The terms indicate that two different rotational movements are being carried out by the grinding wheel spindle: the rotation of the grinding wheel around its own axis and the rotation of the grinding wheel spindle around the axis of the hole being ground.

Planetary internal grinding is not a new invention. It originated many years ago, during the early development phase of internal grinding, and its role—at that period—exceeded the present one. Subsequently, for many of the operations which were once the exclusive field of planetary internal grinding, substitute methods have been developed, such as precision boring, followed, perhaps, by honing. The blocks of multiple-cylinder type internal combustion engines are an important example of that technological change. However, there are still various applications for planetary internal grinding which are significant enough to warrant the regular production of this type of internal grinding machine, as well as its discussion, which follows.

Figure 3-5.1 illustrates a representative model of a large planetary internal grinding machine whose work-holding surface, bridging over the table, is designed to accept very large and heavy workpieces. The 36-inch (about 92-cm) transverse adjustment of the cross slide and the 40-inch (about 102-cm) table travel indicates the range of work sizes, bore distances, and depths which this particular model can handle.

The orbital radius of the grinding spindle can be adjusted from 1/16 inch (about 1.6 mm) to 1-9/16 inches (about 39.5 mm) and the gradual increase of that radius during the grinding process accomplishes the wheel infeed. By combining the planetary radius range with the use of an appropriate internal grinding spindle the machine is capable of grinding holes within a diameter range of 1¼ inches (about 32 mm) to 14½ inches (about 368 mm) and to a maximum depth of 34 inches (about 860 mm).

The machine table has a hydraulically actuated reciprocating movement for the purpose of traverse grinding, with stroke lengths and positions controlled by adjustable table dogs. The table speed is infinitely variable to a maximum of 14 fpm (about 4.25 m per minute).

Of course, planetary internal grinding machines are built in larger sizes as well as smaller sizes than the described model, which has only been selected to illustrate the application of the planetary grinding principle in the design of internal grinding machines.

Jig Grinding Machines *(See Fig. 3-5.2)*

The operational movements of these machines are similar to those of the previously discussed planetary grinders, the grinding wheel spindle carrying out an orbital movement around the center of the hole being ground in a stationary workpiece. The principal reasons for the reversal of the conventional principles with regard to the member which carries out the rotational movement are, however, somewhat different and can be related to two specific conditions:

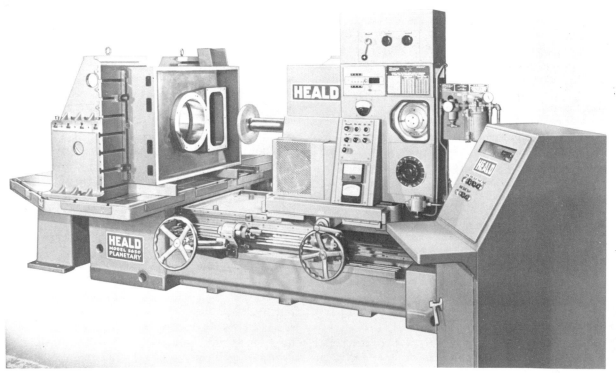

Courtesy of Cincinnati Milacron, Inc., Heald Machine Div.

Fig. 3-5.1 Large size planetary type internal grinding machine with worktable designed for mounting heavy and bulky workpieces.

(a) The workpieces of jig grinding, mainly tool and fixture elements, usually have several holes in very tightly toleranced mutual locations. It is a more convenient, and also much more accurate method, to position the workpiece which is firmly mounted on a table with movements along two coordinate axes, than to mount the work separately for each hole after having aligned its required center with the axis of a rotatable work spindle.

(b) The holes in workpieces which are commonly processed on jig grinding machines are, as a rule,

much smaller in diameter than the part itself. Consequently, extremely high peripheral speed would result on the outside of the part, if it had to be rotated to produce an adequate surface speed for the hole being ground. That drawback of a rotating workpiece is further aggravated by the usually eccentric location of the holes in relation to the center of gravity of the workpiece. The compromises in speed selection, which such conditions require when grinding with rotating workpieces, can be avoided by applying the principles of planetary grinding.

Here, only the principles and some of the justifications of jig grinding, as a variety of internal grinding, are mentioned for the purpose of a reasonably complete review of internal grinding methods.

Vertical Internal Grinding Machines

The term "vertical," indicates the orientation of the work's axis of rotation which, in internal grinding, is coincident with the axis of the hole being ground. This arrangement of the machine which affects the orientation of other elements, such as the direction of the reciprocating motion in traverse type internal grinding, differs from the common design around a horizontal axis.

There are several reasons for using the vertical arrangement in certain types of special-purpose internal grinding machines. The more important reasons are pointed out in the following, and in the case of some workpieces, several of these conditions for which the vertical arrangement is preferable, may be present at one time.

(a) Workpieces of medium size, e.g., in the 6- to 10-inch (150- to 250-mm) diameter range, although well within the capacity of the regular, horizontal-spindle type internal grinding machines, can be handled much more easily for automatic feeding and loading by lying flat, instead of in an erect position, particularly when the external work surface does not permit the rolling of the part in a chute. Typical examples are automotive ring gears with about 6 inches (150 mm) hole diameter. The vertical internal grinder shown in Fig. 3-5.3 can be equipped with work-handling devices comprising a conveyor on which the parts are lying flat. From there the parts are transferred automatically, by means of a hydraulic transfer arm, into the vertically mounted pitch control chuck on the work head of the grinding machine. This system of chuck, which is also used on the regular horizontal type internal grinding machines, locates, by the action of its jaws, parts

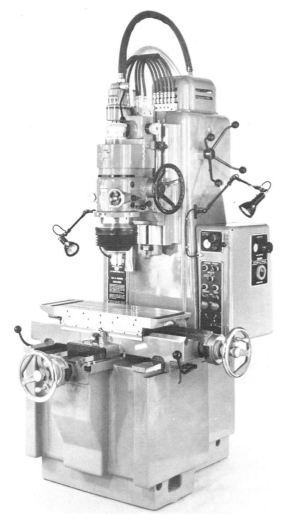

Courtesy of Moore Special Tool Co.

Fig. 3-5.2 Jig grinding machine, a particular type of planetary internal grinder in vertical arrangement, equipped with high-precision coordinate table for the accurate positioning of the work according to the required center locations of the holes being ground.

Courtesy of Bryant Grinder Corp.

Fig. 3-5.3 Vertical internal grinding machine specially adapted for the grinding of internal cams and other contoured bores.

with external gear teeth on the functionally significant tooth pitch line.

(b) Very large and heavy parts of essentially annular shape, that is, with short length in relation to the diameter, can be mounted, including positioning and clamping, much more easily when lying flat. Typical capacity data for a standard model of vertical internal grinding machine designed to handle extra-large workpieces (see Fig. 3-5.4) are the following:

Swing,   max. 62 inches (about 158 cm)
Face Plate diam.   64 inches (about 162 cm)
Clearance over face plate,

                        max. 24, 36, or 52 inches
                        (about 61, 92, or 132 cm)

(c) Multiple operations to be carried out on a workpiece in the same setting, both for reasons of operational efficiency and assured concentricity. Such operations may produce surfaces which are cylindrical or tapered, internal or external, the latter using the same spindles which were designed for internal grinding. The vertical arrangement permits mounting the grinding head on a compound slide which is supported and displaced along a cross rail guide above the face plate of the work head (see Fig. 3-5.4).

Courtesy of Cincinnati Milacron, Inc., Heald Machine Div.

Fig. 3-5.4 Work area of a heavy internal grinding machine with cross-rail-mounted grinding heads. (Machine is shown with protective covers removed.)

In other models, the cross rail guide design permits mounting two compound slides with two wheel heads, or even an additional side head grinding compound slide, the latter for carrying out, e.g., rotary face grinding on parts whose cylindrical and tapered surfaces have been ground in the same setup with the heads mounted on the cross rail (see Fig. 3-5.5).

Multiple Spindle Internal Grinding

Workpieces of small or medium size with holes to be ground in closely toleranced geometric relationship with a face or an external surface, can be finished efficiently by grinding the interrelated surfaces in a single setup, either simultaneously or sequentially.

For such operations multiple-head internal grinding machines, in designs adapted to the work characteristics, are being used effectively. The machines are equipped with two wheel heads (see Fig.

Courtesy of Cincinnati Milacron, Inc., Heald Machine Div.

Fig. 3-5.5 A different setup of the vertical internal grinding machine shown in Fig. 3-5.4, with the cross-rail-mounted spindles in retracted position and the side head grinding compound in operation.

3-5.6) which are, in general, arranged side-by-side, although in specially designed multiple-spindle internal grinding machines, the spindles may be positioned in mutually inclined or even perpen-

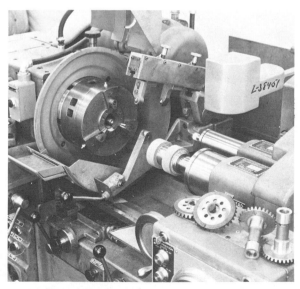

Courtesy of Cincinnati Milacron, Inc., Heald Machine Div.

Fig. 3-5.6 Work area of an internal grinding machine with two spindles, designed for the sequential grinding in a single operation of several internal and external work surfaces.

dicular locations, in the latter case, mounted on a swinging type spindle base.

The side-by-side arrangement may permit grinding of the work surfaces with both spindles without changing the position of the work head, except for the infeed movement (see Fig. 3-5.7). In other cases, the sequential grinding will require the indexing of the work head in a transverse direction (see Fig. 3-5.8).

Courtesy of Cincinnati Milacron, Inc., Heald Machine Div.

Fig. 3-5.7 Diagrams showing the operational steps of a two-spindle internal grinder used for the sequential grinding of the face and the bore on a workpiece processed in a single operation: 1. Grinding the face with sensitive manual facing attachment; 2. Grinding the bore with reciprocating wheel movement.

When the radial locations of the work surfaces to be ground do not interfere with the fixed lateral positions of the two grinding spindles, such as in the case of a relatively large bore with adjoining face, the transverse indexing of the work head is not needed; the sequential operation is accomplished by the independent retraction of the facing head, while the reciprocating movement of the table imparts the traverse required for the grinding of the hole.

The principles of sequential grinding, an operation commonly carried out with two spindles, can be applied to bore and face grinding on machines equipped with a single spindle, yet with two independent

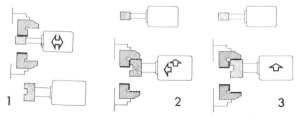

Courtesy of Cincinnati Milacron, Inc., Heald Machine Div.

Fig. 3-5.8 Diagrams showing the sequential grinding of two bores and a bottom face on a special internal grinding machine equipped with two spaced wheel heads: 1. Grinding the small bore in two phases with reciprocating wheel; 2. Plunge grinding of the large bore to semi-finish size and finish grinding the face; 3. Finish plunge grinding of the large bore.

feedboxes, one for the work-head transverse movement and one for the wheel-head advance and retraction. In a typical application the relieved face type grinding wheel will first plunge grind the recessed face and rough grind the bore, then the wheel retracts from the face and carries out the finished grinding of the bore with a short stroke reciprocation.

The preceding brief, and by far not complete, review of multiple-spindle grinding, should, however, point out still another aspect of the applicational flexibility with which internal grinding operations can be carried out with the aid of special purpose or specially adapted equipment.

Double-End Internal Grinding Machines

Special double-end internal grinding machines also designated as "duplex," have two wheel heads facing each other, and the work head located in the middle. They are designed to grind the workpiece simultaneously from both ends. A typical representative of this type of internal grinding machine is shown in Fig. 3-5.9.

Workpieces which have coaxial holes at two opposed ends can be finished efficiently and with a high degree of alignment accuracy when both holes are ground simultaneously on duplex type internal grinding machines. The same method and equipment can be applied for bores of different diameters. The capacity range of the illustrated model is related to the size of the hole in the work spindle, which has about a 12-inch (300-mm) diameter. The machine is designed for automatic cycling, including speeds, feeds, wheel dress, and size control.

Double-end grinding with two simultaneously operating wheel heads in opposed positions is not limited to workpieces with holes at two ends. A characteristic example of the flexible application of the described machine design principle is shown in the operating diagram of Fig. 3-5.10, which indicates the arrangement of the functional elements involved in the simultaneous grinding of two workpieces. The workpiece has a single hole with a straight and a tapered section, each of which must be ground to a close mutual alignment by using reciprocating wheel spindles. The work table is of the indexing type and holds two workpieces which are laterally displaced, the same way as the wheel slides. The wheel slide at the right carries the spindle for grinding the straight section of the hole, which is mounted on a cross slide for approach, infeed, and dress compensating motions. The compound slide on the left-hand side has axial movement in its lower portion, while the upper slide, in an angular position, carries out the reciprocating movement for the grinding and the wheel dress. The feed is accomplished by the axial movement of the lower slide, and is controlled by the radial withdrawal of the feed wedge.

Courtesy of Cincinnati Milacron, Inc., Heald Machine Div.

Fig. 3-5.9 Double-end internal grinding machine with two opposed wheel heads for the simultaneous grinding of coaxial internal surfaces at both ends of a workpiece.

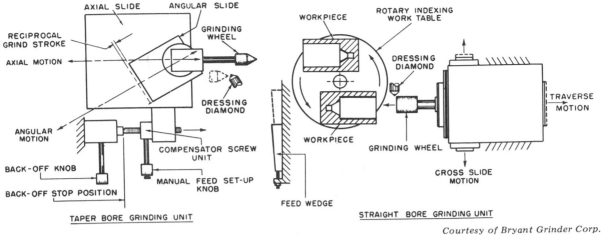

Courtesy of Bryant Grinder Corp.

Fig. 3-5.10 Operational diagram of a double-end internal grinding machine with two heads, opposed, parallel yet laterally displaced, for the simultaneous grinding of two workpieces, finishing sequentially different elements of the same internal surface.

Still another example for the application of the double-end principle is the internal grinding machine designed for the simultaneous grinding of the raceways in the outer rings of double-row tapered roller bearings. That particular bearing design also comprises straight sections at both ends, which are ground in the same setup by swinging the angular slides into axial positions for finishing the cylindrical sections of the opposed position bores. Double-end internal grinding machines built for the described operation are designed to function in an entirely automatic cycle, except for the loading and unloading of the large, about 10-inch (250-mm)-diameter workpieces.

Center Hole Grinders

Center holes, usually made with an included angle of 60 degrees, are important design elements for locating and holding rotating parts either during a production process or, less frequently, in service. The critical characteristics of center holes are the angle, the alignment of the axes, as well as the roundness and finish. In order to assure the required accuracy the center holes of heat-treated parts are often finished by grinding on special machines. These represent, in principle, particular types of internal grinding machines, using mounted wheels held in the chuck of an internal grinding spindle.

Center hole grinding machines may operate either as plunge grinders, producing the angular center holes by the axial advance of the wheel spindle carrying a properly dressed wheel, or they may be designed to operate as planetary grinders. This latter type

combines the orbital motion of the wheel spindle with its reciprocating movement and infeed.

Both types of center hole grinders are discussed in greater detail in Section 1, because their most important applications are in conjunction with precision type cylindrical grinding.

Internal Form Grinding with Crush-Dressed Wheels

Crush dressing is an efficient method for developing irregular or even intricate wheel profiles, which are then transferred by plunge grinding to produce a corresponding form on the work surface. The crush-dressing process exerts a very substantial pressure on the wheel, which is transmitted to the wheel spindle. For that reason the application of crush dressing is usually reserved to grinding machines with adequately dimensioned wheel spindles and spindle bearings or, more appropriately, to grinding machines expressly designed for crush dressing.

In principle, that method is not adaptable to internal grinding machines of regular design. However, there are special internal grinding machines available which have been developed for the application of crush dressing and used for production operations by which internal surfaces of intricate form must be produced in quantities sufficiently large to warrant the special tooling required in that process.

Grinding wheel profiles similar to those produced by crush dressing can be developed with the aid of rotary diamond rolls. This, however, is not an unconditionally equivalent alternative, due to economical and technological limitations. On the other hand, in the application of rotary diamond rolls, the high pressures created by crush dressing are avoided and consequently, wheel profiling by rotary diamond dressing is applicable to most types of regular internal grinders, except for those operating with spindles having a very low stiffness.

Auxiliary Devices for Internal Grinding

The operational elements of internal grinding with regard to both the equipment and the movements, can be duplicated on certain types of machine tools not expressly designed for internal grinding but serving other primary purposes. As supplementary equipment for permitting internal grinding, an internal grinding spindle with an appropriate drive is needed, preferably supplied as a unit to be mounted on an existing machine tool member, which either has in itself, or can be supplied with a longitudinal and transverse movement.

Machine tools well adapted to carry out internal grinding operations as a supplemental capability, with the aid of easily installed accessories, are: cylindrical grinders and engine lathes. Internal grinding on cylindrical grinding machines by means of a unit which is often a permanently installed accessory of universal type cylindrical grinders, was discussed in Chapter 1-5. Such units, commonly mounted on a base to be swung out of the operational area when not in use, can produce work quite comparable to that of regular internal grinding machines. The similarity of the operating principles of external and internal grinding of round surfaces explains such an essentially equivalent performance.

Carrying out internal grinding on a lathe is feasible and often applied in small shops and in toolroom practice, although it is simply an expediency and is not recommended as a regular production method. Toolpost type grinding machines with internal spindles (see Fig. 3-5.11), may be mounted on the cross slide of a lathe, but for accurate work, the dresser diamond should be held in a bracket mounted on the machine bed, or headstock. Because lathes are not built for grinding operations, the guideways and other exposed members should be protected against the penetration of harmful grinding dust. Such internal grinding operations must usually be carried out dry, a conditions which calls for light cuts and results in low performance.

Extra-Large Special Internal Grinders

For the purpose of visualizing the highest capacity ranges of internal grinding machines, examples of such equipment, built to handle particularly large and heavy workpieces, are described briefly in the following:

Figure 3-5.12 shows the working end of a very large size internal grinding machine built for a special purpose, namely, the finish grinding of the powder

Courtesy of Century-Detroit/Babcock & Wilcox

Fig. 3-5-12   Extra-large-size internal grinding machine, with its working end in the foreground; visible also is the multiplicity of hydraulic equipment and piping, which are required by the particular design system of this grinder.

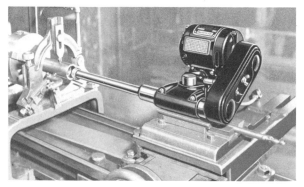

Courtesy of The Dumore Co.

Fig. 3-5.11 Toolpost grinding machine with long quill type internal spindle for deep-hole grinding, mounted on the cross slide of a lathe.

chamber of heavy artillery cannons. This particular grinding machine, weighing 55 tons, can accept work up to 120 inches (over 3 m) long, has a useful stroke length of 82 inches (over 2 m), and uses a 6-3/4-inch (about 170-mm)-diameter spindle. Hydrostatic bearings are used in the wheel head, cross slide, traverse slide and work head; the combined effect on the overall stiffness resulting from this advanced type of support makes possible the attainment of an extremely fine work finish, in the order of 3 microinches (.08 micrometer R_a). The runout of the work head is held to 50 microinches (1.25 micrometers). Another contributing factor to the high quality of performance is the drive of the spindle by a hydraulic motor, in place of the conventional outboard electric motor with belt and pulley drive.

This grinding machine is also equipped with a tracer control system, which automatically traces the complex powder-chamber form from a two-dimensional flat template. The diagram, Fig. 3-5.13, illustrates the general shape of the powder chamber being ground, as well as the arrangement of the templet which also comprises a section utilized for truing the grinding wheel.

Another example for the use of the giant internal grinding machines is the bore grinding of the main landing gear for the largest types of airplanes. One type of such workpieces, which weighs 1800 pounds (about 820 kg), is 9 feet (about 2.75 m) long, has a bore diameter of 11 inches (about 280 mm), and a bore depth of 80 inches (over 2 m). The attainable

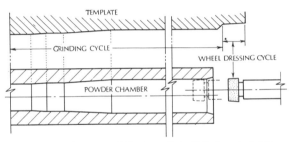

Courtesy of Century-Detroit/Babcock & Wilcox

Fig. 3-5.13 Diagram of the tracer control system showing the relative arrangements of the workpiece and of the two-dimensional template, one section of the latter also serving as a guide for the wheel truing.

bore diameter accuracy in this operation is said to be 50 microinches (about 1.25 micrometers).

Supplementary Functions and Equipment of Internal Grinding Machines

While the principal elements and movement discussed in Chapter 3-2 characterize the essential features of the internal grinding process, the actual operation of internal grinding machines also requires a considerable number of supplementary functions. Some of these are common to grinding machines in general, however, others are distinctly different and are peculiar to internal grinding machines.

A complete survey of supplementary functions would, of course, not be a feasible objective. However, a few of the more important ones will be discussed in the following, with emphasis on functions and equipment which are essential and/or characteristic elements of internal grinding operations.

Work-Holding for Internal Grinding

Holding the workpiece in internal grinding differs from work-holding in other methods of grinding because of the unique conditions which must be satisfied for assuring access to, and the proper position of, the internal surface to be ground. The particular conditions of work-holding for internal grinding result from the following characteristics which are common to all internal grinding operations:

(a) The axis of rotation of the workpiece is not physically present; it must either be derived from an external reference surface, e.g., the outside diameter of ring shaped parts, or it will have to be found as the center of the unfinished hole, using some adequate method, such as indicating.

(b) The workpiece has to be held in position by means of clamping elements which must not inter-

fere with the free access of the grinding spindle to the hole to be ground.

(c) The orientation of the hole being ground will be controlled by its alignment with the work-head spindle of the grinding machine. When mounting the work for internal grinding such alignment must be assured either by a dependable locating surface on the workpiece (e.g., a face perpendicular to the required ground hole) or it will have to be defined by indication or some similar procedure.

(d) The grinding forces in internal grinding usually are not too high, consequently, a moderate clamping pressure is sufficient in the majority of cases. This condition permits the application of work-clamping methods which may be inadequate for most other machining processes.

The observation of the preceding conditions and requirements can avoid excessive reliance on operator's skill and reduce the non-productive time which results from the mounting and positioning of the work for internal grinding. The implementation of these requirements must start in the design of the part, by providing dependable locating surfaces, followed by the appropriate choice or design of the applicable work-holding devices.

Another aspect which is often a prime consideration in the design of work-holding devices for internal grinding machines is their adaptability to automatic operations. That requirement involves both the unobstructed introduction and withdrawal of the workpiece, as well as the instantaneously operated clamping and releasing action, whose controls are tied in with other elements of the automatic grinding cycle.

Courtesy of Cincinnati Milacron, Inc., Heald Machine Div.

Fig. 3-6.1 For grinding the inner tapered clutch surface of a steel gear, the workpiece is held by using a draw-bar operated, expanding stake fixture which locates and clamps the part.

Some of the work-holding devices of internal grinders have been mentioned in other chapters of this section; some others are similar to those used in various types of external grinding machines. However, in order to present a rather comprehensive survey of work-holding systems used in internal grinding, Table 3-6.1 enumerates many of the devices which serve different types of internal grinding machines and are applied to a wide variety of workpieces processed by internal grinding.

Courtesy of Bryant Grinder Corp.

Fig. 3-6.2 Workhead of an internal grinding machine equipped with double four-jaw chucks for holding long workpieces such as machine tool spindles.

In addition, the devices shown in Figs. 3-6.1 to 3-6.3 are examples of the design flexibility which permits the holding of workpieces with widely different characteristics, while still meeting the important functional requirements which are listed in the preceding discussion as items (a) to (d).

Related to and complementing the operation of holding devices are the equipment of work feeding and loading, the availability of which is a mandatory requirement of fully automatic processing. Most of the feeding devices used for internal grinders rely in their operation on gravity, particularly convenient for ring-shaped parts which roll along chutes of properly designed cross section, although other methods of feeding, such as power driven conveyors are also used, when needed.

Courtesy of Cincinnati Milacron, Inc., Heald Machine Div.

Fig. 3-6.3 Special work-holding fixture for the internal grinding of a non-symmetrical part. The crank bore of a forged steel connecting rod is being finish-ground to very tight tolerances, locating the part by its pin end and using three twist fingers for clamping.

Dressing and Truing on Internal Grinding Machines

As a result of the particular conditions which characterize the internal grinding process, the dressing and truing of the wheel is much more frequently needed than in most of the external grinding methods. The necessarily small wheel diameter causes a more frequent contact of the individual elements on the wheel surface with the work being ground, hence, a more rapid wheel wear. At the same time, the extended contact area due to the wraparound effect of the work surface and the confined position of the wheel

Table 3-6.1 Work-holding Systems and Devices Used in Internal Grinding

SYSTEMS		DEVICES	
DESIGNATION	CHARACTERISTICS	EXAMPLES	DISCUSSION
Centering of work guided by the unfinished hole	Parts of irregular shape, or when the hole being ground is not concentric with the OD of the part, are mounted by referencing from the unfinished hole, either by means of a mechanical element (plug) or by tramming (with an indicator). Applications of this system are: (a) individual parts, such as in toolmaking; (b) irregularly shaped parts, generally held in special fixtures.	Face plate with T-slots Face plate with independently operated jaws. Magnetic plate Special fixtures	Not a complete holding device, but provides adequate mounting base for special work locating and clamping members. Used for parts with irregular shape. For holding parts whose outside surface is not of circular shape, or out-of-round to a degree excluding the use of concentrically acting jaws. Used for parts which have a flat mounting surface normal to the hole and require centering in relation to the work spindle of the grinder. Intermediate elements which transmit the magnetic force may be needed to provide clearance between the work and the plate. Parts of irregular shape require special holding fixtures when internal grinding is carried out in continuous production. Such fixtures must provide means for the rapid and precise locating of the part assuring the proper position of the hole being ground with respect to other functionally critical elements of the part. Clamping is either radial or axial and often power operated.
Concentric OD clamping—applicable over wide size range	Annular (ring shaped) parts with bores essentially concentric with the OD, can be clamped on the outside surface which serves also for locating the work. Concentrically closing jaws perform both the clamping and locating over substantial diameter ranges, although with only fair locating accuracy. Applied for parts (a) with limited accuracy requirements, or (b) whose	Three-jaw chucks Four-jaw chucks	The versatile applications of these universal work holding devices also include internal grinding, particularly in cases where the wide capacity range is a desirable property. Drawbacks are the limited centering accuracy, the danger of part distortion due to excessive clamping force and the time consuming manual operation. Used for parts with configuration requiring four-point clamping, otherwise similar to the preceding. Some models also permit independent jaw adjustment.

Table 3-6.1 *(Cont.)* **Work-holding Systems and Devices Used in Internal Grinding**

SYSTEMS		DEVICES	
DESIGNATION	CHARACTERISTICS	EXAMPLES	DISCUSSION
	OD will be finished subsequently by referencing from the ground hole.	Interchangeable-jaw chuck Face clamping chucks	Mechanism is similar to preceding types, however the jaws are interchangeable and usually soft for being adapted to the outside shape of the non-cylindrical part. Selected for parts with clamping surfaces at right angles to the axis of the hole; also for assuring the locating from the back face of the workpiece.
Concentric clamping on finished OD	Devices in this category provide excellent centering accuracy with respect to the finished OD of essentially ring shaped parts. The short .04—.08 in. (1 to 2 mm) opening movement of the clamping elements for releasing the part necessitates the use of inserts or jaws precisely adapted to the diameter and form of the outside work surface. Extensively used system for the production type internal grinding of precision parts.	Diaphragm chucks with three or six jaws. Multi-jaw diaphragm chucks. Collet chucks. Diaphragm chucks for particular work shapes. Sliding jaw collet chucks	The diaphragm actuated clamping system provides sufficient clamping force for most grinding operations, it is quick acting, easily adapted for power operation, and retains the accuracy of the originally set centering. The base elements hold adjustable jaws thus providing wide clamping range, but the actual opening movement is very small. These are made with 12 or more jaws for distributing the clamping force over a wide area, to avoid distortions. The wide area of the jaws covers a large portion of the work surface, producing a minimum of clamping pressure. Can be actuated by diaphragm or other mechanical systems, usually power controlled. Typical examples are diaphragm chucks with jaw tips contacting the flanks of gear teeth thus referencing the bore being ground from the pitch circle of the gear. Such devices, designed especially for internal grinders are made in many varieties, adjusted to the configuration, locating elements and the method of clamping required by the workpiece.

(Continued on next page)

Table 3-6.1 (*Cont.*) Work-holding Systems and Devices Used in Internal Grinding

SYSTEMS		DEVICES	
DESIGNATION	CHARACTERISTICS	EXAMPLES	DISCUSSION
Locating on the periphery of the rotating work.	The entire periphery of the rotating part is used for locating it in relation to fixed position support elements, thus producing a hole separated from the OD by a wall of controlled thickness. The system permits quick loading and unloading, often automated. Its application is limited to annular parts with precisely finished OD, a condition which exists in a very large number of components requiring internal grinding.	Supporting rolls — (a set of three, one of them the driver.)	The system comprises three rolls: (1) a large regulating roll for driving the work; and (2) a support roll below the work forming a V-block; into which (3) the pressure roll forces the workpiece. The system is used for parts with finished OD which must be concentric with the bore, but squareness of the face to the bore is not necessary. Relative position of the rolls to the work center can be: (a) work and regulating roll centers in a common plane for max. support; and (b) work center raised for improved roundness.
		Fixed shoes — (Two shoes in conjunction with a rotating backing plate which drives the work)	The non-rotating shoes assure the highest degree of work stability, however, the face of the workpiece must be flat and square to the OD by which the work rests on the shoes. Small amount of eccentricity between the centers of the backing plate and the work provides a radial force holding the work firmly against the shoes, which are adjustable and interchangeable to suit the size and the OD form of the workpiece.

inside the work, impede the discharge of the detached chips which thus tend to load the wheel surface. The described conditions necessitate frequent dressing for maintaining the proper cutting ability of the grinding wheel.

The need for wheel dressing within short intervals, a functionally required procedure, is often utilized for the benefit of size control. Dressing the wheel by removing a surface layer of specific thickness produces a known, although gradually decreasing, wheel diameter. By compensating for the reduction of the wheel diameter with a corresponding adjustment of the cross slide position, the operating surface of the wheel can be retained in a constant position with respect to the work head which, in turn, determines the position of the work surface.

Frequently, the workpiece size is controlled by the fixed position, in relation to the axis of the hole, of the diamond on the dresser. When the grinding wheel is dressed with the diamond in this position, a precise relationship is established between the position of the grinding wheel and the hole. From this position, the hole can be ground to size by a controlled amount of infeed and by utilizing the spark-out technique at the end of the cycle.

Frequent wheel dressing causes an accelerated wheel-size reduction, a condition which in internal grinding has a proportionately greater effect on useful wheel life than in external grinding where substantially larger wheels are commonly used. For that reason, the choice and the maintenance of the proper layer thickness which has to be removed in dressing, termed the amount of "dress-off," deserve particular attention in internal grinding.

The devices and tools of wheel dressing. As a general practice, wheel dressing by single-point diamond is used in internal grinding which, in the majority of cases, involves surfaces with straight elements, cylindrical, or tapered. Dressing devices used in internal grinding machines may be either of the rigidly mounted design (see Fig. 3-6.4), or of the "lift-up" type. The latter can be swung out of the way to permit the unobstructed traverse of the grinding wheel slide, thereby making possible the use of the shortest possible quills, particularly when grinding small holes. For automatic operations the lift-up type wheel dressers are hydraulically operated, with a positive cam interlock to prevent an untimely descent (see Fig. 3-6.5).

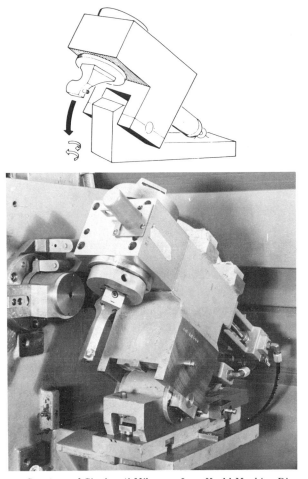

Courtesy of Cincinnati Milacron, Inc., Heald Machine Div.

Fig. 3 6.5 Diagram (top) and close-up view (bottom) of a lift-up type radius dresser such as is used for wheels grinding ball tracks in bearing rings.

Courtesy of Cincinnati Milacron, Heald Machine Div.

Fig. 3-6.4 Work area of a shoe-type internal-grinding machine equipped with dresser diamond mounted in a fixed bracket, and with single-jet air gage.

For truing curved wheel surfaces having a circular arc for contour, which may be either convex or concave, special radius dresser devices are used; these are designed to permit the sensitive adjustment of the required radius length. Typical capacity values for the radius range are 0 to 1 1/8 inches (about 28.5 mm) concave, and 0 to 2½ inches (about 63.5 mm) convex. Another type of special truing unit, mainly used on universal type internal grinding machines,

serves the angular truing of the wheel from 0 to 90 degrees, with a typical stroke length of 1¼ inches (about 32 mm). For operations on internal grinding machines which involve the grinding of flat surfaces with the face of the wheel, special face-truing attachments, in addition to the regular truing devices must be used.

In the grinding of internal surfaces with contours other than straight, or perhaps with a short radius circular arc, special profile truing devices are required. Substantially improved truing efficiency may be achieved when truing complex contours with the aid of diamond studded rotary dresser rolls (see Fig. 3-6.6) when the volume of the work to be produced warrants their rather high initial cost. In specially engineered multiple-wheel internal grinding operations the rotary diamond and the single-point truing methods may be combined: rotary diamond for roughing the entire contour and single-point for the second, the finishing wheel, which grinds only a section of the profile.

For completing the enumeration of internal wheel dressing methods, reference should also be made to crush dressing, applied on internal grinding machines expressly designed for operating with this system of wheel dressing.

In single-piece or small volume production, wheels with diameters which exactly correspond to the size required for the effective grinding of small holes, are not always available. In such cases, the closest larger wheel diameter is selected and .the installed wheel is dressed off to the required size. For such wheel shaping, which precedes the regular diamond truing, the use of an abrasive stick is recommended as a means to conserve the diamond.

Setting the position of the truing diamond is a critical element of the machine setup. For assuring a dependable control of the form and size of the ground surface independently of the progress of the wheel wear, the diamond must contact the grinding wheel in the same plane that contains the initial line of contact between the wheel and the work. In order to retain that setting in a consistent manner, even after the periodic turning of the diamond nib, well-centered diamonds of essentially symmetrical form should be selected for truing in precision type internal grinding.

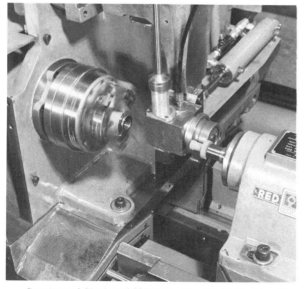

Courtesy of Cincinnati Milacron, Inc., Heald Machine Div.

Fig. 3-6.6 Diamond studded rotary dresser used for shaping the internal grinding wheel to a complex contour.

Automatic new wheel dressing is a system supplied with certain models of production type internal grinding machines to accelerate the actions required in connection with wheel change and make them independent of the operator's attention. When the wheel is worn down to a pre-established size, the operation comes to a halt and the machine slide resets to the new wheel position. After the operator has changed the wheel, he presses a particular button which starts the machine dressing at a coarse rate for the rapid elimination of the wheel runout, prior to the regular truing which is a part of the automatic grinding cycle.

Controlling the Frequency of Wheel Dressing in Internal Grinding

The following discussion applies to the periodic dressing of the wheel in the course of the internal grinding process and is independent of the need for regular wheel dress-off following the installation of a new grinding wheel.

In manually controlled operations the need for wheel dressing is usually determined by the operator relying on such guides as: (a) past experience, according to which wheel dressing is needed after a specific number of parts has been ground; (b) dressing-off for each single part, generally prior to the finishing phase of the operation; (c) observing the grinding performance for indications pointing to the degradation of the wheel surface, revealing itself by a different sound, the development of heat in the workpiece beyond the usual level, vibrations, chatter on the ground surface, etc.

In automatic internal grinding, wheel dressing for each workpiece is frequently applied as a safeguard against premature wheel deterioration or as the integral element of the process when controlling the work size by automatic compensation and wheel dress. However, when the work size control is accomplished by direct gaging—either by mechanical plug or continuously monitoring indicator, dressing the wheel for the purpose of size control is not necessary and should be applied only as required by the actual condition of the grinding wheel, whose useful life can thus be extended. Unnecessary wheel dressing is wasteful, causing excessive wheel and diamond usage, it often extends the cycle time and requires more frequent wheel changes, which always involve the interruption of production.

Internal grinding machines used in automatic operations are designed with different systems for maintaining the wheel dress frequency at the functionally necessary level. In the following, a few examples of such systems are discussed:

Skip-dress, a term indicating the skipping of cycles before the wheel dresser is operated, applying it only, for example, in every fourth cycle, thus producing four consecutive parts with the same dress-off. The skip-dress frequency is usually determined empirically, relying on experience gathered by grinding a sufficient number of parts with different skip-dress periods. The grinding machines designed for the application of that method are equipped with adjustable cycle counters and switches for initiating the dress-off operation.

Dress initiation by work size monitoring. A system, which, designated by its developer, the Bryant Grinder Corp., as "Dress on Demand," is based on the concept that a dull or loaded wheel, by cutting less efficiently, will increase the pressure on the wheel spindle causing its deflection. The wheel supported in a deflected spindle will not penetrate into the work surface to a depth which corresponds to the position of the wheel head in relation to the workpiece. In the Dress on Demand system, the cross feed is set to produce a specific worksize, as based on the performance of a sharp grinding wheel; the actual attainment of that size is checked by an in-process gage which monitors the grinding operation. Should the worksize not reach the expected value at the end of the pre-established cross-feed travel, that condition, termed "size-late," will be detected by the gage which then initiates the dressing cycle, consisting of slide position compensation and wheel dress.

This system represents an adaptive control of the wheel-dressing frequency, triggering it only when the cutting ability of the wheel declined due to regular wear, as well as to uncontrolled variables such as non-uniform stock allowance, hard work material, inconsistent wheel structure, etc.

Gaging and Size Control in Internal Grinding

For assuring the specified finish worksize regular gaging of the work surface is more often required during internal grinding than in most other grinding methods which operate with a wheel periphery generally larger than the work surface. In internal grinding that relationship is reversed, accelerating the breakdown rate of the wheel, thus resulting in a less favorable grinding ratio than is attainable in the external grinding of the same work material.

The need for regular gaging applies to all precision type internal grinding operations excepting those which substitute for actual gaging of a system of size control by means of fixed-position wheel dressing. Such methods, however, are considered as functional equivalents of in-process gaging and will be included in the following discussions.

Post-Process Gaging

Post-process gaging, which is extensively used in continuous production on external type grinding machines, is applied much less frequently to internal grinding. The successful use of post-process gaging in internal grinding as a method of size control, is predicated on very consistent grinding performance, causing only a slow and gradual size shift. Only under such conditions can the size changes detected by post-process gaging serve to indicate the need for machine adjustment, either manual or automatic. Furthermore, size control based on trend indications by post-process gaging is only applicable to work size tolerances which are wide enough to permit the regular occurrence of size degradation and still produce parts which are within the specified dimensional limits.

In order to satisfy the progressively tighter work tolerances, the formerly widely used size control systems based on post-process gaging are rapidly being replaced by the more responsive in-process gaging.

The use of post-process gages for checking the actual size of the ground internal surfaces, without using the indications for initiating manual or automatic machine adjustments, however, a still widely applied process, is a quality control function, which may support internal grinding process control, but is not considered a part of it.

In-Process Gaging

Each commonly used method of in-process gaging for the purpose of size control in internal gaging, belongs to one of the following four major categories:

(a) *Direct manual gaging* at different stages of the process, by interrupting the grinding, retracting the wheel, and measuring the dimensional conditions of the hole being ground. This method is time consuming and dependent on the skill of the operator, but requires no special equipment besides a hand-operated gage and involves no particular machine setup operations. For these reasons its use is preferred in single-piece and low-volume production, such as in the toolroom or when manufacturing small lots of identical parts.

(b) *Finished size detection* is a method comparable to the so-called attribute-gaging in dimensional inspection, which signals the attainment of the finish size of the hole being ground. Because the penetration of a physical element, functionally a gage plug, into the hole is used for determining whether the required size has been reached, this method of size control is termed "plug-gaging." It is only applicable to through-holes of cylindrical shape.

(c) *Continuous in-process gaging* is carried on while the grinding is in progress, by monitoring the size changes of the hole being ground and either indicating the momentary work size, generally in relation to the required finish size, or controlling the operation directly, by stopping the feed when the finish size has been reached, for example. This category comprises many varieties of internal gages, several of which will be discussed in greater detail.

(d) *Size control by wheel dressing* is based on the principle that the operating surface of the grinding wheel must be in a definite relationship to the axis of the hole for producing a specific work diameter. That relationship is affected by the unavoidable wear of the wheel during the grinding process, that wear—for a given wheel diameter—being generally proportionate to the volume of stock removed from the work surface. For that reason the controlled dressing is usually carried out in the final phase of the internal grinding operation, when the extent of wheel wear in removing a small amount of work material can be precisely anticipated and counterbalanced by a preestablished setting value.

A general review of the gaging methods which are commonly used in internal grinding is presented in Table 3-6.2, complemented with Fig. 3-6.7, showing four diagrams of the most frequently applied systems of in-process internal gaging. A more detailed discussion of several characteristics gaging processes and various equipment follows.

Plug Gaging

The system (see Fig. 3-6.8) uses a solid gage plug attached to the end of a shaft, called the "gage rod," which extends through the hollow work-head spindle and rotates concentrically with, and at the same speed as, the workpiece. The gage rod also moves axially at a rate and direction identical with the reciprocating movement of the grinding spindle. The gage plug is mounted in a spring holder which permits it to slip back on the shaft when hitting an obstacle, such as the face of a hole having a diameter smaller than the plug.

During the grinding operation the reciprocating movement of the shaft will, at each stroke of the wheel spindle in a direction away from the work head, cause the plug to advance toward the rear end of the workpiece, where it attempts to enter the hole. As long as the diameter of the hole is smaller than the plug diameter, that attempt will be frustrated and the plug will have to slip back. When the grinding action has increased the hole diameter to the finish size, the gage plug, made to that diameter, will succeed in entering the hole, thereby actuating a limit switch which then terminates the wheel infeed.

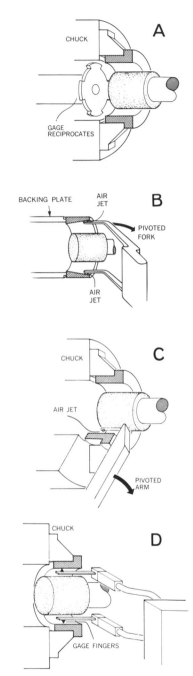

Courtesy of Cincinnati Milacron, Inc., Heald Machine Div.

Fig. 3-6.7 In-process gaging in internal grinding; frequently used systems. A. Plug gaging; B. Double-jet air gaging with nozzles on pivoted fork; C. Single-jet air gaging with nozzle on pivoted arm; D. Electronic gaging with contact points mounted on retractable gage fingers.

The capability of plug gaging with respect to the accuracy of the attainable size control has been considerably improved in recent years by various design changes, which reduced the mass of the gage and of

Table 3-6.2 Frequently Used Systems of Work Size Measurement on Internal Grinding Machines

SYSTEM DESIGNATION	INSTRUMENT TYPE	DIAGRAM	DISCUSSION
Manual Rigid or Adaptive	Gage Plugs		Gage plugs of the solid or adjustable type are very seldom used in modern production for gaging on internal grinding machines, particularly because of their lack of capacity to indicate the difference between the actual condition of the work in progress and the nominal dimension. Their advantages are: relatively low cost and the failsafe accuracy of the inspection, provided that the gage plugs are regularly verified.
	Micrometers		Inside micrometers, whether of the regular spindle type or the self-centering three-jaw type, have the advantage of supplying information on the actual size of the hole, thus guiding the operator with regard to the stock still to be removed. The drawbacks are the skill required for its proper handling and the possibility of errors when reading the vernier scale. The use of inside micrometers on internal grinding machines is generally limited to toolroom or small batch production.
	Mechanical Indicators		Mechanical inside indicators are available in several different designs, as a single-spindle type with centering paws, fork-gage type, or indicating plug gages with expanding split plug as the sensing member. Indicator gages always require some kind of setting master, the most common and reliable being a gage ring of appropriate diameter. A special type is the swing-in indicator which compares the position of the hole surface to a fixed element on the grinder. When hand gaging is preferred, the indicator type instruments are the most frequently used on internal grinding machines.

(Continued on next page)

Table 3-6.2 (*Cont.*) **Frequently Used Systems of Work Size Measurement on Internal Grinding Machines**

SYSTEM DESIGNATION	INSTRUMENT TYPE	DIAGRAM	DISCUSSION
Rigid Mechanized	Power Operated Plug Gage	 CHUCK GAGE RECIPROCATES WITH WHEEL PLUG GAGING	Widely used system for the automatic precision grinding of cylindrical through holes by applying reciprocating wheel movement. A plug at the end of a rotating shank extending through the bore of the workhead spindle, alternates synchronously with the traversing wheel. When the wheel retracts, the plug attempts to enter, from the rear, the hole being ground. Until the bore reaches its finish size the plug cannot enter it and is pushed back against a light spring pressure. When the hole reaches the finish size, the entry of the plug actuates a switch for terminating the infeed.
Air or Electronic Indicating Instruments	Single Point Probe	 SHOE CENTERLESS FIXTURE AIR JET ARM PIVOTS SINGLE JET AIR GAGING	Single-point probes relate the position of the rotating internal surface to a fixed element of the machine. That position should reflect the size of the internal feature without significant interference by inaccurate work location. Single probes are also applicable for small bores with very little space between the wheel and the bore wall, and can be used for either straight or tapered bores.
	Fork Probe From The Front	 CHUCK GAGE FINGERS	A fork probe introduced from the front measures the actual diameter of the internal surface in a specific plane along its axis. The indications are independent of the locational accuracy of the work holding, but sensitive to the tipping or shifting of the probes which must be located in a common plane normal to the work axis and in coincidence with a diameter of the bore. Unless the grinding wheel leaves sufficient free space inside the bore, the application is not feasible, or requires very thin and delicate contact fingers.

(*Continued on next page*)

Table 3-6.2 (*Cont.*) **Frequently Used Systems of Work Size Measurement on Internal Grinding Machines.**

SYSTEM DESIGNATION	INSTRUMENT TYPE	DIAGRAM	DISCUSSION
Air or Electronic Indicating Instruments	Fork Probe From The Rear	SHOE CENTERLESS FIXTURE · GAGE FINGERS · ELECTRONIC GAGING	Fork probe introduced from the rear is used when direct diametrical measurement is needed yet the small size of the bore prohibits the use of a regular gage fork. Such sensing heads are applicable for very small through holes, as long as they are cylindrical and are ground with a reciprocating wheel whose repetitive withdrawals permit the entry of the probe by following the wheel. Used on grinders designed for plug gaging and equipped with reciprocating gage rods.
Indirect Size Control	Diamond Controlled Work Sizing		Diamond sizing is a system of size control frequently used in internal grinding. It operates based on the experience that the variations in the rate of wheel wear in finish grinding with very small stock removal are insignificant to a degree permitting the control of the ground size by a specific amount of infeed movement following the truing of the wheel with a fixed position diamond.

the rod tube, made its guidance more accurate, reduced friction and shortened the response time, now in the order of about 0.01 second. Concurrently with the improvement of the plug gage, the number of oscillating strokes has been increased to about 300 or even more, per minute, permitting the reduction to very small increments, of the rate of the finishing feed, a now practical minimum being 0.000 010 inch(0.00025 mm) per pass. As the result of these improvements, size control by plug gaging in the order of 0.000050/0.000100 inch (0.00125/0.0025 mm) is now possible for small holes of about 0.5 to 1 inch (13 to 25 mm) diameter.

A few aspects related to plug gaging deserve mention:

(a) The repeated attempts of the plug to enter the work hole before it reaches the finish size, results in a large number of mechanical contacts which have to overcome a spring pressure and thus tend to cause wear. To reduce the progress of that wear, carbide-tipped gage plugs are generally used.

(b) The plug wear is further reduced by limiting the attempts for entering the hole to the period of finish grinding, keeping the gage rod in retracted position during the roughing phase of the cycle. A retraction air cylinder is often used for this purpose and for withdrawing the plug gage from the finished holes.

(c) On chuck type internal grinding machines the gage rod should rotate concentrically with the work spindle. On centerless type internal grinding machines which operate by a drive plate, the work center is

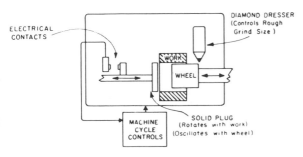

Courtesy of Bryant Grinder Corp.

Fig. 3-6.8 Block diagram of the essential movements and actions in "plug gaging."

displaced by a small amount, with respect to the axis of the work-head spindle; in this application the gage rod will have to be adjusted to run concentrically with the hole in the workpiece.

Continuously Monitoring In-Process Gages

These in-process gages operate by having the sensing member, known as the "probe" of the gage, in continuous contact with the work surface being ground, for the purpose of registering and indicating the change in size which takes place by removing stock from the surface. The amplifier and indicator of the gage are located remotely and are connected with the probe by a hose, or a cable. There are two systems permitting the amplifying and indicating member of the gage to be located remotely with respect to the probe which has to operate in the grinding area: air gages and electronic gages. Both systems are used extensively in internal grinding machines.

Air gages have the advantage of using probes which are not in direct mechanical contact with the rotating work surface, consequently, they are not subjected to frictional wear. Modern air gages with differential circuits are not sensitive to variations in the pressure of the air supply or environment, and these operate with relatively simple sensing heads which, however, require an area of free space between the wheel and the work surface, more than is available in very small holes. The use of air-electrical transducers permits the utilization of the signal resulting from air gaging, for the automatic control of the grinding cycle.

Electronic gages permit very high ratios of amplification, and various types of electronic gages used in grinding machines provide indications over different ranges with corresponding sensitivities: coarser, but embracing a wider range, for roughing; and finer, over a reduced range, for finishing. For that purpose some of the double-range gages have two indicator dials, side-by-side (Fig. 3-6.9), or a single indicator that operates over different ranges, automatically signaling the momentarily operative range on the dial face. Typical values for the least graduations on electronic gages used on internal grinding machines are 0.0005 inch (0.0127 mm) for coarse, and 0.000 050 inch (0.00127 mm) for fine indications.

When using diamond tipped gaging fingers the wear due to continuous mechanical contact with the work surface can be held at a level which, over a workday period, does not interfere with even the very fine incremental indications often required for high precision grinding operations. Of course, a daily checking of the gages with masters and readjustment, when needed, are usually part of the regular quality control procedures for high-precision type automatic grinding.

Courtesy of Marposs Gauges Corp.

Fig. 3-6.9 Remotely located amplifier instrument of an in-process electronic grinding gage with two side-by-side indicators having, respectively, coarse and fine graduations, covering overlapping dimensional ranges with common zero (finish size) base.

Signals originating from the gage at different stages of the stock removal process can be used for initiating cycle control actions, such as for ending the rapid approach, the coarse feed, the finishing feed, and the spark-out, respectively, with, perhaps, intermediate wheel dressing and cross slide compensation.

The sensing members used in conjunction with any of these gage systems, the air probes or the electronic contact fingers, can be of the single-point or of the double-point type. The single-point probe (see Fig. 3-6.10) measures the position of the surface being ground in relation to a fixed point on the machine. In shoe type internal grinding, the measured dimension is the wall thickness of the workpiece, which will decrease as the hole diameter grows, in the course of the grinding operation. Single-point gage probes are applicable for small holes which do not provide enough space for introducing, together with the grinding wheel, a double-point probe. The single-point probes can be supplied in various designs for gaging cylindrical, tapered, or curved surfaces; are simpler to main-

Courtesy of Bryant Grinder Corp.

Fig. 3-6.10 Block diagram of air gaging on an internal-grinding machine, with feed-back control, using a single-point probe (finger).

tain than the double-point type; and are preferred when the hole diameter measurement (referenced from an external surface) is found to be sufficiently reliable.

Courtesy of Marposs Gauges Corp.

Fig. 3-6.11 Measuring head with two interchangeable and adjustable contact fingers for the continuous size monitoring of cylindrical hole diameters. The head is moved in and out of the hole in synchrony with the reciprocating wheel, with automatic locking of the fingers before each withdrawal to permit re-entering the hole.

Courtesy of Marposs Gauges Corp.

Fig. 3-6.12 Double head, recessed hole measuring unit mounted on a swivel bracket for bringing the contact fingers attached to each head into and out of the measuring position. The fingers retract automatically for passing the narrower hole mouth such as that of ball bearing outer rings, and are released when in gaging position inside the ball track.

Double-point sensing members, often termed "forks," are also used for both air and electronic gage systems, and measure the actual diameter of the hole in a specific axial plane. Gage heads with two sensing fingers are of a rigid-type (e.g., airforks for tapered internal surfaces), or are adjustable for accommodating different hole diameters within a specific measuring range (Fig. 3-6.11). Gage heads with retractable fingers (Fig. 3-6.12), are used for measuring recessed internal surfaces, such as the tracks of ball-bearing outer rings.

Double point gage heads entering from the rear, through the hollow work-head spindle, are used (Figs. 3-6.13 and 3-6.14) for gaging the actual diameter of very small diameter holes which do not provide sufficient space even for the slender contact fingers of electronic gages to be introduced alongside the grinding wheel. Such gages are commonly installed on the rod tube which otherwise is used for the solid plug gage and oscillates synchronously with the grinding spindle. Because the gage fingers enter the work hole by following the traverse movement of the grinding wheel, the space limitations that restrict the use of fingers introduced from the front (parallel with the wheel) do not affect the use of probe forks which penetrate into the hole from the back of the work-piece. Although limited in their use to through holes, such gage probes present the further advantage of avoiding interference with the loading and unloading of the work.

While interrupted internal surfaces cannot be measured with air gages, special types of electronic gage heads are available which lock the position of

Courtesy of Marposs Gauges Corp.

Fig. 3-6.13 Measuring head of an electronic bore gage for entering the workpiece hole from the rear by passing through the hollow work head spindle. The gage head, which has two contact fingers, is shown in a totally withdrawn position while being set to the required bore size by means of a bracket-mounted micrometer head.

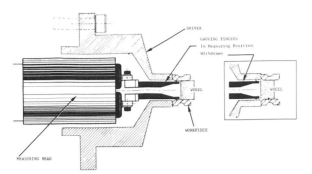

Courtesy of Marposs Gauges Corp.

Fig. 3-6.14 Diagrams illustrating the operation of the electronic gage head shown in Fig. 3-6.13, which permits the two-finger type gaging of very-small-diameter holes without interfering with the automatic loading of the workpiece. The gage operates by moving in synchronism with the reciprocating grinding wheel and is shown in measuring position (left), respectively withdrawn (right).

the sensing fingers when over the ribs and retain that locked position while passing across the axial grooves.

Work size control by wheel dressing, in common shop language known as "diamond sizing," is a relatively simple, yet dependable method of controlling the work size in the automatic internal grinding process. Diamond sizing is applicable to practically any shape of hole, cylindrical; tapered; or profiled; irrespective of whether a through hole or a blind hole must be ground.

Although not a new development, the increased stability and alignment accuracy of modern internal grinding machines and the improved consistency of the grinding wheels are important factors in making diamond sizing applicable for holes with very tight dimensional tolerances. The advantage of diamond sizing is that it provides dependable size control without the use of special in-process gages, which are rather expensive to install as well as to maintain, particularly for small hole sizes which require very delicate gage probes.

Diamond sizing in automatic operations is only applicable on internal grinders designed for that process, i.e., the equipment will perform the different movements which are integral constituents of work sizing with the aid of controlled wheel dressing. Although the elements of the movements and, particularly the mechanisms by which these are carried out, differ in various makes and models, the principles are common enough to consider the diagram in Fig. 3-6.15 as a characteristic illustration of the process.

In essence, the distances and the rates of the infeed movements are set to fixed values which will, after the wheel enters the hole, first produce a rough grinding infeed, then back off and retract the wheel. Then follows a "compensation" movement over a distance which should (a) counterbalance the expected wheel wear during the rough grinding, and (b) provide for the penetration of the dresser diamond to a depth sufficient to produce a geometrically regular and freely cutting wheel face. That is brought about by the wheel passing along the diamond which is set at a specific distance from the finish size of the hole surface. Thereafter, the wheel again enters the hole and advances over a fixed distance for finish grinding the hole to the required diameter.

Beside these basic movements, described for explaining the principles of the process, various other elements may be incorporated into the automatic sequence, such as retraction after dressing, to produce a clearance for avoiding diamond drag on the wheel, compensated by a corresponding make-up addition to the infeed travel.

Grinding of Flat Surfaces on Internal Grinding Machines

The grinding of flat surfaces in positions essentially or exactly normal to the axis of the part, can be carried out on many models of internal grinding machines. The actual application of internal grinders

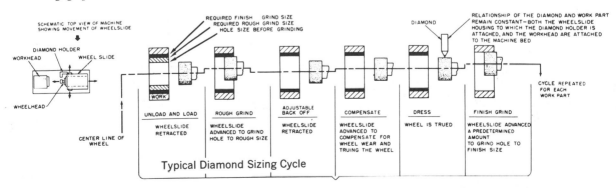

Courtesy of Bryant Grinder Corp.

Fig. 3-6.15 Block diagram describing the operation of a typical diamond sizing process in internal grinding.

for producing flat surfaces is, however, limited to cases in which particular conditions warrant the substitution of internal grinders for the appropriate type of regular surface grinding machines. Examples of such cases are:

(a) In toolroom work for flat grinding operations which are carried out occasionally only, using the internal grinder as an acceptable, although not equivalent, substitute equipment

(b) For work surfaces on which a concentric grinding pattern must be developed, requiring special, and not always available, types of surface grinding machines

(c) When the flat surfaces being ground must be very precisely perpendicular to a hole; a requirement which may be easier to satisfy by grinding the two mutually interrelated surfaces in a single setup

(d) Grinding ring-shaped flat surfaces in recessed locations, such as at the bottom of a hole which cannot be reached by the wheel of any regular type of surface grinding machine.

There are two distinctly different methods by which flat surfaces are produced on internal grinding machines:

(a) *Rotary surface grinding* with the periphery of the grinding wheel mounted on the spindle in the wheel head (see Fig. 3-2.8). For carrying out this operation the work head is swiveled on its base by 90 degrees (on machines designed with such a wide range of adjustment) and the workpiece is mounted in a chuck or similar holding device which properly exposes the work face being ground. The grinding of the surface is carried out by the reciprocating movement of the table over a stroke length sufficient to assure that the periphery of the wheel traverses over the entire radial width of the rotating work surface.

In grinding flat surfaces, the diameter of the applicable wheel is not limited, therefore, heavy wheel spindles of the integral type are used for such operations. Because of the essentially linear contact between the wheel and the work surface, grinding wheels with harder bonds than are usually recommended for internal grinding are used for such operations.

Ring-shaped surfaces which are not exactly perpendicular to the work axis but are inclined to it by a specific angle (wide-angle cones) can also be produced by this method. To produce such conical surfaces, which may be either convex or concave, the work head is swiveled over an arc, less than 90 degrees, equal to the amount of the cone incline angle. For producing a convex (protruding) cone the wheel traverses from the center of the work towards its periphery away from the wheel head, while con-

Courtesy of Voumard Machines Co., Stoffel Grinding Systems

Fig. 3-6.16 Internal-grinding machine equipped with a retractable face-grinding attachment, which has hydraulically actuated swivel movement and can be operated either by manual control or in an automatic cycle.

cave (receding) cones are ground by traversing from the work center toward the wheel head.

(b) *Surface grinding with the face of the wheel.* Various makes of universal internal grinding machines can be equipped with special devices for face grinding as a supplemental operation (see Figs. 3-6.16 and 3-6.17). In its general arrangement, with a base mounted on a major member of the machine and carrying the grinding spindle on a swing arm, the attachment resembles the internal grinding devices with which universal type cylindrical grinding machines are equipped.

However, on internal grinding machines the face grinding attachment is, for operational reasons, mounted on the work head instead of the cross slide

Courtesy of Voumard Machine Co., Stoffel Grinding Systems

Fig. 3-6.17 Close-up view of the face grinding attachment, installed on an internal-grinding machine and equipped with cup wheel for effective face grinding.

of the machine, consequently, it must contain an individual feed device for face grinding with the cup wheel. The illustrated type of attachment is available either with automatic or manual feed control, both for grinding and wheel truing. The swing movement of the attachment and the retraction are controlled hydraulically.

Grinding flat surfaces with the face of the wheel, usually a cup wheel, on internal grinding machines offers several advantages, such as:

(1) Internal and face grinding can be carried out in a single operation, assuring excellent geometric interrelation (squareness) of these two surfaces

(2) Surface grinding with the face of a cup wheel on a rotating workpiece produces a surface pattern (cross hatch) which may be desirable for functional or other reasons

(3) Cup wheels, particularly of the flaring cup type can reach surface areas which are inaccessible to peripheral surface grinding

(4) For grinding narrow, ring-shaped flat surfaces, straight wheels with a recessed or relieved face may be used, often grinding the adjacent hole surface with the periphery of the same wheel which can be mounted on the regular internal spindle without the need for a special attachment.

The Setup and Operation
of Internal Grinding Machines

The particular conditions which characterize the grinding of internal surfaces affect many of the procedures required for the machine setup and operation; the pertinent actions differ in various respects from those generally followed for the grinding of external surfaces. A few examples of such conditions which call for the very rigorous control of the machine setup and of the operational variables are:

(a) The mounting of the grinding wheel at the end of an extending, frequently rather long spindle tends to amplify the effects of misalignments

(b) The confined location of the grinding area usually prevents the operator from observing the progress of the grinding and from becoming aware, during the process, of any corrections which may be needed for assuring proper operations

(c) The proneness to deflection of the generally flexible wheel spindles under the effect of grinding forces, particularly when they are in excess of the anticipated amount due to improper work preparation, grinding data, wheel conditions, and other operational variables.

Because of the importance of the correct setup, the manufacturers of internal grinding machines, particularly of models built for automatic operation, usually supply detailed manuals for the setup and for the operation of such equipment. While there would be no point in excerpting, or even duplicating such information, a few important and rather generally applicable aspects of setup and operation will be discussed in the following. That discussion is intended to indicate some major areas of criticality and also to supply information which is valid for most of the in-

ternal grinding operations, irrespective of whether they are carried out singly or in continuous production.

The order of listing does not reflect the sequence in which these items have to be taken care of, nor should the aspects which are not mentioned be considered of secondary importance.

Work Preparation and Stock Allowances for Internal Grinding

Successful internal grinding operations, particularly in automatic processes without an operator in attendance, are strongly dependent on proper work preparation. The workpieces presented to internal grinding machines must satisfy two major requirements:

(a) Assure correct and repeatable location of the work in relation to the hole to be ground, with respect to both the centering and the orientation (alignment) of the workpiece

(b) Provide a proper amount of stock on the hole surface to avoid interference with the wheel when it enters the hole, and also provide sufficient material for the grinding to produce a clean and sound work surface of the specified form and size.

The stock left on the work surface for removal in the grinding operation, generally termed "grinding allowance," affects several technological aspects of the operation, as well as the required grinding time. The proper planning of the stock to be left in the primary operation for the subsequent grinding is particularly important in the case of internal surfaces in order to reduce the unfavorable effects of this method's inherent process limitations.

The amount of stock to be removed in internal grinding very closely determines the time required for the grinding phase of the operation. This is due to the narrow limits within which the rate of stock removal in internal grinding can be varied without harmful effects on the accuracy (e.g., due to spindle deflections), the integrity, the finish, etc., of the ground work surface.

In the majority of cases it is more economical to remove stock from an internal surface in the preceding operations, usually before hardening of a workpiece made of steel, and to leave only the predetermined minimum amount of stock on the work surface for internal grinding. For the same reason, a pregrinding condition which permits dependable work centering will help to avoid substantial stock allowance as a means of compensating for deficient work location. In high-volume automated production the centering of the work results from the positional and form accuracy of the locating surface, usually the outside diameter of the workpiece. That condition must guide the selection and execution of the operations which precede the internal grinding.

Excessive variations in stock allowance can also be detrimental to the efficiency of the internal grinding operation, particularly when carried out in a preset cycle which has to be established by providing for the worst conditions, that is, the maximum stock allowance. For that reason it may prove advantageous to carry out the semifinishing to close limits even when such requirements add to the cost of the preceding operation, usually executed on an automatic lathe or similar equipment.

In this respect, mention should be made of the effects of technological advances on the optimum processing data for internal grinding. High-speed grinding, stock position sensors, adaptive controls, etc., have substantially raised the efficiency and adaptability of internal grinding. By reducing the sensitivity of the process to variations in stock allowance, the formerly observed rigorous work-preparation requirements can often be relaxed. Consequently, in internal grinding processes using high performance machines with adaptive controls, it may be economically advantageous to permit more liberal tolerances for the semifinished work, to be thus processed by highly efficient primary methods, such as cold, or even warm, forming.

Although the most economical and still technologically satisfactory amount of stock left for internal grinding is controlled by many factors which must be evaluated in their combined effect, general guidelines on stock allowances may be used as a first approach in designing the process, as well as for time estimating purposes. A graph indicating the generally recommended stock allowances for internal grinding is presented in Fig. 3-7.1. The graph contains several curves showing the amount of stock allowance as a function of the hole diameter, plotted for different ratios of hole diameter to hole length. Although the length of the hole has only a minor effect on the amount of the required stock allowance (as indicated by the closeness of the different curves in the diagram), the consequences of spindle quill deflection on the safe amount of stock allowance should not be disregarded.

The Selection of the Applicable Work-Holding Device

The three-jaw chuck is, in principle, the most universally applicable work-holding device for internal grinding. Nevertheless, in actual operations the three-jaw chucks in their basic design, are rather rarely used for holding work on internal grinding machines. A listing of the main conditions which limit the actual use of that essentially universal work-holding device may serve as a review of the considerations which will have to guide the selection of work-holding methods and tools for internal grinding.

(a) Three-jaw chucks are designed for universal applications and provide a wide capacity range but because of their design characteristics, their *centering accuracy* is inadequate for many types of precision work carried out on internal grinders.

(b) Three-jaw chucks, designed originally for lathe operations, can exert substantial pressure for holding the work, but are not adapted to the *sensitive control of the clamping force.* Internal grinding generally requires a small holding force only, while the application of excessive pressure can cause form distortions and surface damage.

(c) The concentrated force applied by a chuck contacting the work at three points only is another condition which may leave marks on the finished work surface and/or cause distortions, whereas devices with *large clamping surfaces* generally avoid these harmful consequences.

(d) *The operational time* required by hand-actuated chucks for clamping and releasing the work is much longer than that needed by holding devices which open by a small amount only, just enough for accepting and permitting the removal of the work which, for internal grinding purposes has an already finished outside surface.

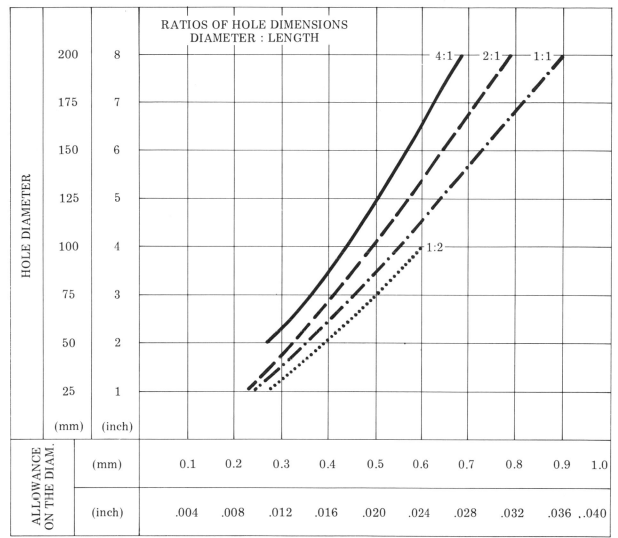

ALLOWANCE ON THE DIAM.	(mm)	0.1	0.2	0.3	0.4	0.5	0.6	0.7	0.8	0.9	1.0
	(inch)	.004	.008	.012	.016	.020	.024	.028	.032	.036	.040

The indicated values are predicated on parts: mounted in well-centered position; out-of-roundness is less than 1/10 of the allowance; free of major distortions (such as from heat treatment); and without significant metallurgical defects (e.g., deep decarb.)
(The presence of unfavorable conditions may require increased stock allowances.)

Fig. 3-7.1 Generally recommended stock allowance on the diameter of holes finished in internal grinding.

(e) The bulk and mass of the three-jaw chucks is not conducive to *vibration free rotation* when running at high speed such as is needed for producing optimum work peripheral speed in grinding small-diameter holes.

(f) *The adaptability to automatic work loading and discharging* is an essential requirement in many types of internal grinding operations, a condition which cannot be satisfied with regular three-jaw chucks.

(g) *The capacity to accept workpieces of irregular configuration* or with special locating requirements is not assured with the basic types of three-jaw chucks.

In view of the above indicated and similar requirements which characterize the mounting of workpieces for internal grinding operations, a wide selection of different holding devices, including three-jaw chucks in special high-precision design, has been developed. Examples of such devices are discussed in the preceding chapter, in the selection of which functional requirements such as listed above, will have to be evaluated.

Alignment of the Operating Elements of Internal Grinding Machines

Alignment of the machine elements and of their movements is not a condition which, in general, can be established or corrected in the course of the machine setup. The proper conditions of alignment must be assured by the design and in the building of internal grinding machines. However, it is possible during the test runs checking the correctness of the setup, to detect the presence of alignment errors by their effect on the performance accuracy of the grinding machine. Furthermore, the effect of alignment errors can be measured and their origin associated with specific machine elements, should such a detailed examination of the machine conditions be needed.

The reason for discussing the alignment requirements of internal grinding machines is the importance of that condition with regard to the attainable accuracy of the internal grinding operation. A further purpose is to clarify some misconceptions still prevalent, concerning the nature, the consequences, and the possible remedies of alignment errors of internal grinding machines.

The basic conditions required for the grinding of a hole of regular cylindrical form are shown in Fig. 3-7.2, which illustrates the following relationships:

(1) A common plane contains: (a) the axis of the grinding wheel and also the position of the dresser diamond, and (b) the axis of the workpiece.

(2) Within that common plane both axes of rotation are parallel with each other.

(3) Also parallel with these axes is the path of the traverse movement of the wheel, both within the grinding area and along the diamond point.

Any deviation from these basic conditions of mutual locations and of the path of movement will interfere with producing an accurately cylindrical shape of the ground hole.

It is evident from this list of requirements that while the positioning of the mutually related elements into a common plane may be a matter of adjustments, some of which can be taken care of in the setup, the parallelism of the paths of movement is a condition inherent in the machine.

The guideways of the moving machine members must be straight and parallel (or perpendicular for the infeed movement) within an accuracy which will produce the required control of the movement. It is for that reason that the manufacturers of internal grinding machines have developed guideway designs capable of maintaining the very accurate positions and orientations of the moving machine members along their entire travel and over a considerable period of service. A few of the guideway systems used in high-precision types of internal grinding machines have been discussed in Chapter 3-2.

Deviations from the described alignment conditions will cause proportional distortions of the produced internal surface, commonly resulting in holes which, instead of being cylindrical:

(a) Will be tapered (when the path of the wheel contact is in the plane of, but not parallel with, the axis of the hole).

(b) Will be bell mouthed (either symmetrical or asymmetrical) when the path of the wheel contact is in a plane intersecting that of the hole axis (see Fig. 3-7.3).

An important requirement of efficient and accurate internal grinding is also the form of the grinding wheel which is generally cylindrical, except for the rare applications of internal plunge grinding of

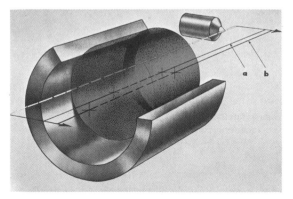

Courtesy of Bryant Grinder Corp.

Fig. 3-7.2 Basic locational relationships in internal grinding, with the wheel axis (a), the work axis (b), and the dresser diamond point contained in a common plane.

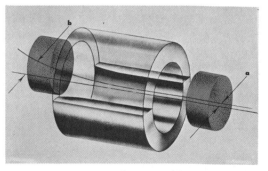

Courtesy of Bryant Grinder Corp.

Fig. 3-7.3 Bell mouthed hole resulting from misalignment between the axis of the work and the path of the reciprocating wheel. The difference between *a* and *b* indicates, in a highly exaggerated manner, the divergence of the wheel path from the hole axis plane.

forms. In order to produce a wheel of cylindrical form by truing, the grinding wheel must contact the diamond along a line which also contains the axis of the wheel. When these two elements are in mutually inclined planes, the produced wheel shape will not be a cylinder but a figure with concave side elements (as shown in Fig. 3-7.4). The actual form of the concave-sided wheel can be symmetrical, or a concave-sided taper with either the leading or the trailing end smaller, depending on the position along the wheel where the two planes—containing the wheel axis and the line of diamond contact, respectively—intersect. Wheels of this shape will not cut along their entire width, and the work surface section contacted only by the concave portion of the wheel surface, will be finished to a smaller diameter than the rest of the hole.

Courtesy of Bryant Grinder Corp.

Fig. 3-7.4 Grinding wheel dressed to concave shape and the work surface it produces when the axis of the wheel *a*, is inclined in relation to the plane containing the path *b* of the wheel, along the diamond point.

The proper height setting of the three participating members—the work axis, the wheel axis, and the diamond point, while not a matter of alignment, is a related requirement. Improper conditions, which may be corrected in the setup, affect the size-holding accuracy of the operation when diamond sizing is used. Figure 3-7.5, A and B, illustrates, diagrammatically, two common errors in the mutual height setting of these members: (A) The plane containing the wheel axis and the diamond is separated from the plane of the work axis; as the wheel diameter decreases, the ground hole will become progressively smaller. (B) The diamond point is displaced with respect to the plane containing the axes of the work and of the wheel; as the wheel diameter decreases, the hole diameter will gradually become larger. These consequences will occur regardless of which direction the mutual displacement of the regularly coinciding planes has taken place.

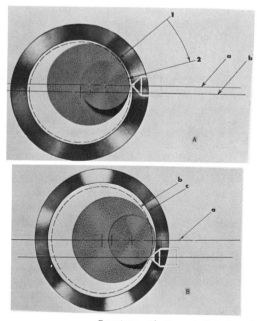

Courtesy of Bryant Grinder Corp.

Fig. 3-7.5 Changing wheel diameter interferes with the control of the bore diameter in diamond sizing when the center planes of the work, the wheel, and the diamond point are not common. A. Wear of the wheel will cause decreasing hole diameter due to the shifting of the point of their mutual contact from 1 to 2. (*a*—Plane of the wheel center, *b*—Plane of the work center.) B. Wear of the wheel causes increasing hole diameter. (*a*—Plane of the work and wheel axes, *b* and *c*—Hole diameters resulting from different stages of wheel wear.)

Essentially, the same conditions of form and size control, to be assured by the proper alignment and positional relationship, apply to the grinding of tapered holes which are produced by swiveling the work head at an angle corresponding to one-half of the included angle of the taper.

Finally, the uncommon case of deliberately created, yet precisely controlled misalignment, as a means of achieving particular form-generating objectives, should be mentioned. Specific types of contructional elements must be manufactured with holes the side elements of which have a very small, but accurately controlled convexity. Such configuration is called "crowning," and its purpose is to assure uniform stress distribution when an anticipated load is applied against the surface. Typical examples are the raceways of outer rings for roller bearings, both cylindrical and tapered. The described contour of the ground inside surface can be produced by utilizing the relationships just discussed, between the alignment of the machine elements and the hole form produced by internal grinding. By a slight, but precisely calculated tipping of the work head, a controlled misalignment can be created which will cause the wheel to grind a bell-mouthed hole (see Fig. 3-6.5), although, with a contour convexity of very small magnitude, e.g., 0.0002 inch over 1 inch bore width (0.005 mm over 25 mm bore width). For assuring the symmetrical location of the convex bore contour, the path of the wheel travel must intersect the axial plane of the workpiece in the middle of the hole width.

Coolant Application in Internal Grinding

The unfavorable conditions under which the internal grinding process must be carried out increase the importance of adequate coolant (grinding fluid) application. The properly directed and dispensed coolant can substantially alleviate the adverse effects of such inherent process characteristics as an enclosed work area, relatively small wheel surface, large area of contact between the wheel and the work, etc.

The purpose of the coolant application in internal grinding is essentially identical with that in other grinding process, that is, cooling; lubrication; washing off the chips; rust inhibition; etc. However the proper coolant application has an added importance due to certain particular conditions of internal grinding, which are here briefly reviewed:

(1) Cooling of the contact surfaces must be more intensive because any point on the wheel surface enters into operational contact at much shorter intervals than in external grinding, where a much lower rotational speed is required for a comparable peripheral speed.

(2) Chip removal action of the coolant from the work area with its wraparound work surface, can only be effective when the coolant pressure and volume are adjusted to these conditions. Unremoved chips and grits will churn in the hole area and may even damage the ground work surface.

(3) The free space through which the coolant can circulate, including also introduction and draining off, is generally restrained, particularly in small holes for which wheels closely approaching the hole diameter are often used.

In order to cope with such adverse conditions by adapting the coolant application the following requirements must be satisfied:

(1) *The distribution* of the coolant in the work area by properly dimensioned and arranged nozzles must assure that:
(a) The coolant flow extends over the whole width of the wheel
(b) The coolant penetrates into the wheel contact zone

(c) In traverse grinding the entire length of the traverse path is covered; also the coolant reaches the work surface ahead of the cut
(d) In plunge grinding, in addition to the contact area, the coolant is also applied to the sides of the wheel.
(2) *The pressure* of the coolant must be sufficient:
(a) To penetrate into the often very narrow wedge-shaped space between the wheel and the work surface
(b) To break the barrier of the high-pressure air surrounding the wheel which is rotating at high speed
(c) To clean out the chips which tend to be lodged in the space between the grains of the wheel.
(3) *The volume* of the coolant supply must be adequate to provide every nozzle of the distribution system with the amount of coolant needed for covering the operating areas, assuring appropriate heat absorption, and also an effective washing away of the chips and the dislodged wheel particles.

While equipping the internal grinding machines with an adequate coolant supply system is generally the responsibility of the machine tool manufacturer, the selection and adaptation of the nozzles to suit the dimensions and configuration of the work, as well as the assurance of the system's proper functioning, are important activities of the setup and operation of internal grinding machines.

Common Faults in Internal Grinding

Precision type metalworking methods, such as internal grinding, which involve a series of coordinated movements, both concurrent and sequential, are prone to operational imperfections, particularly when the performance is appraised in the perspective of rigorous dimensional and geometric tolerances.

Problems in the operation of any machine tool can, of course, arise from circumstances which are unrelated to the particular metalworking method whose implementation is the intended purpose of the equipment. The possibility of a machine malfunctioning increases as its controls become more sophisticated. That trend toward sophisticated design is distinctly present in modern, production type internal grinding machines which are built with an advanced degree of automation, operate at particularly high speeds, and have very sensitive means for size control. To counterbalance the inherently higher incidence of improper function of a more complex system, many technological advances are incorporated into the design of modern internal grinders, thus assuring sustained trouble-free operation, easy defect identification, and quick fault correction.

Table 3-7.1 Examples of Common Faults in Internal Grinding

AREA OF DEFICIENCY	EXPERIENCED DEFECTIVE CONDITION	PROBABLE CAUSES
Size and Form Control	OUT-OF-ROUNDNESS Distorted shape, lobing, circumferential waviness.	Distortion of the work due to excessive or incorrectly applied clamping action of the work-holding device Overheating of the work, causing nonuniform expansion which affects the clamped section to a lesser degree Runout of the workhead spindle Vibrations and/or deflections of the grinding wheel quill
	TAPERED HOLE instead of a straight one, or taper of incorrect angle.	Alignment errors or improper swivel adjustment of the workhead Nonuniform breakdown of a wheel which may be too soft Unbalanced overtravel of the wheel at the two ends of the hole
	BELLMOUTHED HOLE Larger at the end(s) than in the center.	Misalignment (tipping) of the workhead Excessive overtravel of the wheel at the end(s) of the hole.
Work Surface Conditions	CHATTER MARKS High-frequency waviness of the ground surface.	Wheel spindles or quills too long and/or too thin; may have to be made of special rigid material Vibrations in the grinder or transmitted from neighboring equipment Improper belt tension; may be too tight Pulleys not properly aligned Belt in faulty condition or not designed for the speed at which it is operated Out-of-balance condition of the wheel or of the pulleys.
	FEED LINES Marks caused by the reciprocating wheel.	Improperly, too coarsely dressed wheel Too-hard wheel, may need breaking of the corners as an expediency Incorrect alignment due to tipped wheel head.
	THERMAL DAMAGE Burning, heat checks, discoloration of the ground surface.	Too hard wheel, actually or apparently. The latter caused by too low work speed Wheel too fine or with too dense structure Excessive infeed rate Insufficient or inadequately directed coolant Incorrect wheel dressing caused by a worn diamond or by dull or glazed wheel
	DAMAGING OF THE WORK SURFACE Scratches or wheel marks.	Improper dressing, using high feed or faulty diamond Dirty coolant Too soft wheel bond, releasing sharp grains which are caught between the wheel and work Wheel with too coarse grains.

(Continued on next page)

Table 3-7.1 (*Cont.*) Examples of Common Faults in Internal Grinding

AREA OF DEFICIENCY	EXPERIENCED DEFECTIVE CONDITION	PROBABLE CAUSES
Grinding Wheel Condition	LOADING OF THE WHEEL Pores partly filled with lodged chips	Wheel too fine and/or structure too dense Wheel too hard, retains dulled grains; temporary help by increasing the traverse rate Insufficient coolant flow Wheel dressing with too-fine feed or dull diamond; impedes free cutting action.
	GLAZING OF THE WHEEL Detectable by shiny appearance.	Infeed too light, causing insufficient pressure on the wheel Excessive lubricity of the coolant Improper wheel specifications; too fine, hard, or dense Dressing produces inadequate sharpness of the wheel face.
	BREAKAGE OF THE WHEEL	Excessive wheel speed, beyond the specified limit Improper wheel mounting, without or with wrong washer, excessive clamping, uneven locating surface on the quill. Overheating of the wheel by inadequate cooling and too-heavy feed Out-of-balance wheel, loose or improperly centered bore Mechanical damage, wheel jamming into the workpiece.
	IMPROPER TRUING AND DRESSING	Inadequate diamond, dull or chipped Wrong height setting of the diamond Unstable diamond support, loose setting, insufficient rigidity of the dresser Improper truing data with respect to the depth of cut and feed rate.
Operation of the Machine	SLIPPING OF THE DRIVE BELTS	Belts supported by pulleys with substantially different diameters, particularly when with short center distance, are sensitive to the presence of oil and dirt Belts operated at high speed must be made of special material, woven endlessly Multiple Vee belts must be exactly matching in dimensions and condition of wear Belt tension may be (a) too loose, causing: whipping, slippage, vibrations; (b) too tight, causing: accelerated wear of the belt and spindle bearings.

(*Continued on next page*)

Table 3-7.1 *(Cont.)* Examples of Common Faults in Internal Grinding

AREA OF DEFICIENCY	EXPERIENCED DEFECTIVE CONDITION	PROBABLE CAUSES
Operation of the Machine	FAULTY WHEEL HEAD OPERATION Vibrations or overheating	Excessive speed, possibly due to errors in the selection of pulleys Mechanical damage caused in handling Wear in the bearings Insufficient lubrication or use of inadequate oil. *NOTE*: In the case of damage or wear, expert repair under controlled conditions is needed.
	UNEVEN TRAVEL OF HYDRAULIC TABLE	Air pockets in the feed lines when machine was standing idle. May disappear following repeated reversals of travel. Insufficient oil pressure or volume; also dirt in the oil. Leakage, improper sealing. Mechanical damage, such as of the piston rod. Dirty or improperly lubricated table ways.

Typical examples of that development are the use of solid state electronic controls, defect spot-identification light signals, modular control boards, the grouping in easily accessible location of all hydraulic control elements, and many others.

However, the subject of this discussion is these potential faults in the operation or in the resulting workpiece, which are related to the internal grinding process but are essentially independent of the model of grinding machine on which they are carried out. A review of the common faults and of their potential causes can be helpful in identifying and analyzing such conditions as a basis for corrective action. Furthermore, the realization of the causes of potential trouble should permit the institution of preventive measures as a means of avoiding, or at least reducing, the incidence of a defective internal grinding performance.

Of course, it is not possible to present an all-inclusive listing of faults, nor can the actual causes be pointed out in a generally valid manner. The enumeration of potential faults must, therefore, be limited to such as are considered common and the listed causes are only probable ones, serving to assist in the determination of, but not intended to substitute for, an actual diagnosis.

With due regard to these considerations a tabulated listing of common faults and of their probable causes is presented in Table 3-7.1. In order to permit easier identification, the listed examples have been grouped into four categories, reflecting the general nature of the deficiency. That grouping, however, does not represent a rigid delineation of fault sources and its sole purpose is to facilitate the association of problems with probable areas of origin.

The Grinding Wheels and Operational Data of Internal Grinding

Grinding Wheels for Internal Grinding

In internal grinding, due to the wraparound effect of the shape of the work surface, an extended area of contact will result between the wheel and the work, as illustrated in Fig. 3-8.1. This diagram, with proportions exaggerated for reasons of clarity, also shows that generally, the relative speeds of the wheel and the work have a direct effect on the dimensions of the chip removed by a single abrasive grain. The faster the relative work speed, the longer the section of its surface will be presented to the individual grains of the grinding wheel while they are passing along the arc of contact.

The geometric shape and the dimensions of the theoretical arc of contact are the result of the relative diameters of the grinding wheel and of the work surface, as well as of the depth of cut. The chip thickness is also controlled by the depth of cut. In traverse grinding, still another factor affects the grinding conditions—the reciprocating speed of the wheel relative to the rotational speed of the work. This relationship will control the "lead" of the wheel path, extending the effective length of the contact arc, when compared with conditions in the plunge type internal grinding.

The described conditions thus affect the requirements of the grinding wheels used for internal grinding, the composition of which will reflect the following characteristics:

(a) Coarser grain and open structure to provide more space for the retainment of chips before these can be released at the end of the grain's engagement with the work surface

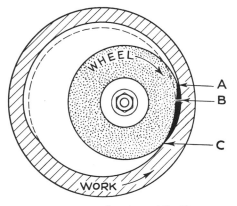

Courtesy of the Norton Company

Fig. 3-8.1 The principles of grinding wheel penetration into the internal work surface, shown diagrammatically as the chip removed by a single grain. The dimensions are highly exaggerated and the proportions distorted for explaining the concept. While the grain on the wheel surface moves from A to C, the work surface, rotating at a slower speed, will advance from C to B, thus bringing additional surface areas into engagement with the grain.

(b) Softer bond to release the dulled grains more easily, because the extended area of contact reduces the natural dressing effect of the work surface. Exceptions in this respect are interrupted surfaces, such as bores with keyways, which act as wheel dresser tools. The long arc of contact, particularly in small holes ground with relatively large wheels, compounded by the difficulty of introducing a sufficient amount of coolant into the restrained work area, add to the importance of a cool cutting action, accomplished by replacing the dulled grains by new, sharp ones.

Sequential Steps in the Selection of Grinding Wheels for Internal Grinding

Wheel diameter and width. Extending wheel life is desirable for reasons of economy as well as productivity which is affected by the frequency of wheel changes. Therefore, wheel diameters close to that of the hole are selected, particularly for the grinding of small-size holes. A larger wheel diameter, with proportionately larger periphery, is also beneficial with regard to the size and form-holding performance of the operation. On the other hand, a gap between the wheel and the work surface which is too narrow hampers the introduction of the coolant and can also prevent the penetration of the probes or sensing fingers of in-process gages. Therefore, in certain applications, the need for sufficient clearance space between the surfaces of the wheel and the work may override the advantages derived from the use of the largest possible wheel diameter.

According to currently accepted recommendations, the following relationships between the diameters of the hole and the wheel are often used as guidelines:

(1) Up to 7/8 inch (about 22 mm) hole diameter: grinding wheel diameters equal to that of the nominal hole size are chosen, and then the wheels are cut down by dressing to a size smaller .040 to 0.120 inch (about 1 to 3 mm) to fit into the hole which, at this stage, is still smaller by the amount of the grinding allowance than its nominal diameter.

(2) From 1 to 2½ inches (about 25 to 60 mm) hole diameter the new wheel is smaller by .150 to .250 inch (about 4 to 6 mm) than the nominal diameter of the hole.

(3) For holes up to 10 inches (about 250 mm) wheels in sizes equal to about 3/4 of the hole diameter are chosen.

(4) For still larger holes, the controlling factor is usually the capacity of the grinding wheel spindle which limits the applicable wheel diameter.

The *width* of the grinding wheel usually has the following values for range limits:

(1) Wheels 1 inch (about 25 mm) diameter: 3/4 to 1¼ times the wheel diameter.

(2) Wheels to 2 inches (about 50 mm) diameter: ½ to ¾ times the wheel diameter.

(3) Wheels to 3 inches (about 75 mm) diameter: ¼ to ½ times the wheel diameter.

(4) For larger wheels the relationship between diameter and width becomes less distinct, the capacity of the spindle and of the drive motor become the controlling factors.

As an example: for wheels in the diameter range of 6 to 7 inches (about 150 to 175 mm), the recommended wheel width range is 1 to 2 inches (about 25 to 50 mm).

The abrasives of internal grinding wheels, similarly to those used in other grinding methods, are chosen to suit the work material. Because internal grinding is applied, in the majority of cases, to workpieces made of steel and ground in hardened states, aluminum oxide wheels are generally recommended. The relatively rare cases in which silicon carbide wheels are used for internal grinding, include workpieces made of nonferrous metals, nitriding type steel prior to nitriding, and glass and sometimes, the rough grinding of cemented carbide. One application for silicon carbide for internal grinding of steel is the fine finishing of gages with very fine grain (500) wheels. For grinding holes in cemented carbide dies, diamond wheels are generally considered the best choice.

Aluminum oxide is used in different grades of friability, selected according to the work material and the operational objectives, balancing the sometimes conflicting requirements of substantial stock removal, cool cutting action, good size and form holding properties, etc.

The bond of aluminum oxide and silicon carbide wheels used for internal grinding is usually a vitrified material which effectively withstands the high temperatures often generated in that process. Organic bonded wheels, such as resinoid or rubber, are only used when the accomplishment of very high finishes is one of the major process requirements, even at the cost of having to grind with very light cuts. The relatively large area of contact between the wheel and the internal work surface calls for a soft *grade* bond. For hardened steel the generally recommended grades are "K" for small bores and "L" for larger holes. Even softer grades, such as "J," are used for very hard materials, while for soft work materials internal grinding wheels with grade "M" may provide the best performance. These grades are intended to serve as approximate mean values, considering that both the grading scales of different wheel manufacturers and the operational conditions of internal grinding operations vary.

The grain size is adjusted primarily to the rate of intended stock removal and the required work surface finish. A higher rate of stock removal calls for coarser grain, varying from 46 for soft, to 60 for hardened steel. Fine finishes commonly require fine grains in the 80 to 100 range, such as for blanking and drawing dies. For finishing cemented carbide dies with diamond wheels, 220 is the recommended grain size.

On the other hand, in general operations on hardened steel, high grade commercial finishes can be obtained with medium (54 or 60) grain size wheels, by dressing with slow diamond advance and taking light cuts in the final passes of the operation.

Chemically Treated Grinding Wheels for Internal Grinding

Sulfur treated grinding wheels are sometimes used for internal form grinding. By providing improved grinding lubricity, preventing the chemical welding reaction between the metal and the abrasive, and reducing the loading of the wheel pores, the treated wheels improve the chip formation, assure cooler cut and prolong the form holding of the wheel face. Sulfur treatment, however, besides adding to the cost of the grinding wheel, also has other drawbacks, such as contamination of the coolant, unpleasant odor, and reduced life of the dressing diamond. Consequently, a judicious evaluation of the benefits and the possible disadvantages of using grinding wheels treated with sulfur impregnation, or by some other processes: e.g., wax or a proprietary chemical, should precede the pertinent selection.

Precision Mounted Wheels

For the internal grinding of parts with hole diameters of about ½ inch (about 12 mm) or smaller, mounted wheels are generally used. These wheels are cemented on the knurled end of a mandrel which, for the precision type of grinding operations, is made with shank diameters held to a tolerance of 0.003 inch (0.0075 mm), or better. Popular mandrel diameters are 1/8 − .0001 in. = .1249 in. and 3/16 − .0001 in = .1874 in. The wheel diameters are held to 0.001 inch and the runout must not exceed 0.001 inch. Equivalent metric shaft diameters are 3 and 5 mm.

Various styles of mandrels are used, such as (a) tapered shoulder type, (b) shoulder type, and (c) double shoulder type, all with cylindrical shank (see Fig. 3-8.2). For larger wheel sizes, threaded flange type mandrels are often selected which offer some cost savings. For small diameter wheels, usually those less than 0.095 in. (2.5 mm), the knurled portion of the mandrel should support the wheel over its full length, to avoid flaking, while for larger diameters partial wheel support is admissible.

For manufacturing reasons a minimum wall thickness must be maintained for mounted wheels, a dimension determined by the simple formula of $\frac{D-B}{2}$, where D = the outside diameter of the wheel

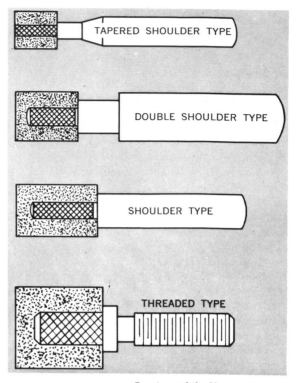

Fig. 3-8.2 Frequently used types of cylindrical precision mandrels for mounted wheels.

and B = its mounting hole or recess diameter. The minimum wall thickness varies according to the grit size. Values are listed in the following table which may be used as a guide. For mounted wheels the letter M is added to the wheel specification symbol.

Table 3-8.1 lists the recommendations of the Norton Company for the grit sizes and grades of mounted wheels in various common applications.

Operational Data for Internal Grinding

Internal grinding machines are often used to handle a wide range of work sizes, a condition which would require a comparable range of work speeds. Actually that requirement is seldom met, and often not even approached, as indicated by the example of a widely used model of internal grinding machine. That grinder will accommodate work with hole diameters from 1/8 to 5 inches, representing a ratio of 1 to 40, whereas the lowest and the highest speed of the work head result in a spread which can be expressed by a ratio of 1 to 3.

Similar, although not such extreme lack of adaptability, may often be found in the commonly used internal grinding wheel spindles of such machines

Table 3-8.1 Grit Sizes and Grade Ranges for Precision Mounted Wheels

GRIT SIZE	USED	GRADE
60 70 80	Toolroom and general purpose work, Parts with hardness R_c .50-60	I, J, and K
90 100 120	Production grinding parts with hardness R_c 58-64	L and M
150 180 220	Smaller wheels Precision mounted wheels in the 0.096 to .500-inch (2.44-to 12.7-mm) diameter range	M to Q
240 280 320	Miniature mounted wheels 0.096 inch (2.44 mm) and smaller (see the following notes regarding grit size limits)	Q to T

NOTE: Grit size limits of the miniature mounted wheels are controlled by the wall thickness of the wheel, which is made with a hole or recess for mounting on the mandrel.

$$\text{Wall thickness} = \frac{D - d}{2} \qquad \text{where} \quad \begin{array}{l} D = \text{outside diameter} \\ d = \text{hole or recess diameter} \end{array}$$

Min. wall thickness	mm	0.53	0.43	0.38	0.33	0.23	0.15	0.13
	inch	.021	.017	.015	.013	.009	.006	.005
Coarsest grit size		100	120	150	180	220	240	320

Recommendation of the Norton Company

with speed ranges often much narrower than the range of the used wheel diameters. (*NOTE:* Such differences must never be permitted to result in higher than recommended wheel speeds.) The discrepancy between the desirable and the actually produced wheel speeds will be further accentuated by the gradual wear of the wheel, a relatively rapid process in internal grinding due to the generally small starting diameters of the wheels.

These are a few highlights of the conditions which tend to reduce the efficiency of internal grinding, unless equipment designed for, or adaptable to the operational conditions is available. For that reason the informative values regarding the optimum grinding data are often difficult to implement, and deviations—even quite significant ones—may have to be accepted. In such cases, variations in the composition of the grinding wheels and of the applied feed data can be used to partially compensate for the unfavorable speed values.

However, in modern equipment, primarily in internal grinding machines built for high volume production, work and wheel speeds which represent optimum values for particular workpieces, are generally provided by means of appropriate equipment design. Other types of modern internal grinding machines serving more varied applications, are equipped with steplessly variable work-head speeds, covering a range which is comparable to the work capacity range of the machine. This development toward assuring optimum speed values is a major factor in promoting certain internal grinding operations into the category of high production methods.

Surface Speed of the Grinding Wheel

Until recently, internal grinding wheels which are commonly made with vitrified bond, have been operated with a maximum surface speed of 6000 to 6500 sfpm (33 to 36 m/sec). Most of the currently used internal grinding machines and the commonly available grinding wheels, are designed to run at that maximum speed level, not to be exceeded in view of the hazards which such procedure may involve.

In the past few years, substantial advances have been made toward applying the principles of *high-speed grinding* to the grinding of internal surfaces as well. Internal grinding machines, specially designed for and actually operating with wheel surface speeds about twice the formerly used conventional values, are receiving increasing acceptance in industries using internal grinding in high-volume production, such as for the precision grinding of the bores and raceways of antifriction bearing rings.

To make the high-speed internal grinding effective and economically warranted, a corresponding increase of the work speed and the reciprocating speed was also needed, with feed rates adjusted to the raised stock removal capacity of the process. Finally, to reap the full benefits of the accelerated grinding process, the speed of the supplementary operations also had to be raised, at least, proportionately.

The Work Speed

The peripheral speed of the work surface should be in a specific relationship to the surface speed of the grinding wheel, in order to assure optimum cutting action. To satisfy that requirement in internal grinding, the speed of the external work surface must always be higher by an amount determined by the wall thickness of the workpiece.

That condition, while barely significant in thin-walled parts of relatively large diameter, may be a limiting factor in obtaining optimum hole surface speed in parts with outside diameters substantially larger than the bore. In view of these conditions, it is advisable to mention here the maximum (external) surface speed of the work at which grinding machines offering a wide range of work-head speeds may be safely operated. A typical value for such external surface speed limit is 400 sfpm (120 m/min).

As a first approach in determining the grinding data, the assurance of a basic ratio of 1 to 80 between the work and the wheel surface speeds, can be selected. When relating this ratio to the 6500 sfpm (33 m/sec) wheel speed, a work surface speed of about 80 fpm (25 m/min) will result. Applying this basic value to a few frequently ground hole diameters, the following *work rotational speeds* can be determined:

Hole diam (mm)	6	12	25	40	60	100	150	200
Equal to about (inches)	¼	½	1	1⅝	2½	4	6	8
Work speed (rpm)	1250	625	310	190	125	80	55	40

Modern, general-purpose internal grinding machines with an infinitely variable work-head speed over a range, e.g., 0 to 500 rpm, will be adequate to provide the optimum work speed for most of the common hole sizes listed above, except for those under ½-inch diameter. When the regular wheel speed is maintained for small-hole diameters, while the surface speed of the work area being ground drops significantly below the optimum value, the grinding wheel will tend to act harder, even to the extent of affecting its performance because of loading of the wheel surface. The harmful effects of such conditions can often be reduced, or even eliminated by selecting a wheel one or two grades softer than the operation would otherwise require, such as using grade K, instead of L.

The Reciprocating Speed in Traverse Type Internal Grinding

When grinding the holes of narrow workpieces, wheel widths are used which permit operation with short strokes, in the order of 1/8 to 1/4 inch (about 3 to 6 mm).

In order for the wheel to cover a distance in excess of the entire length of the hole during each revolution of the part, very high reciprocating speeds are needed in grinding small hole diameters. This is particularly the case when high-speed grinding is applied and the work is rotated at a speed adequate to approximate the 1 to 80 surface speed ratio of the work and the wheel, mentioned above. Accordingly, modern production type internal grinders have maximum reciprocating speeds as high as 400 to 500 strokes per minute.

For longer holes, having lengths which are a multiple of the wheel width, the general principles of traverse grinding will be observed; the traverse speed per work revolution may be kept equal to 1/4 to 1/3 of the wheel width, slower for finishing, and sometimes faster for roughing. To provide the optimum traverse speed, modern, general-purpose internal grinding machines have hydraulically actuated and steplessly variable table speeds in a range of, e.g., 0 to 18 fpm (0 to 5.5 m/min).

The Wheel Infeed

In internal grinding it is a commonly applied practice to remove the bulk of the stock allowance, about 90 to 95 per cent of the total, in the rough grinding phase of the process and then, often preceded by wheel dressing and size compensation, to follow with the finish grinding, terminating the process with an adequate spark-out period.

For traverse grinding the infeed is actuated at the end of each single or double stroke. In plunge grinding, or very short-stroke type traverse grinding, the feed is continuous and its rate is usually determined in relation to the rotational speed of the work.

The incremental infeed values in roughing are from 0.00025 to 0.0006 inch (about 0.006 to 0.015 mm) per pass or per work revolution; less for sensitive, thin-walled parts, and larger for heavy workpieces. In finish grinding these rates are reduced by a factor varying from 2 to 4. In operations where the wheel is being dressed for each piece the dressing is usually applied following the roughing and, in automatic operations, the cycle includes the adjust-

ment of the cross slide to compensate for the wheel layer removed in the dress-off.

The spark-out period, during which no cross-slide advance takes place, will usually extend over 2 to 6 strokes or work revolutions, depending on the amount of the wheel spindle deflection whose effect should thereby be corrected. Longer spark-out periods are used when needed for improving the produced surface finish.

The values mentioned for feed increments apply to the diameter of the workpiece and are obtained by cross-slide movements equal to one-half of the respective amounts.

SURFACE GRINDING

General Purpose and Systems of Surface Grinding

The Role of Flat Surfaces in Design and Manufacture

A flat surface is the most convenient and dependable reference plane for designing either simple or complex figures. In manufacturing, the flat surface serves as a locating base in relation to which essential dimensions and positional conditions can be produced. In the actual manufacturing process a flat surface offers the advantages of positive workpiece location and usually assures dependable work holding.

For these reasons many mechanical parts, both basic structures and component elements, are designed with one or more flat surfaces which serve as dimensional and positional reference planes and are also used as the locating surfaces of parts mounted on machine tools.

This essential role in design and manufacture which flat surfaces frequently perform implies that the accuracy of those part characteristics which are related to these reference planes is dependent on the physical condition of the flat surfaces representing such planes. Consequently, most flat surfaces must be produced with an accuracy consistent with that expected from the derived product parameters.

Flatness of a technical surface expresses a condition in which all elements of that surface lie between two imaginary parallel planes at a distance defined by the specified limits of the required flatness. This condition applies equally to solid and to interrupted surfaces, in the latter case considering the continuous areas of the surface only. While interruptions, as long as they do not interfere with the predominant characteristics of the flat surface, can generally be tolerated; protruding elements, unless

designed to be located outside the functionally effective area, can defeat the essential purpose of a nominally flat surface.

The Favorable Characteristics of Surface Grinding

Having considered the important role of flat surfaces as reference planes for dimensioning and as locating planes for manufacturing, the selection of the most suitable methods for producing such surfaces is an important function of process design.

Many methods present themselves and are actually used in manufacturing practice. The most widely employed methods are planing or shaping with single-point cutting tools, milling with multiple-point cutters, and grinding. Sometimes, more than one of these methods are combined, applying them sequentially to produce a particular surface, such as milling for maximum stock removal, followed by grinding for improved dimensional and geometric accuracy, as well as surface finish.

Flat surface lapping, another abrasive method, has limited usage only and is generally applied for the ultimate refinement of a previously fine-ground surface.

Recent progress in grinding methods has made possible very substantial stock removal rates in the roughing phase of the grinding operation, and is often referred to as *abrasive machining*, Such effective removal of large amounts of work material had been previously accomplished only by methods which utilized metallic cutting tools.

There are workpiece or material conditions, and also accuracy specifications which make surface grinding the only applicable, or by far the most economical, method of producing the required surface. In some other case it may simply be a matter of convenience such as with regard to work loading and holding, the availability of equipment, or the desire to save on special tooling, etc., which causes surface grinding to be selected as the preferred method. On the other hand, the favorable position of surface grinding, in comparison to other processes, may be predicated on the use of special equipment, such as heavy-duty or highly automated grinding machines.

In order to aid the process engineer in making the proper choice, particularly when closely equivalent alternative methods may also have to be considered, a review of the favorable characteristics of surface grinding is presented in Table 4-1.1. The purpose of that listing is to indicate the various aspects in which the properties of grinding may be superior to those of other methods used in producing flat surfaces. The table discusses these aspects in a general sense only. Actual values with regard to either the relative capabilities of alternate methods, or the economic advantages of one against the other, will have to be developed for specific cases. With respect to surface grinding, some informative values that are useful for such comparison will be found in subsequent chapters of this section.

The Systems of Surface Grinding

The geometric shape of surfaces produced by methods included in this category is essentially flat. While the designation "flat surface grinding" would be more appropriate, expressing more precisely the purpose of the process, the shop usage of the term is so widely accepted that "surface grinding" is generally considered to imply the grinding of flat surfaces.

A flat surface can be produced in grinding either: (a) by *traversing movements* carried out in a common plane, or (b) by *rotating motion* around an axis normal to the plane of the surface being ground. These two types of movements, generally imparted to the workpiece, are considered to comprise two major categories in the classification of surface grinding methods.

The grinding of flat surfaces may be carried out either (a) with *the periphery* or (b) with *the face* of the grinding wheel. Depending on the process objectives, both methods find application for either traversing or rotating workpieces.

In face grinding it is generally not the entire face of the grinding wheel which is engaged in the actual cutting process; the grinding is carried out with an annular, that is, a ring-shaped portion of the wheel face, with the rest of the wheel face recessed, or nonexistent. The purpose of this arrangement is to limit the area of contact between the wheel and the work in order to reduce the development of heat, to provide better coolant flow, to facilitate the removal of chips, etc.

In some operations that area of the workpiece which is presented to the wheel during the grinding process is relatively small, thus the shape and the dimensions of the work provide that limiting control of the wheel and work contact area which, for the reasons just indicated, is a desirable operational condition. Therefore, it is technically possible and, for increased productivity, preferable to keep the entire or at least major portion of the wheel face operative.

Surface grinders which operate by utilizing the full wheel face, or most of it, are known by the distinguishing designation of *disc grinders*. In disc grinders, too, the progress of the workpiece along the face of the wheel can be a straight line traverse across the wheel face, or a circular movement producing an arc-shaped contact path with respect to the disc wheel.

Another important characteristic of disc grinders is the possibility of simultaneously grinding opposite sides of the work, by passing the workpiece between the essentially parallel faces of two grinding discs which are facing each other. This method, which is utilized in the majority of disc grinding operations, is carried out on equipment known as double-disc grinders.

A survey of these major methods of surface grinding, with a brief discussion of each method's general characteristics, is presented in Table 4-1.2.

The methods considered in this section may also be designated as precision surface grinding. This is in distinction to other grinding methods and equipment which, while also producing substantially flat surfaces, have for their primary purpose the removal of undesirable material from the work surface without, however, producing an accurate surface intended to represent a plane and to constitute the boundary of a tightly toleranced dimension.

The Relative Advantages of Different Surface Grinding Methods

The differences between various methods of surface grinding are, in some cases, quite substantial, with only a few significant characteristics in common, including, of course, the final objective. The question could be raised, "Why are such widely different systems offered, and actually used, when any one of them seems to be aimed at, and accomplishes the same basic objective, that is, to generate a flat surface by grinding?"

**Table 4-1.1 Favorable Characteristics of Surface Grinding Methods
(Examples for comparison with alternate methods)**

Aspect	Condition	Discussion
Workpiece Characteristics	Work material and hardness	Grinding is applicable to hardened steel, castings with inclusions, very hard materials, etc., even to those not adapted to other methods of metalcutting.
	Surface preparation	Grinding can be applied to surfaces with scale, high degree of unevenness, etc., without work preparation.
	Interrupted surfaces	Interrupted surfaces which often cause chipping of cutting tools, and also of workpiece edges, are ground with ease when using proper grinding wheels.
	Machining allowance	Grinding is practicable with very small surface allowance, while cutting tools usually require a specific minimum depth of tool edge penetration.
Properties of the ground Surface	Flatness	Flatness of the ground surface, a major objective of the process, generally exceeds, by far, those produced by other machining methods.
	Dimensional accuracy	Grinding generally, has size-control capabilities which are superior to other metalcutting processes applied for comparable stock removal.
	Surface finish	Because of the very large number of cutting edges which operate in grinding, the produced surface is much smoother than one attainable by any other standard metalcutting method.
	Distinct profile definition	When profiled straight element surfaces are produced by surface grinding a contour accuracy unattainable by other methods of metalworking results.
Flexibility of Adaptation	Selective application of surface grinding	Surfaces with precisely delimited boundaries, even very close to shoulders, can be produced by several surface grinding methods.
	Location of the ground surface on the workpiece	The surface ground can be in precisely controlled linear or angular relation to defined reference surfaces or points on the workpiece.
	Angle and form grinding	Surfaces other than flat, such as tapered or profiled, can be produced by various surface grinding methods, particularly the peripheral type (see Table 4-2.1).
	Process automation	Certain types of surface-grinding methods are adaptable to various degrees of automation, including material handling and size control.
Work Mounting and Holding	Magnetic plates or chucks	In the majority of surface-grinding operations the workpieces are mounted on magnetic chucks, avoiding expensive fixtures by the use of an easily operated and precise device which is highly versatile in its applications.
	Multiple part grinding	Magnetic plates, occasionally with inexpensive auxiliary devices, replaced with special fixtures in rare cases only, make surface grinding applicable simultaneously to a varying, often large number of workpieces.

(Continued on next page)

Table 4-1.1 (*Cont.*) **Favorable Characteristics of Surface Grinding Methods**
(Examples for comparison with alternate methods)

Aspect	Condition	Discussion
Work Mounting and Holding	Mounting nonmagnetic workpieces	For surface grinding such workpieces often can be mounted on a magnetic plate or chuck with the aid of auxiliary devices which are held or even caused to be operated by magnetic force.
	Light clamping force for mechanical holding devices.	Surface grinding in general, exerts on the workpiece a force which is much smaller than that of alternate metalworking methods, consequently, a light clamping force, causing little workpiece distortion, is frequently sufficient.
Economy and Productivity	Simple tooling for work holding	The versatile applicability of magnetic retention simplifies the tooling for work holding, reducing cost and lead time when changing over to different parts.
	Off-machine tool sharpening eliminated	Cutting tools used in most types of metalworking methods require regular resharpening on special equipment, away from the production machine. Surface grinding requires only dressing of the wheel without its removal or disturbing the setup. In face grinding even the wheel dressing is required at long intarvals only.
	High rate of stock removal	Surface grinding, as a method, is not limited to precision work, but is also suitable, with proper equipment, for substantial stock removal, the method being well adapted to abrasive machining.
	Combining roughing and finishing in a single operation	Grinding, in general, is adapted to operate with changing rates of stock removal, such as by reducing infeed toward the end of the operation. Certain types of surface-grinding machines are also designed to comprise two or more grinding stations through which the work is made to pass in a sequential or continuous progression.

The reasons for the diversity of the methods and equipment are the specific advantages of each system, which will be discussed in detail in subsequent chapters. Such advantages, which in many respects are quite obvious and significant, apply to particular combinations of workpiece dimensions, operational characteristics, production quantities, quality specifications and many others.

From the following discussions of various methods of surface grinding a general outline of preferential areas of use will emerge; in some cases defining exclusively applicable methods. The resulting guidelines should prove useful for selecting the method which is the most appropriate.

It must be pointed out, however, that in a field as dynamic as grinding technology, where progress in methods and equipment is continuous and often rapid, particular advantages which once may have assured the positive superiority of one method in specific applications, are being challenged by progress in other methods. Consequently, the advantages of the individual methods, as analyzed later in this section, should be considered as the first indications for utilizing a given grinding process without, however, excluding the possibility of applying an alternate method.

In order to demonstrate in which respects traditional method selection can be superseded due to progress in machine and tool development, a few randomly selected examples will be pointed out in the following. In these cases, alternate methods different from those which generally may be selected, proved to be superior and increasingly are being adopted throughout industry.

In general: Peripheral grinding with a relatively small area of contact is the preferred method for precision type grinding operations which can be executed with excellent control of size and geometry, while the rate of stock removal is not one of the prime considerations.

Typical exception: Recent developments in peripheral surface grinding methods and tools, involving the application of high-speed grinding on powerful

Table 4-1.2 The Systems of Surface Grinding

Operative Wheel Surface	Work Movement and Equipment Designation	Diagram	Discussion
Periphery (Peripheral surface grinders with horizontal spindle)	Traversing ——— Peripheral surface grinders (with reciprocating table)		Most commonly used system, applicable over a very wide range of models with drive motors varying from about 1 hp to 100 hp. It is one of the basic types of toolroom equipment and offers high flexibility of adaptation for other than plain flat surfaces, as well. Also capable, with appropriate equipment, of substantial stock removal in heavy machining operations.
	Rotating ——— Rotary surface grinders (with horizontal spindle)		Although the wheel and work relationship is similar to the preceding system, rotation of the work, in contrast to the traversing, limits the field of application to workpieces which are adapted to, or actually require the continual rotational movement instead of the straight line reciprocation.
Annular face portion (Face grinding machines)	Rotating ——— Rotating table type face grinders (with vertical spindle)		The large contact area, resulting from operating with the face of the wheel generally assures a higher rate of stock removal when grinding essentially plain flat surfaces. Productivity being one of the prime objectives of this system, the rotary work movement, because of its continuity, is commonly preferred over reciprocation which involves constant reversals of the work movement. The system is adaptable to a wide variety of work surface configurations and locations.

(Continued on next page.)

Table 4-1.2 *(Cont.)* **The Systems of Surface Grinding**

Operative Wheel Surface	Work Movement and Equipment Designation	Diagram	Discussion
Annular face portion (Face grinding machines)	Traversing ——— Reciprocating type face grinders (mostly with vertical spindle)		Workpieces, which due to general configuration, are less or not adapted to mounting on a rotating table may still be ground efficiently with the face of the grinding wheel. Face grinding machines with reciprocating work movement commonly have vertical wheel spindles and horizontal mounting surfaces of the machine table, but are also built with these mutual positions of the main operating members reversed.
Substantially the total face of the wheel (Disc grinders)	Workpiece individually introduced into grinding zone ——— Gun or indexing type double disc grinders		A large portion of the opposing faces of two disc-shaped wheels are simultaneously grinding opposite sides of the workpiece; which is introduced into and held in the grinding zone until the wheels, in a gradual axial infeed, produce the required work thickness.
	Workpiece continuously passing through the grinding zone ——— Single or double disc grinder with rotating carrier		The workpieces retained in a rotating carrier plate are passing across the faces of two opposed wheels (or a single wheel and a support plate); the thickness of the ground work is determined by the gap between the grinding wheels which are advanced only to compensate for wear and dress-off.

(Continued on next page.)

Table 4-1.2 *(Cont.)* **The Systems of Surface Grinding**

Operative Wheel Surface	Work Movement and Equipment Designation	Diagram	Discussion
Substantially the total face of the wheel (Disc grinders)	Workpiece is caused to traverse the entire disc wheel faces ——— Feed-thru type double disc grinder		Similar to the preceding, however, the entire wheel face participates in the grinding of a continuous row of workpieces which are made to traverse essentially along the diameters of two disc wheels, the faces of which are set apart by a distance corresponding to the required work thickness.

peripheral surface grinding machines with wheel sizes in the order of 32 to 36 inches (about 800 to 900 mm) diameter and drive motors to about 125 hp, resulted in achieving substantial stock removal rates; comparable with those of high performance face grinders.

In general: Face grinding, due to the large area of contact, generally requires grinding wheels with coarse grains, a relatively soft bond, and open structure and is primarily intended for heavy stock-removal rates, with lesser emphasis on accurate size control and high-quality finish.

Typical exception: Developments in work-holding and loading devices now permit for small size workpieces, the efficient use and often automated operation of face type grinding methods, particularly in the disc grinder system. These again, by their size, limit the wheel and work contact area, thus permitting the use of the grinding wheel in compositions well suited for accurate size and finish control.

In general: Rotary surface grinding, a system of peripheral grinding with a rotating work table, unless implemented with machines having continuously variable speed control, has the inherent drawback of nonuniform work speed, a condition which would put the system at a disadvantage as compared with peripheral grinding using straight line reciprocated work traverse.

Typical exception: The lay of the ground surface on disc-shape elements used as sealing surfaces should have an essentially concentric pattern to prevent the escape of fluid or gases. Such regular grinding

patterns on flat surfaces may only be produced by rotary surface grinding.

The purpose of this brief discussion of unconventional selection of the applied surface grinding method is to indicate the need for an analysis of the pertinent operational conditions and the importance of up-to-date information regarding the capabilities of modern grinding equipment. Such circumspect procedure may, in some cases, reveal rewarding alternatives and contribute to achieving optimum method selection.

Abrasive Machining in Surface-Grinding Operations

Abrasive machining, in the current interpretation of the term, refers to a metal-removal system in which grinding is substituted for machining operations previously performed by cutting tools of more or less conventional design.

In order to qualify for the designation "abrasive machining," the grinding process must have the capacity for stock removal rates which are comparable to those achieved in the conventional metal cutting process it may replace. Grinding in general, disregarding rough grinding such as is used in foundry and steel mill operations, is a precision process and has been used, until recently, essentially as a finishing operation only. Advances in machine design and dimensioning, power input and grinding wheel composition permitted a substantial increase of stock removal by grinding, which thus has become, in particular applications, a potential substitute for conventional machining methods.

Substitution of a new technological process for one used previously is, in general, warranted only when the technical results are, at least equivalent, while other benefits, such as increased productivity, reduced cost, etc., are also realized.

Producing flat surfaces with good form accuracy and surface finish, by maintaining specific dimensional limits, is the area where abrasive machining is showing, perhaps, the most remarkable results and has found the widest acceptance.

In comparison to conventional methods of metal removal, such as milling, planing, or face turning, the use of surface grinding can produce substantial benefits, which will result as the sum of savings in different cost areas. A few potential advantages of high performance surface grinding are listed in the following. In some cases several, or even most of these advantages can be utilized:

(1) The rate of stock removal can actually exceed that of machining the same workpiece with cutting tools.

(2) The functionally adequate surface area can be produced with a lesser amount of material converted to chips; this is achieved by designing parts expressly for abrasive machining (e.g., with reduced machining allowances or with a noncontinuous surface which would have to be avoided in machining with cutting tools).

(3) Holding the workpiece in grinding is often feasible on a magnetic chuck, but even if mechanical work-holding fixtures are used, those for abrasive machining may be much simpler because of the lower clamping forces required for grinding.

(4) Multiple operations are often needed in machining with cutting tools, such as roughing followed by finishing, in order to produce the required geometry, finish, and size. Surface grinding generally has the capability of producing in a single operation, surfaces which are equivalent to those resulting from finish machining.

(5) Grinding, the process of abrasive machining, requires for tool reconditioning only an occasional wheel dressing, frequently obtained automatically and with simple initiation by the operator, while cutting tool reconditioning involves expansive sharpening or the expert exchange of expendable tool inserts.

It should, however, be pointed out that abrasive machining, in order to be efficient, is not simply a method involving the operation of high powered equipment. Abrasive machining, whether used for producing substantially accurate flat surfaces, or any other shapes, must be handled as a system, involving such items as:

(a) The proper design of the part for utilizing the advantages of grinding (e.g., designing pads instead of a continuous surface for a part area which must represent a plane)

(b) The selection of grinding machine equipment capable of removing substantial amounts of stock without interfering with the production of the required work quality

(c) Designing the process by specifying appropriate grinding data for the purpose of highest efficiency and also for avoiding metallurgical damage to the workpiece

(d) The use of grinding wheels, the assurance of coolant supply, etc., adapted for the efficient implementation of the process.

A few other characteristic conditions of abrasive machining, which can constitute disadvantages in comparison with the more conventional machining methods, may also be mentioned:

(a) Substantially higher unit horsepower (hp/cu in. /min) is required for abrasive machining as compared to other machining processes.

(b) Unless the work is carefully planned there is a possibility of metallurgical damage to the surface that has been ground.

(c) Usually more time is required to change the grinding wheel than to change a cutting tool such as a milling cutter.

It must be remembered that most surface grinding machines are not designed for performing abrasive machining. The majority of work which is carried out on surface grinders does not require such a highly productive process. In the subsequent discussion of surface grinding equipment (both peripheral and face grinding types) references will be made to models which are either adaptable or have been expressly designed for abrasive machining.

Peripheral Surface Grinding – Principles, Applications, and Basic Equipment

Characteristics of Peripheral Surface Grinding

In peripheral surface grinding the work traverses in a tangential plane along the grinding wheel's periphery, which in this method, is the operative wheel surface. The traversing movement of the work is generally in a straight line at right angles to the grinding wheel axis. The movement of the work in that plane can also be rotational; this arrangement, applied much less frequently, has the distinctive designation of rotary surface grinding and will be discussed separately in Chapter 4-6.

Surface grinding with the periphery of the grinding wheel, in comparison to using the face of the wheel, has several major characteristics which will generally determine the choice between these two alternate methods. In many cases such a choice is obvious, simply because the limitations of the alternate method eliminate its use. In some other cases, however, the selection of the most suitable method would be a matter of conjecture, unless the ways in which the characteristics of alternate methods differ, and the specific advantages which each of these methods offer, are understood and considered from the viewpoint of a particular operation. Such advantages may be related to process performance, work quality, or equipment adaptability, etc., several aspects of which may apply to a particular case.

A review of a few important characteristics of peripheral surface grinding which may be definitely advantageous, or actually required in specific applications, follows:

(1) *The narrow area of contact* extending over a relatively short distance in the direction of work travel, precisely defined by the selected depth of cut

and wheel diameter, permits easy chip disposal and favorable coolant access to the work surface. This results in a cooler cut, the avoidance of work surface injuries by chips retained in the contact area, and also permits the use of wheels with finer grains, harder bond and denser structure, thus achieving better size control and surface finish (see Fig. 4-2.1).

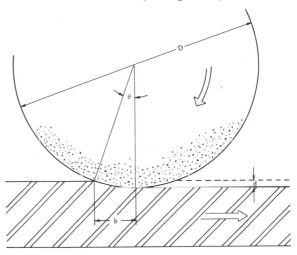

Fig. 4-2.1 Diagram showing (in exaggerated proportions) the contact area between the wheel and the work in peripheral surface grinding. The breadth of the momentary contact area b is the function of the wheel diameter D and the depth of cut d. As an example: for $D = 12$ inches (300 mm) and $d = 0.001$ inch (0.025 mm), the b breadth is theoretically 0.136 inch (about 3.45 mm). The actual length of the grain path is somewhat longer or shorter, due to the traverse movement of the work in a direction either opposite to or in the same direction as the wheel rotation. The work traverse is, however, performed at a speed usually less than one per cent of the peripheral wheel speed.

269

(2) *Angular surface grinding*, which produces flat surfaces, usually along the entire length of the workpiece, parallel with, but inclined to a vertical plane along the direction of the table traverse, can be accomplished either by mounting the workpiece in a tilted position on the machine table, or by dressing the wheel to a tapered periphery and applying plunge grinding. Heavy workpieces, such as the guideways of machine tool beds, are ground on special way grinders with inclinable grinding wheel heads.

(3) *Taper surface grinding* which produces a work surface inclined to a vertical plane normal to the direction of traverse can be achieved with dependable accuracy by raising one end of the workpiece with respect to the plane of the table travel, such as by mounting the work on a sine plate.

(4) *Profile grinding* by shaping the wheel periphery to a simple or complex profile with conveniently applicable wheel-truing devices. Plain angular elements, also radii, both convex and concave, singularly or with adjacent straight sections, can be produced with accessories which are precisely adjustable over rather wide ranges. Intricate profiles may be shaped with the most suitable type of several alternative truing systems, such as profile bars, pantographs, crush rolls, diamond blocks and rotary diamond rolls. The peripheral surface grinding process will then generate a work profile, representing the inverse shape of the wheel surface and extending over the entire ground length of the workpiece.

(5) *Easy observation of the work area.* In certain types of surface grinding operations, most particularly in tool grinding, the observation of the work area during actual grinding is a desirable, or even essential condition. In tool sharpening, for example, the objective of surface grinding is to eliminate the damaged surface, or a layer which contains the dulled portion of the cutting edge. As soon as a sufficient amount of stock has been ground off the surface, the objective of the operation is accomplished, and continued grinding would only unnecessarily weaken the workpiece and reduce its service life.

In actual application it is possible to determine the required depth of grinding by means of a preceding measuring operation and then controlling the grinding process for removing the corresponding surface layer. However, such measurements may be difficult to accomplish, as in the case of complex blanking dies or milling cutters with intricate profiles and, perhaps, nonuniform tooth wear. Therefore, it is frequently more convenient, as well as more economical, to continue the surface grinding while visually observing the workpiece and to terminate the grinding as soon as the visible marks of tool edge wear have disappeared.

Peripheral surface grinding with reciprocating table movement resulting in disengagement of the work at the end of each stroke, is a method particularly well suited for such operations.

The Applications of Peripheral Surface Grinding

The basic characteristics of peripheral surface grinding, some of which were pointed out in the preceding discussion, determine in the first place, the major application areas of this grinding method. A further characteristic, not discussed in detail because it is more a potential than a concretely observable property, is the flexible adoption of a great variety of work-holding, wheel-truing and dimensional-measuring accessories, which are opening up still further areas for the successful application of this method.

Peripheral surface grinders in their basic form are relatively simple and inexpensive machines, which can be afforded even by small shops. As a matter of fact, hardly any toolroom or jobbing shop operates without a peripheral surface grinder, and many tool shops use a considerable number of such equipment.

While toolroom work, both manufacturing and repair, are very important areas of its application, there are also many uses for peripheral surface grinding in other fields of production. This method is applied frequently in manufacturing where flat surfaces of considerable length or in confined locations must be produced by grinding, for reasons of accuracy, surface finish, or other specifications calling for a ground surface. Other important uses are short or extended straight surfaces with profiled cross section, which can only be ground by peripheral surface grinding.

The preceding listing of but a few preferred or required uses indicates how varied are the applications of peripheral surface grinding. While in several such operations peripheral surface grinding may not be the only method by which the required surfaces can be produced, it should still be considered and evaluated with respect to operational economy, including tooling costs, the quality of the produced surface, and the complementary operations which each of the alternative methods may require.

As a guide for giving due consideration to this method when designing a process which involves flat or straight profiled surfaces, a survey of the most frequent uses of peripheral surface grinding is presented in Table 4-2.1.

The Principal Design Types of Peripheral Surface Grinding Machines

The four generally used design types are reviewed in Table 4-4.2. These types have the common charac-

Table 4-2.1 Examples of Work Surface Configurations Adaptable for Peripheral Surface Grinding

General Category	Type of Work Surface	Diagram	Discussion of Process Conditions
Open flat surfaces in different positions	Plain flat surfaces, continuous or interrupted		Most common application of peripheral surface grinding on surfaces whose width is not limited by that of the wheel. By transverse movement at the ends of the reciprocating table strokes the entire work surface is gradually presented to the wheel in overlapping passes. The infeed is operated prior to the directional change of a series of transverse movements, or at the ends of the reciprocating strokes when infeed grinding is applied.
	Workpiece supported at an angle to produce an inclined flat surface (taper grinding)		By using appropriate work-holding devices the ground part surface can be in an inclined position with respect to the selected reference element on the workpiece. The applied work-holding device, such as, e.g., a vise with angular adjustment, a sine plate, or a special fixture is selected to conform to the required angular accuracy of the ground surface. Compound angles may also be produced by the same method.
	Parallel flat surfaces at different angles of inclination (angular grinding)		Several flat surfaces with parallel elements can be ground in subsequent phases of an operation, either by changing the incline angles in which the work is supported for grinding the individual surfaces, or by changing the angle of the wheel spindle inclination, such as on special way grinding machines.

(Continued on next page.)

Table 4-2.1 *(Cont.)* **Examples of Work Surface Configurations Adaptable for Peripheral Surface Grinding**

General Category	Type of Work Surface	Diagram	Discussion of Process Conditions
Bounded flat surfaces in different orientations	Flat surface bounded by a shoulder		By precisely limiting the reversal position of the transverse movement the ground surface can be controlled to reach very close to an extending shoulder which is parallel with the reciprocating table movement. Using a face-dressed (concave) wheel and sensitive transverse adjustment, often in combination with a positive stop, a light cut may also be applied to a low shoulder for improving the squareness of the adjoining surfaces.
	Flat, single, or parallel surfaces in recessed locations (slots in a common plane)		The choice of grinding wheel widths, combined with the possibility of narrowing the wheel edge by truing, permits grinding the bottom surface inside shoulders, or even by using an appropriate wheel truing device, the complete contour of slots parallel with the direction of table traverse. Several parallel slots, of equal or different dimensions, may be ground in the same setup, either by relying on the graduations of the hand wheel for the transverse adjustment or supplementing it with a position indicator for improved spacing accuracy.
	Flat recessed surfaces spaced around the periphery of the part		Mounting the workpiece into an indexing fixture for incremental rotation around the part axis, the slots around the periphery of the workpiece (either in radial positions or off-center) can be ground in precisely controlled angular spacing and dimensions. The operation may be carried out as infeed grinding (finishing the slots successively), or by indexing the work at each table reciprocation, for sequential grinding with infeed applied following each completed turn of the workpiece.

Table 4-2.1 *(Cont.)* **Examples of Work Surface Configurations Adaptable for Peripheral Surface Grinding**

General Category	Type of Work Surface	Diagram	Discussion of Process Conditions
Bounded flat surfaces in different orientations	Flat recessed surfaces in different orientations on the face of the part		Vertical axis indexing devices mounted on the reciprocating table of peripheral surface grinders equipped with appropriate truing devices, are used to grind slots of different configurations and with the required angular spacings on the face of a workpiece (crenelated configuration). Such operations may be carried out by selecting the most suitable process: grinding the surfaces individually in succession, or by continuous indexing combined with incremental infeed, also as an automatic operation by the coordinated control of reciprocation, indexing, and infeed.
Profiled straight element surfaces	Straight surfaces with several cross-sectional elements of regular form		Straight surfaces composed of several adjacent sections with profiles which may be straight lines mutually inclined at different angles, circular arcs producing convex or concave surfaces, or comprising a combination of such elements, including the straight line tangent to the circular arc, can be ground precisely on most types of peripheral surface grinding machines which are equipped with the appropriate wheel truing devices (see Table 4-2.4). Such devices are generally available as optional, but regularly manufactured accessories.

(Continued on next page.)

Table 4-2.1 *(Cont.)* **Examples of Work Surface Configurations Adaptable for Peripheral Surface Grinding**

General Category	Type of Work Surface	Diagram	Discussion of Process Conditions
Profiled straight element surfaces	Straight flat surfaces with special (intricate) profiles		Surfaces of the described type occur frequently in toolmaking, both for sheet-metal dies and for profiled cutting tools, as well as in the production of parts with complex profiles which must be finished by grinding because of accuracy requirements or work material properties. For grinding such profiles in widths not exceeding that of the grinding wheel, regular peripheral surface grinders are suitable when equipped with the appropriate truing device, such as single-point diamond duplicator, crush dresser, form diamond block or rotary diamond dresser (see Table 4-3.1).
Cutting tool sharpening	Single plane flat surfaces in controlled locations		The sharpening of many types of cutting tools and dies is carried out on regular peripheral surface grinding machines, as long as the sharpening consists essentially in producing new flat surfaces in precisely controlled location and orientation. For such operations the workpiece may be held simply on the magnetic chuck such as in the case of dies, on an arbor, in an adjustable vise or in special workholding fixtures for assuring the proper location of the workpiece.
	Flat surfaces in several parallel or related planes		Cutting tools whose edges are restored by grinding individual flat surfaces which are mutually parallel but located in displaced planes, such as short flat broaches, or even located in different planes, can be processed on peripheral surface grinders equipped with appropriate holding devices. The basic design of the reciprocating table surface grinder may be retained for developing specialized equipment, such as for producing long flat broaches, for which purpose the machine will be equipped with a wheel head operating with a small cup wheel (see Fig. 4-4.12).

teristic of horizontal spindle position and reciprocating table movement but vary with regard to the design of the members which carry out:

(a) The transverse or crosswise movement; and

(b) The downfeed or vertical adjustment of the wheel spindle in relation to the fixed table level.

Very light machines are frequently built with vertically adjustable columns (type A). For large, heavy machines, the design in which cross movement is obtained by sliding of the wheelhead, sometimes referred to as the *outrigger design* (type D), is preferred and even necessary in many instances.

Types B and C differ in the member to which the crosswise movement is assigned: the table supported on a compound slide or the column which, for that purpose, is mounted on a slide. Different reputable manufacturers use one or the other of these two design systems for peripheral surface grinders in the same size ranges, achieving comparable results with respect to operational accuracy and service life.

This survey of design types would be incomplete without mentioning the traveling column type surface grinders, built for very long parts which require a corresponding length of traverse movement. By designing the grinding machine with a traveling column along a fixed position table, considerable floor space can be saved. This objective is the main reason for the occasional building and operation of peripheral surface grinders with traveling column.

Capacity Ranges and Typical Models of Peripheral Surface Grinding Machines

The method of grinding straight, generally flat surfaces with the periphery of the grinding wheel, while the work is traversed along the operating surface of the wheel, is applicable to workpieces of widely varying sizes. The required capacities and capabilities may also vary extensively, the latter aspect including substantially different performance rates and quality specifications. Grinding machines designed to implement this method may also be required to offer different degrees of adaptability for performing other than plain surface grinding operations.

To permit a survey of such an extended range of functionally similar production equipment, a classification is used with categories based primarily on work capacity, a characteristic which in this context is to some extent interrelated with productivity and often with operational accuracy as well. This classification is not based on any standards or industrial practices, but has been adopted only to provide a convenient system for reviewing that broad terrain

which different models of peripheral surface grinders cover.

As a first step in that survey a comprehensive listing of the various categories of peripheral surface grinding machines is presented in Table 4-2.3, indicating also the general ranges over which the listed characteristics of different models within the individual categories may vary. In this table only basic machine types have been considered, while special models and equipment serving limited applications will be discussed in Chapter 4-4 of this section.

Specific models of peripheral surface grinding machines will now be reviewed briefly, as representatives of the different categories. While the table refers to each of the discussed models, these are not unique and others, possibly several different makes and types of peripheral surface grinding machines, essentially equivalent to the presented examples, could have been chosen for visualizing the characteristics of the individual categories.

Representative Models for the Major Categories of Peripheral Surface Grinding Machines

The differences between the categories, which were defined in Table 4-2.3, can best be appraised by comparing representative models of each group.

The models chosen for the following discussion, while products of reputable machine tool manufacturers and considered well suited for illustrating the dominant characteristics of the individual categories, are not necessarily outstanding in their group which may comprise several other, essentially equivalent makes and models. The purpose of this discussion, however, is not to evaluate and to recommend specific models, but to demonstrate principles on the basis of which such appraisals and subsequent selections can be made by the user, by giving due regard to the prevailing particular aspects and circumstances.

Light Shop Type Surface Grinding Machines

Machines of the type shown in Fig. 4-2.2 make surface grinding machines accessible even for small shops with limited financial means. While productivity with regard to both stock removal rates and operational times may not be important considerations in small shop practice, precise work is an essential requirement for surface grinding jobs in general. Accordingly, most types of small peripheral surface grinders, including the illustrated model, are designed for accurate work, producing excellent flatness, good finish, with a size control capacity in the order of 0.0002 inch (0.005 mm), or better.

The table is traversed by hand-operated lever, acting on a drum which transmits its rotation to the

Table 4-2.2 Design Systems of Horizontal Spindle Peripheral Surface Grinders

Type Specifications	Diagram	Discussion of Characteristics
Compound table and vertical column adjustment		Used for lighter types of grinding machines in which the combined weight of the generally cylindrical column and of the wheel head still permits a sensitive adjustment of the vertical position. The system is frequently applied also for universal tool grinders, in that case, with plain traversing table and with the transverse movement carried out by the column base.
Compound table and vertical wheel-head adjustment		The machine base and the upright constitute a single-piece casting for assuring constant relation between the ways for the table saddle and the vertically adjustable wheel head. The saddle carries out the transverse movement and contains the ways for the longitudinal reciprocation of the table. This design is frequently used for small and medium-size machines, including those intended for toolroom applications.
Plain reciprocating table, transverse base for the upright and vertical wheel-head adjustment		An alternative version of the preceeding type, yet eliminating the saddle for the table which thus has the guideways of its longitudinal movement directly in the base, the latter containing also the guideways for the transverse adjustment of the upright. The mutual perpendicularity of the two movements in the horizontal plane is assured by virtue of the inherent rigidity of the base casting. This design system is used for both the light and medium heavy types of grinders.
Plain reciprocating table, compound saddle for the wheel head		The upright supports and provides the guideways for the vertical adjustment of the saddle, which has horizontal guideways for the wheel-head slide. Using a single piece of casting for the base and the upright, assures constant relation between two mutually perpendicular movements. Carrying out the transverse movement with the slide of the wheel head avoids the displacement of the much heavier machine members. This design system is preferred for medium and particularly for heavy machine models.

ies include various work-holding and special wheel-dressing devices.

Toolroom Type Surface Grinding Machine — Light

Accuracy and versatility are major attributes of toolroom equipment, properties which also apply to peripheral surface grinders, such as that shown in Fig. 4-2.3. The illustrated model is available in three versions, representing different degrees of power actuation for the table reciprocating and cross-wise feed movements. These are, in increasing order of power actuation:

(a) Manual operation of both movements

(b) Hydraulic longitudinal movement combined with hand cross feed

(c) Both movements equipped with hydraulic drive.

Toolroom grinding may require a wide range of different fixtures for work holding, many types of which are available with the basic machine as optional accessories. The vertical capacity of this model—the distance between the table and the new wheel in its

Courtesy of Brown & Sharpe Mfg. Co.

Fig. 4-2.2 Light surface grinding machine for general shop applications, shown with basic equipment for table reciprocation by hand lever.

table movement by means of a steel tape with adjustable tension, a design which practically eliminates backlash. For easy movement and accurate guidance the table is supported on precision steel rolls and a device providing adjustable drag assures table actuation with the feel required for precision work.

These machines can be equipped, as optional accessories, with exhaust attachment for dry grinding or with wet grinding equipment, comprising supply tank with motor-driven pump, adjustable nozzle, splash guard and table guard, as well as the necessary tubing and control elements. Other optional accessor-

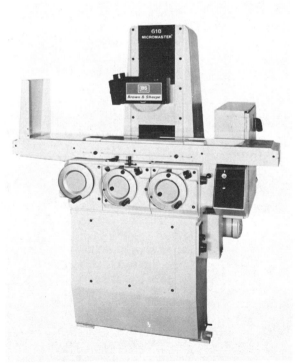

Courtesy of Brown & Sharpe Mfg. Co.

Fig. 4-2.3 Toolroom type surface grinding machine for light precision work, with optional power movements, adaptable for varied operations with a wide range of special accessories.

Table 4-2.3 Principal Types of Peripheral Surface Grinding Machines Grouped into Categories

Designation	Characteristics and Principal Applications	Capacity Range		Grinding Wheel Diam. × Width	Spindle Drive Motor hp	Approx. Machine Weight	Example
		Table Surface Width × Length	Max. Work Height				
Light shop type	Simple machine, often with manual cross feed, but not always. May be inexpensive equipment for occasional flat surface grinding, while other models serve for precision work on small workpieces.	5 x 10 to 6 x 12 inches (125 x 250 to 150 x 300 mm)	12 inches (300 mm)	8 x 1/2 inch (200 x 13 mm)	1	1200 to 1400 lbs (550 to 650 kgs)	Fig. 4-2.2
Toolroom type — light	Toolroom work requires high operational accuracy and comprises many varieties of the basic surface grinding, which call for a wide choice of special accessories.	8 x 18 to 12 x 14 in. (200 x 450 to 300 x 600 mm)	15 in. (375 mm)	8 x 1 to 12 x 1½ in. (200 x 25 to 300 x 38 mm)	1½ to 3	2,500 to 5,000 lbs (1100 to 2200 kgs)	Fig. 4-2.3
Toolroom type — heavy	Toolroom grinding on large parts, such as fixture elements, may also involve substantial stock removal, requiring large wheels and powerful drive motors, while still versatile enough for varied work.	10 x 24 to 16 x 42 in. (250 x 600 to 400 x 1050 mm)	15 to 18 in. (375 to 450 mm)	12 x 1 to 14 x 1½ in. (300 x 25 to 350 x 38 mm)	3 to 7½	5,000 to 10,000 lbs (2200 to 4500 kgs)	Fig. 4-2.4
High clearance type tool grinder	Making and maintaining heavy and bulky die blocks by grinding the whole operating surface, often without the removal of the long guide pins, may require toolroom grinders with extra-wide bed and high throat clearance.	16 x 48 to 24 x 96 in. (400 x 1200 to 600 x 2400 mm)	24 to 32 in. (600 to 800 mm)	20 x 3 to 28 x 3 in. (500 x 75 to 700 x 75 mm)	10 to 15	20,000 to 26,000 lbs (9,000 to 12,000 kgs)	Fig. 4-2.5

(Metric Equivalents are Approximate.)   *(Continued on next page.)*

Table 4-2.3 (*Cont.*) **Principal Types of Peripheral Surface Grinding Machines Grouped into Categories**

Designation	Characteristics and Principal Applications	Capacity Range		Grinding Wheel Diam. × Width	Spindle Drive Motor hp	Approx. Machine Weight	Example
		Table Surface Width × Length	Max. Work Height				
Production type — light	Production surface grinding, while demanding limited versatility only, may have emphasis on accuracy in preference to productivity. The designation "light" applies to the rate of stock removal.	14 x 36 to 18 x 48 in. (350 x 900 to 450 x 1200 mm)	17 to 24 in. (425 to 600 mm)	14 x 1½ to 16 x 2 in. (350 x 38 to 400 x 50 mm)	7½ to 10	8,000 to 12,000 lbs (3600 to 5400 kgs)	Fig. 4-2.6
Production type — medium	General production type surface grinding covers a broad area with indistinct boundaries. Machines of this category comprise work size and stock removal capacities ranging from small to large.	18 x 48 to 24 x 72 in. (450 x 1200 to 600 x 1800 mm)	24 to 28 in. (600 to 700 mm)	20 x 3 to 20 x 6 in. (500 x 75 to 500 x 150 mm)	10 to 15	18,000 to 28,000 lbs (8,000 to 12,500 kgs)	Fig. 4-2.7
Production type — heavy	Surface grinding machines built for substantial cutting performance are of particularly sturdy construction, operate with wide wheels and have workpiece capacities from medium to large. Powerful motors drive the main spindle and table.	24 x 60 to 36 x 120 in. (600 x 1500 to 900 x 3000 mm)	25 to 36 in. (625 to 900 mm)	20 x 6 in (500 x 150 mm)	30	33,000 to 65,000 lbs (15,000 to 30,000 kgs)	Fig. 4-2.8
Production type — extra large	Surface grinding is also required on particularly large workpieces, representing considerable weight and often processed by abrasive machining, stimulating still further expansion of this category of peripheral surface-grinding machines.	36 x 72 to 42 x 240 in. (900 x 1800 to 1050 x 6000 mm)	30 to 40 in. (750 to 1000 mm)	20 x 6 to 28 x 6 in. (500 x 150 to 700 x 150 mm)	30 to 125	70,000 to 120,000 lbs. (32,000 to 55,000 kgs)	Fig. 4-2.9

(Metric Equivalents are Approximate.)

highest position—is, in relation to the general dimensions of the machine, rather high in order to accommodate tall workpieces held in fixtures which are mounted either directly on the machine table, or held on a previously installed magnetic chuck.

Tool grinding generally requires dependable positional controls, which, in this model, consist of a vertical adjustment of the grinding wheel and the crosswise movement with the aid of hand wheels having regular graduations of 0.0002 inch (about 0.005 mm), but with the use of an auxiliary knob, 0.0001 inch (about 0.0025 mm) graduations.

Work surface configurations other than flat, must often be ground on toolroom surface grinders, and usually this is accomplished by the appropriate truing of the operating face of the grinding wheel. The long list of optional accessories for these models of surface grinding machines includes truing attachments for radii, angles, and devices with combined movements for producing continuous radius and tangent shapes. The "over-the-wheel" truing attachment (see Fig. 4-2.4) makes the process easier to perform than by using a device mounted on the machine table and it assures both the controlled parallelism of the trued wheel face with the machine table, as well as the dependable size compensation corresponding to the thickness of the dressed-off wheel layer.

Courtesy of Gallmeyer & Livingston Co.

Fig. 4-2.4 Toolroom type surface grinding machine for heavy duty precision work, equipped with wet grinding attachment, special wheel truing device, and magnetic chuck.

For grinding complex profiles the same toolroom type surface grinder can be equipped with an auxiliary device, projecting the workpiece profile on a screen in high magnification. The progress of the grinding can be observed by the operator by comparing the work surface to a contour chart mounted on the screen. The chart shows the required finished work profile in the same magnification as that used for projecting the contour of the actual workpiece.

Toolroom Type Surface Grinding Machines — Heavy

Toolroom operations may need fixture elements or other precision components which, because of their large size and heavy weight, require grinding machines designed to accommodate them. Such work, in order to be efficient, will often have to be carried out with substantial stock removal rates, necessitating adequate wheel drives, as well as wheel spindles of commensurate stiffness and strength. All these operational characteristics which point toward a generally heavier grinding machine must not diminish the capacity of the equipment for assuring the high performance accuracy, the sensitivity of controls, and the adaptability of the machine to perform the varied types of work commonly involved in toolroom operations.

The toolroom type surface grinder shown in Fig. 4-2.4 represents a series which includes models with table surfaces of 36 inches (about 900 mm) length and with widths of 12, 14, and 16 inches (about 300, 350, and 400 mm). These machines are of the saddle type design, a one-piece casting of great rigidity supporting the table and also carrying the traverse mechanism.

The powerful, 7½ hp spindle drive motor is mounted on an adjustable bracket and drives the grinding wheel spindle through V-belts. The purpose of this arrangement is to provide two different wheel speeds, the regular one for full size wheels and a higher speed for worn wheels. Separate motors are provided for the hydraulic pump, the coolant pump, and the power rapid vertical travel of the wheel head.

The wide range of traverse and feed movements satisfies most of the varied operations which characterize toolroom work:

(a) Table speeds are infinitely variable, from 3 inches to 150 feet (about 75 mm to about 45.75 m) per minute.

(b) Cross-feed is hand or hydraulically operated; the latter method is variable up to a maximum of 0.75 inch (about 19 mm) per table reversal.

(c) Downfeed by hand using a twin handwheel with the outer one for coarse adjustment and the inner one graduated in ten thousandths of an inch (0.0025 mm) for fine adjustment. The machine can also be equipped with automatic downfeed with incremental settings from 0.0001 to 0.001 inch (about 0.0025 to 0.025 mm) per table reversal.

High Clearance Tool Grinding Machines

Die grinding is one of the regular operations in the toolroom; and in plants producing large stampings using correspondingly heavy dies, surface grinders of somewhat special design, commonly designated as *die grinders*, may be required. The reason for that requirement is the practice of restoring the sharpness of the tool by grinding the die in its partially assembled state, that is, without removing the die (leader) pins. Due to the frequently extensive length of these pins, regular surface grinding machines generally do not have sufficient vertical space, (or throat clearance), these being available only on special equipment. The operational area of a special die grinder is illustrated in Fig. 4-2.5.

Courtesy of Mattison Machine Works

Fig. 4-2.5 Work area of a high clearance type surface grinding machine, showing a die block with leader pins mounted on a magnetic chuck for grinding the operating surface without disturbing the installed pins.

Other characteristics of die grinders are the uncommonly wide table surface compared to the table length, and the large-diameter grinding wheels which are needed for the high throat clearance. Although the width of the wheels used on these machines is generally not more than 3 inches (about 75 mm), their large diameter calls for drive motors of higher power than are commonly used on other types of peripheral surface grinders designed for toolroom operations.

Light Production Type Surface Grinding Machines

Peripheral surface grinding machines of this category may be considered to represent a transition between toolroom and production types. Work generally carried out in production operations does not require that high degree of versatility which characterizes toolroom work, although the performance accuracy expected from this category of production grinders should in no way be inferior to that required in toolroom operations.

High accuracy with respect to size, geometry, and finish can usually be obtained in grinding at reduced stock removal rates, particularly in the finishing passes of the operation. Consequently, it would serve no useful purpose to design grinding machines for high precision performance with powerful drive motors which would hardly be utilized. On the other hand, advanced models of light-type production surface grinding machines are designed to assure, in addition to the expected high quality of the grind, a relatively elevated productivity as well. However, in this case the high productivity is not accomplished by substantial stock removal, which is generally not even needed in view of the normally small stock allowances of such workpieces, but by maintaining steady stock removal rates down to the finishing stage, so that a reduced overall operational time is attained. That performance is predicated on machine design details which assure the highest degree of operational accuracy at consistent stock removal rates, such as by means of wheel spindles of exceptional stiffness, very smooth table movements, and precise downfeed control.

A typical representative of the above characterized category of peripheral surface grinding machines is shown in Fig. 4-2.6, for which model the following performance values are indicated by the manufacturers:

Surface finish:
　3 microinches AA (0.075 micrometers R_a)

Flatness over the entire 14 × 36 inches (350 × 900 mm) working area:
　0.0005 inch (0.0125 mm) TIR

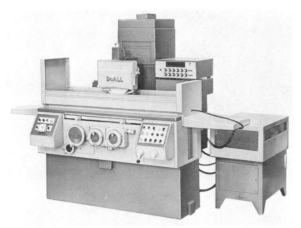

Courtesy of DoALL Company

Fig. 4-2.6 Production type surface grinding machine for precision work, shown equipped with digital readout type automatic downfeed control and externally mounted hydraulic pumps.

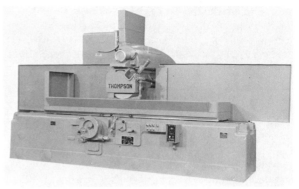

Courtesy of Thompson Grinder, Div. of Waterbury-Farrel

Fig. 4-2.7 General production purpose surface grinding machine with horizontal spindle representing a basic type manufactured in a wide range of different table widths and length capacities. A massive freestanding column supports and provides vertical guides for the up and down moving saddle with its dovetailed cross-slide ways for the wheel head.

Size holding with direct readout device (optional): 0.0001 inch (0.0025 mm).

A few of the design features considered characteristic are the grinding wheel spindle supported on hydrostatic bearings, the hydraulic column counterbalance, the (optional) availability of automatic downfeed with a sizing accuracy of ±0.000 050 inch (0.001 25 mm) when operating the downfeed in increments of 0.0002 inch (about 0.005 mm).

Medium Size Production Type Peripheral Surface Grinders

General purpose surface grinding with an accuracy satisfying the majority of specifications, performed with substantial stock removal rates, represents a very important sector of peripheral surface grinding operations, which may involve workpieces over a wide range of sizes and weight groups. Machines of this type, a representative model of which is illustrated in Fig. 4-2.7, are manufactured in sizes characterized by the following approximate work-table capacity data:

> *width:* from 12 to 24 in. (about 300 to 600 mm)
> *length:* from 40 to 168 in. (about 1 to 4.25 m)
> *height:* from 16 to 36 in. (about 400 to 900 mm).

The maximum wheel diameter is 20 inches (about 500 mm) with widths varying from 3 to 6 inches (about 75 to 150 mm). Correspondingly, drive motors of 15 or 30 hp may be used. Basically similar surface grinders are also finding applications as the base model of special purpose grinding machines, such as: crush dress form grinders, die-block grinders, etc.

Heavy Production Type Peripheral Surface Grinding Machines

The model illustrated in Fig. 4-2.8, one of the smallest of its series, while it is not representative of the heaviest types of surface grinders regarding general dimensions and stock removal performance, still belongs to the category of high productivity equipment. Other models of the same series are manufactured with table widths to a maximum of 36 inches (about 900 mm) and table lengths to a maximum of 192 inches (about 4.9 m) and weighing about 65,000 lbs (about 29,000 kg).

The reason for selecting this example to illustrate the category is to stress the point that peripheral surface grinding is essentially a precision grinding process permitting a high degree of work accuracy to be retained even while accomplishing substantial stock removal on generally heavy workpieces.

The base of the machine contains the wide table ways which are well protected by belt-type way covers. Attached to the base is the heavy one-piece upright column which contains the guideways for the vertical slide, also designated as the *wheel head*. This latter part carries the cross traveling slide, permitting a cross-feed movement equal to the width of the table surface. The wheel spindle, which is designed to accommodate 20-inch diameter by 6-inch-wide (about 500 × 150 mm) grinding wheels, is supported on four large ball bearings which are grease sealed, and has an integral drive motor in the standard 30 hp size. The standard motor of the hydraulic drive is 20 hp, while a 3/4 hp motor operates the power vertical movement of the wheel head.

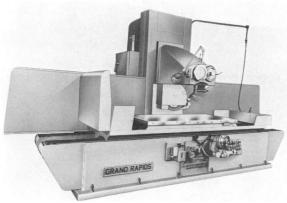

Fig. 4-2.8 Heavy duty production type surface grinding machine designed for precision work on heavy workpieces which often require substantial stock removal to be carried out with large size grinding wheels on machines wich may be equipped with controls offering different degrees of automation.

The high power rating and the general construction of commensurate strength notwithstanding, these machines are intended for very accurate work as is reflected in the general design and execution, such as hand-scraped guideways of the table, the column and the head slides, as well as in their adaptability to diverse workpieces calling for widely different grinding data. Typical values in this latter respect are the following:

(a) The table speeds are infinitely variable to a maximum of 150 feet (about 45 m) per minute, enabling the machine to operate efficiently at the highest speed which is compatible with any specific set of operational conditions.

(b) The cross feed which normally operates at each reversal of the table can have increments set from 1/32 inch (about 0.8 mm) to 3 inches (about 76 mm). The adjustment of the cross-feed rate with a conveniently located handwheel, the setting of its travel length by adjustable dogs which actuate the reversal of the feed direction and the availability of continuous power cross movement at a speed needed for wheel dressing, are a few of the features which also contribute to efficient operation.

(c) The vertical movement whose control sensitivity is highly improved by using an anti-friction ball elevating screw, can be operated with manually controlled downfeed, guided by handwheel graduations in 0.0001 inch (about 0.0025 mm) increments or, optionally, by automatic controls which may also include visual display and automatic wheel size compensation.

Extra-Large Peripheral Surface Grinding Machines

Figure 4-2.9 shows a representative of the extra large grinders which are, in this series, manufactured to a maximum table width of 42 inches (about 1.1 m) and a maximum table length of 240 inches (about 6 m), with main drive motors up to 125 hp. One of the design characteristics of this series of heavy grinders is the box-type upright which consists of two extra-heavy columns, bolted and doweled to the base and connected at the top by a heavy cross bracket, the whole resulting in a unit which is comparable in rigidity to a one-piece rectangular casting. The two mutually facing sides of the columns contain the positive ways for the primary location of the slide and the gibs (main, secondary, and auxiliary) for adjusting the clearances in the major directions.

Fig. 4-2.9 Extra large horizontal spindle surface grinding machine with reciprocating table, adaptable for abrasive machining with high-power drive motor. The double column design and the generally sturdy construction assure high grinding accuracy in the regular, and particularly in the superprecision, versions of these heavy machines.

The illustrated machine is a representative example of those capable of abrasive machining as well as precise grinding. The precision of the grinding performance is assured by the accuracy of workmanship applied to building the machines and by the dependable sensitivity of the controls. In this latter respect various optional systems are also available for the downfeed, which can be semiautomatic, fully automatic, and also complemented with wheel dress compensation. For the convenient and efficient operation of very large grinding machines, pendant controls can be provided. The control of the cross-feed movement reversal points as well as the incremental steps of these movements can be adjusted remotely, from the operator's position, by an electronic device.

Peripheral Surface Grinding— Major and Supporting Functions

Machine Design Characteristics

Peripheral surface grinding machines, although manufactured in sizes and with capacities which extend over a very large range, have many common characteristics with respect to their operating methods, their major and supporting functions, as well as the design principles of their related machine elements. While the similarity of the actual machine characteristics applies to the fundamentals of design and operational concepts only, it is possible to point out and review aspects of machine and auxiliary equipment design, together with their pertinent functions, which are applicable to many different models of peripheral surface grinders.

In the following, examples of aspects related to machine design characteristics, to auxiliary functions and their equipment, and to operational methods, will be reviewed.

The Elements of Transverse Movement

Peripheral surface grinding is frequently applied to work surfaces which are wider than the width of the grinding wheel. Consequently, the work area can be covered only by changing the relative positions of the work and the wheel in a direction normal to the traverse movement of the reciprocating work table so that several consecutive passes are applied to adjacent sections of the work surface. This transverse movement is commonly referred to as the *cross feed*. It can be carried out either by hand adjustment, particularly on smaller machines, or in automatic operation by preset increments, at each reversal of the table traverse, the transverse movement being carried out when the wheel is out of engagement with the work.

It is obvious that such a relative positional change between the wheel and the work can be brought about by displacing either member relative to the other. The transverse movement of the grinding wheel relative to the workpiece, which reciprocates in a constant transverse plane, can be accomplished either by moving the wheel head only, which for that purpose, is supported in a crosswise slide, or by moving the entire column which holds the wheel head on its vertical guideways.

Wheel heads with transverse movements are preferred for large and heavy machines, such as are shown in Figs. 4-2.8 and 4-2.9. These machines require a very high rigidity in the design of the upright column, whose dimensions and weight would make the support on guideways for transverse movement impractical. Carrying out the cross feed by the transverse movement of the work table would be even more ill-advised in such heavy equipment. For these reasons the heavy duty and large size, peripheral surface grinding machines are designed with transverse moving wheel heads supported on a crosswise slide, a category referred to in Table 4-2.2 as the *outrigger* type.

Figures 4-3.1 and 4-3.2 illustrate the machine members involved in the cross-feed movement of two different types of surface grinders, each incorporating a particular design concept applied for assuring a relative transverse motion between the table and the grinding wheel. Manufacturers applying the cross-wise table movement, a design designated as the *saddle type* because of the member which supports the table,

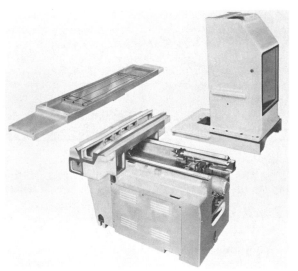

Fig. 4-3.2 Principal structural members of a horizontal spindle surface grinder with crosswise movement of the column. The machine base has integral guideways for the transverse movement of the upright column. The same base supports the table bed with its guideways for the table traverse.

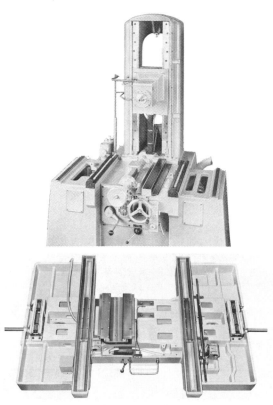

Fig. 4-3.1 Design principles of the saddle type table support in horizontal spindle surface grinding machines. (*Top*) The machine base holding the rigidly bolted column with its needle roller guideways for the grinding wheel head. The box shaped bed has roller guideways for the crosswise movement of the saddle. (*Bottom*) The underside of the saddle; the upper side contains the guideways for the longitudinal table movement.

point to the inherently accurate perpendicularity of the table traverse and cross feed, the two basic movements required to be at right angles to each other. While the assurance of exact perpendicularity is not needed for grinding of plain flat surfaces, the reliance on a high degree of accuracy in this relationship is of significance in precise grinding operations involving shoulders, grooves, or similar multi-plane surfaces. The consistently accurate perpendicularity of the guideways for the vertical wheel head movement, with respect to the table surface (assured by bolting the column solidly to the machine base), is claimed to be another advantage of the saddle-type design system.

For the moving column, displaced along guideways which are solidly attached to the machine base, the rigidity of that support system, using fixed ways, is stressed. As a further advantage, from an operational point of view, the constant distance of the workpiece

from the operator standing in the front of the machine, deserves consideration.

Actually both design systems find applications in peripheral surface grinding machines of different manufacturers, producing equipment of comparable performance. While operational conditions can influence the choice of the user giving preference to one or the other design, other factors such as the details of the acutally applied design with respect to dimensioning and configurations, as well as the applied workmanship, may amply compensate for any conceptual limitations which are inherent in either of these two basic design systems.

Guideways

Guideways for the table movement and for the transverse movement are commonly of the sliding type, combining a V-way and a flat-way. Power lubrication is applied, and a pressure switch is frequently used for protection, which will automatically cut off the machine operation when the pressure drops below a safe level. The larger machines are equipped with tape guards to prevent the penetration of grit and to protect the ways against accidental damage. The table ways may either be integral with the machine bed and table, or may be separate pieces (castings or hardened steel) bolted to the base surface, thus permitting replacement when needed. In exceptional cases, particularly for smaller machines which may be operated by manual table movement, roller supported tables may be used, the guideways consisting of precision ground steel rollers arranged in two rows, one with horizontal axes, and the other having rollers with axes tilted in alternate directions, thus producing a virtual V-way (see Fig. 4-3.3).

Pre-loaded ball ways are also found in some models, and are definitely preferred for operations in which precisely controlled slow table movements are applied, in distinction to the generally fast reciprocating movements of regular surface grinding. Pre-loaded needle roller ways are often used for guiding the vertical adjustment of the wheel head; anti-friction guideways are of advantage here particularly for very low incremental downfeed movements.

Grinding Wheel Spindles and Drives

In most types of peripheral grinding machines the grinding wheel spindles have integral drive motors, a design system well suited for the displacement of

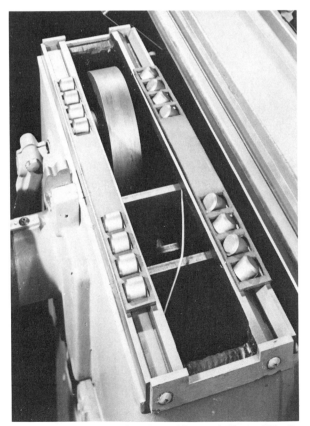

Courtesy of Brown & Sharpe Mfg. Co.

Fig. 4-3.3 Table support system consisting of two rows of steel rollers, with horizontal axes in one row, and with alternately inclined axes in the other row, producing two guideways which are functionally flat and V-type, respectively.

the wheel head, required in this category of grinding machines. Displacement movements of the wheel head are generally applied in the vertical direction for raising and downfeeding the head, and also in a crosswise direction on heavy duty grinding machines. Less frequently, particularly on smaller machines, drives through V-belts are used, also as a means for achieving two different spindle speeds, thereby permitting the use of worn-down smaller wheels at a still efficient peripheral speed. An exceptional case, also requiring belt drive, is the use of high speed spindles with very small diameter wheels, such as may be needed for grinding slots in recessed locations.

An exceptional drive system for grinding wheel spindles with integral motors is the variable speed, DC-motor drive, used in cases where wheel speed variations are needed to adjust the peripheral speed to a particular set of operational conditions, such as in creep grinding.

The bearings of the wheel spindles are most frequently of the antifriction type, either multiple-row ball bearings or sets of tapered roller bearings. In machines intended for toolroom applications or similar quality production work, some models are designed with plain bearings. More sophisticated versions of the plain bearings, are also used in a few models and are sometimes supplied as optional equipment.

The power of the drive motors used in peripheral surface grinders varies, of course, over an extremely wide range, depending on the general size, the applicational purpose, and the wheel dimensions of the different types of machines. The smallest grinders have drive motors of about 3/4 hp, while large, heavy-duty surface grinders, intended for abrasive machining type operations, may have spindle drive motors in the order of 125 hp.

Hydraulic Drive Systems of Peripheral Surface Grinders

The reciprocating movement of the machine table is generally accomplished by a hydraulic cylinder which provides dependable service in most types of small and medium-size surface-grinding machines. Hydraulic drive systems are also used for the crossfeed movement, but not as commonly as for the table reciprocation.

Using hydraulic cylinders and pistons for the reciprocating table movement has many advantages: a basically simple and relatively inexpensive construction, dependable operation, convenient stepless speed variations, and smooth travel and reversals, to name a few.

Besides the many advantages of the hydraulic drive systems a few potentially harmful characteristics of such systems must also be considered. Examples of these are mentioned together with the measures directed at reducing or eliminating the undesirable effects:

(a) Vibrations originating from the high pressure hydraulic oil pumps; these are reduced by special vane shapes, and also by mounting the hydraulic pump unit apart from the grinding machine.

(b) The development of heat in the hydraulic drive system can harmfully affect the accurate positioning of the moving machine members and the maintenance of uniform speed rates. A substantial rise in temperature is prevented by means of cooling devices and, in some systems, by using variable delivery rate pumps instead of throttles installed in the oil lines.

Automatic Downfeed

Automatic downfeed is usually available as standard or optional equipment, in most types of modern peripheral surface grinding machines, except those intended for manual operation only. The degree of automation, however, will vary depending on make, model, and special requirements. In general, feeds at different incremental values, such as those varying over a range of 0.0001 to 0.001 inch (0.0025 to 0.025 mm), can be selected and made operative either at the reversal of the table traverse movement (for "plunge" grinding), or at the reversal of the transverse stroke (for "surfacing"). Further degrees of automation, offered by some manufacturers, may include the following control functions:

(1) A depth limitation by controlling the distance of the feed advance from its starting point.

(2) Depth limitations combined with finishing passes. The feed advance stops, even when that involves a fractional increment, as soon as the finish size has been approached by 0.0001 inch (0.0025 mm), then it produces a final advance of the same value, and subsequently leaves the machine running without downfeed for a pre-set number of sparkout passes.

(3) The grinding wheel head returns to a fixed starting point, which can be set according to the intended amount of stock removal, commonly, within a range of 0.050 inch (about 1.25 mm), although a substantial extension of the retraction range, e.g., to a maximum of 4 inches (about 100 mm), may also be required and supplied, such as when grinding at the bottom of a deep slot.

(4) Visual displays by digital readout of the momentary distance of the wheel from the pre-set finish dimension is also made available by some manufacturers.

(5) Automatic wheel wear compensation is an important complement of the automatic downfeed operation, and may be also useful for manual feed control, using fixed handwheel dial setting in repetitive work. Essentially the feed advance of the truing diamond will cause an equal downfeed of the wheelhead, thus retaining the wheel–work relationship without disturbing the dial setting of the downfeed movement. The operation of the diamond adjustment, such as in the case of a wheel change, is independent of this compensation-coupled truing advance.

(6) Automatic cycling, in conjunction with the automatic downfeed, may also comprise controls for traversing the work table to its extreme position after the wheel is retracted from the work surface. When the table reaches that position, always on the same side in relation to the wheel, the hydraulic pump actuating the table reciprocation stops and the work is presented in an easily accessible location for loading and unloading.

Work-holding in Peripheral Surface Grinding

The grinding of flat surfaces, the most common objective of peripheral surface grinding, generally involves workpieces with two mutually parallel flat surfaces on opposite sides of the part. In such cases one of the flat sides serves as the locating surface which, for parts made of steel, can be retained by magnetic force, such as that exerted by a *magnetic chuck*. Combining the location and the holding of the part in a single operational step and applying these actions to the same surface area, has several significant advantages:

(1) Reduces time and effort spent on loading the work for the grinding operation

(2) Assures a consistently accurate work location

(3) Uniform distribution of the holding force is obtained over a wide area of the work surface (thus avoiding distortions of the workpiece due to insufficient support or excessive clamping pressure)

(4) Inherent accuracy of the grinding machine movements is dependably transferred into the resulting flatness and parallelism of the ground surface, in relation to the locating work surface

(5) The surface to be worked is fully exposed without the interference of clamping elements

(6) Multiple part loading, permitting the concurrent grinding of several small parts, is feasible.

For these and similar reasons, the mounting of the work on a magnetic chuck is the most generally applied method of work holding in surface grinding.

The mounting surface of the part does not have to be continuous for the effective use of a magnetic chuck; interrupted surfaces, as long as they present a flat area of sufficient size and with balanced distribution, can be effectively held by magnetic chucks with appropriate pole spacing.

The two flat surfaces involved, namely, the locating and the worked surface, may be in other than parallel relationship to each other, specifically, at an angle which may be formed by two planes which:

(a) Intersect along a line perpendicular to the direction of the table traverse, such as a wedge

(b) Intersect along a line parallel with the table traverse, such as a prism

(c) Intersect in a single point only, when the mutual positions are at a compound angle.

Magnetic chucks can be incorporated into fixtures of different varieties which are designed to produce, with great accuracy, such diverse relationships between the locating and the ground work surface. The magnetic force originated by a regular magnetic chuck can also be diverted by laminated auxiliary elements, to be applied at any angle including, of course, right angles to the chuck surface. Such elements are also made in V-block form, with the flat base resting on the magnetic chuck, while the V may locate and hold a cylindrical part along two elements of its surface.

Finally, magnetic chucks may also be used to hold nonmagnetic work materials, by means of properly shaped auxiliary members which are retained by the magnetic force and then hold the workpiece by mechanical action.

Vacuum chucks are work-holding devices functionally comparable to magnetic chucks relying, however, on a vacuum (actually the atmospheric pressure) for the retaining force. These are used effectively for holding on surface grinders, nonmagnetic workpieces made of austenitic steels, nonferrous metals, or nonmetallic materials.

Notwithstanding the operational advantages and the adaptability of magnetic and vacuum chucks which hold the workpiece by a pulling action, not all workpieces processed on peripheral surface grinding machines can be located and held by these devices. In such cases, mechanical clamping may be the best method of holding in combination with appropriate locating elements. The most commonly used mechanical clamping device is the vise, which is used on surface grinders in different design varieties, such as plain, tiltable, swivelable, or a type which combines these adjustments.

Peripheral surface grinding machines are also well adapted for grinding slots which may be located in different, usually equally spaced polar positions with respect to the axis of a symmetrical part. These slots may be located parallel to the part axis, such as in a splined shaft, or in radial directions, such as notches on the face of a clutch plate. For this kind of grinding operation the workpiece is held in a horizontal or vertical dividing head, generally of the indexing type. Such work-holding devices are also available with automatic control of the indexing movement, which may be tied in with the operation of the basic machine control system, resulting, e.g., in an automatic spline grinder.

Besides those indicated, there is also an extensive variety of special work-holding fixtures, either obtained by the adaptation of the regular work-holding devices or developed in designs satisfying particular locating and holding requirements of workpieces processed on peripheral surface grinding machines.

In the following, a few types of work-holding devices which are frequently applied on peripheral surface grinding machines will be discussed.

Magnetic Chucks for Peripheral Surface Grinding Applications

Magnetic chucks used on peripheral surface grinding machines are generally rectangular in shape and are manufactured in many different sizes with work-holding surfaces extending over a range from about 4×8 inches (about 100×200 mm) to 42×96 inches (about 105×245 cm).

Magnetic chucks operate most frequently as regular electromagnets, a system simple in design and providing a high degree of holding power.

Another system (see Fig. 4-3.4) with designations "Electroperm" and "Magna-perm," etc., represents a combination of electromagnets and permanent magnets. The operating element is a permanent magnet which, however, is energized by an electric current flowing in a coil which surrounds the magnet. In this system the electric current has to pass through the

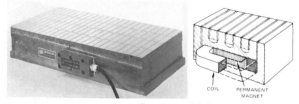

Courtesy of O.S. Walker Co., Inc.

Fig. 4-3.4 (*Left*) Magnetic chuck in a combination design, which operates as a permanent magnetic chuck, except for being energized and de-energized electrically. The diagram (*Right*) illustrates the essential design principles.

coil momentarily only and then the energized magnet will exert the holding force needed to retain the work. For releasing the work, electric current with reversed polarity is passed through the coils. Electrically energized permanent magnet chucks offer several advantages, such as cool operation, superior holding power and the assurance that accidental interruption in the power supply will not cause the unintended release of the work; once charged, the permanent magnet elements will maintain the magnetic field which attracts the workpiece.

Both the electromagnet and the electrically energized permanent magnet chucks operate on direct current which can conveniently be provided by a solid state rectifier, usually supplied for that purpose with a reversing switch for releasing the part at the end of the grinding operation.

Permanent magnets which operate entirely as self-contained units, independently of the electric supply, are also available in small and medium sizes to a maximum work-holding area of 12 × 48 inches (about 300 × 1200 mm), and have metallic or ceramic magnets as active elements. Permanent magnet chucks are supplied with solid base or as swivel chucks which can provide an inclined surface for supporting the work during grinding. (See Fig. 4-3.5). The permanent magnet chuck is turned on and off by means of manually displaced elements; the sliding of a bottom pack composed of steel poles and magnets along the operating elements, causing either an augmentation of the forces in the "on" position, or a slightly reversed polarity in the "off" position.

Workpieces which have been held by magnetic force during the grinding must be demagnetized after the operation. For this purpose AC demagnetizers in one of the three basic types are used, each available in different sizes and output, measured as flux density:

(1) Plate type demagnetizers are preferred in toolroom use and have a flat operating surface along which the part to be demagnetized must be slid by hand, passing it off the demagnetizing field.

(2) Aperture type demagnetizers, applied in operations involving small parts processed by automatic handling, have the shape of a frame through which the workpiece must pass, e.g., rolling down a sloping track through the aperture.

(3) Portable demagnetizers are used for large and heavy workpieces which are difficult to move.

The pole pattern and distribution of the magnetic chucks may be selected to best suit the overall dimensions and configuration of the workpiece, although for common surface grinding applications in the toolroom, a single type, usually laminated with transverse

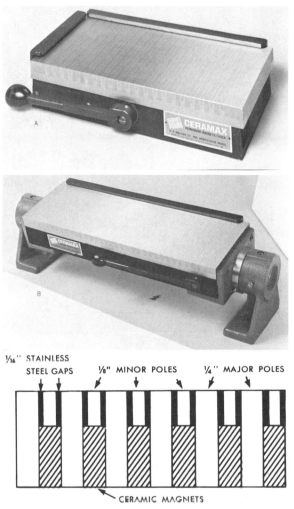

1/16" STAINLESS STEEL GAPS   1/8" MINOR POLES   1/4" MAJOR POLES

CERAMIC MAGNETS

Courtesy of O.S. Walker Co., Inc.

Fig. 4-3.5 Permanent magnetic chucks. A. With solid base; B. With swivel base. The diagram shows the design principles: ceramic magnetic elements are used from which steel poles, separated by stainless steel gaps, conduct the force to the work-holding surface of the chuck.

poles is generally satisfactory. For other uses the poles may be of the grid type or of the laminar (bar pole) type, the latter with transverse or longitudinal laminations. The pole gap dimensions may be of the general purpose, the widely spaced, or the densely spaced variety. For temporarily adapting a magnetic chuck with wide pole spacing for the holding of small workpieces, auxiliary top plates with close pole spacing may be used.

For adapting regular magnetic plates to work configurations which cannot be mounted directly on a flat plate in the horizontal position, auxiliary laminated blocks are used. These are made in rectangular shapes, e.g., for mounting the part on a locating surface at right angles to the worked surface, or of V-block shape (see Fig. 4-3.6) for locating and holding cylindrical workpieces.

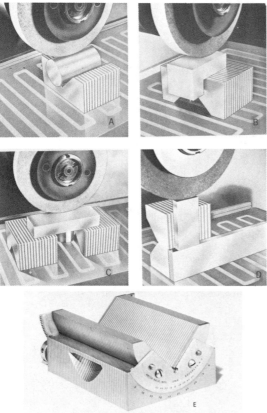

Courtesy of Anton Machine Works

Fig. 4-3.6 Laminated blocks used on magnetic chucks as auxiliary elements for positioning, orienting and holding workpieces of cylindrical and other shapes.
A. V-block holding a cylindrical workpiece
B and C. Laminated blocks holding workpieces provide clearance for protruding elements.
D. Laminated Block used as angle plate.
E. Laminated block fixture which provides an adjustable swivel base for the top member with V-shaped work-holding surface.

Magnetic sine plates, often used in various toolroom operations requiring high angular accuracy, generally have permanent magnets solidly attached or temporarily mounted on a hinged base which can have a single, or two mutually perpendicular pivot axes (see Fig. 4-3.7), the latter is used for producing compound angles. Both types are made in different sizes; typical working surface dimensions are 6 × 6 inches (about 150 × 150 mm) and 6 × 12 inches (about 150 × 300 mm).

Another type of adjustable incline magnetic plate for surface grinding operates by rotating around a centrally located axis and the tilt angle can be set by

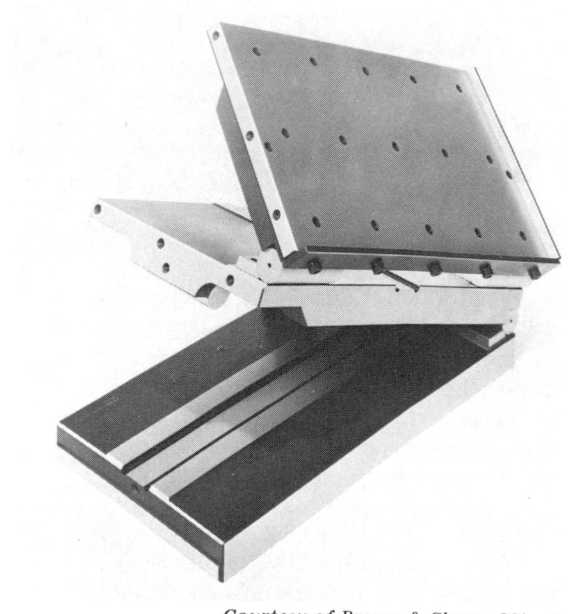

Courtesy of Brown & Sharpe Mfg. Co.

Fig. 4-3.7 Compound sine plate on the top of which a permanent magnetic chuck can be mounted.

means of the attached vernier scale, graduated in degrees and increments of 5 minutes. Such fixtures can be rotated over a wide arc and permit very fast setting without the aid of gage blocks; however, the attainable angular accuracy is less than in the case of regular sine plates.

Nonmagnetic Work-holding on Peripheral Surface Grinders

Vacuum chucks operate by a vacuum supply system and may be used in a manner similar to the magnetic chucks for holding workpieces made of nonmagnetic materials. The workpiece rests on the steel face of the chuck which, in one system, has crosswise grooves containing the vacuum ports. The area of the chuck which is covered by the workpiece can be separated from the surrounding surface area by the insertion of channel seals in order to assure maximum holding power.

Other systems have vacuum channels interconnected by slots milled into the cross ribs. These chucks can be supplied either in basic form without openings, which are then drilled to suit the shape of the work, or with different vacuum opening patterns, such as holes or veins. Such vacuum chucks are conveniently used in conjunction with magnetic chucks on which the vacuum chuck is mounted by bolts, thus combining the effect of the magnetic force used for retaining auxiliary members, with the vacuum effect which actually holds the nonmagnetic workpiece. The auxiliary members used in conjunction with vacuum chucks comprise the masking plate which covers the vacuum openings (not needed) on the chuck surface and the steel locating blocks which, by securing the position of the workpiece against lateral movements, support the retaining action of the vacuum. Figure 4-3.8 illustrates such a setup in combination with a magnetic chuck; it also shows the design principles of the basic vacuum chuck of the type just described.

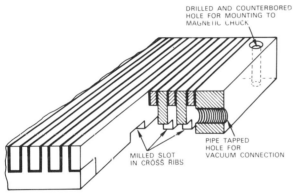

Courtesy of O.S. Walker Co., Inc.

Fig. 4-3.8 Photograph of a vacuum chuck for holding non-magnetic workpieces. It is shown in use on a surface grinder; the vacuum chuck is mounted on a magnetic chuck and the work is held laterally secured by steel blocks, while the rest of the chuck surface is covered by a thin steel masking plate. The diagram is a design concept of the basic vacuum chuck, but without the vacuum openings in the top plate.

Vacuum chucks are usually supplied in the size range of 5 × 10 inches (about 125 × 250 mm) to 12 × 48 inches (about 300 × 1200 mm) surface area. The vacuum needed for operating the chuck can be produced by a vacuum pump unit which also includes the necessary fittings and reservoir tank, all in a compact arrangement to be installed beside the surface grinding machine.

Machine vises used on peripheral surface grinders must be of the precision type, made with an accuracy

corresponding to that expected from the grinding operation. Such devices, also called grinding vises, are made of tool steel, completely hardened and ground with a typical accuracy of 0.0002 inch (about 0.005 mm) with regard to the parallelism and squareness of the operating surfaces and movements. Precision grinding vises may be mounted directly on the machine table, parallel or at right angles to the traverse movement, or they can have an intermediate swivel base. Similar vises are also made for tilt adjustment (see Fig. 4-3.9), mounted on a hinged base which contains a dial graduated in degrees to indicate the incline angle of the vise setting.

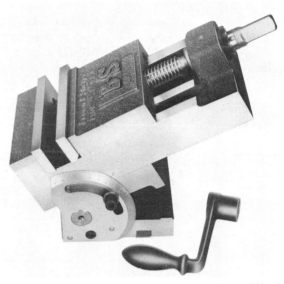

Courtesy of Brown & Sharpe Mfg. Co.

Fig. 4-3.9 Precision grinding vise with tilt adjustment.

Dividing heads and *indexing devices* for grinding surface elements at the required circumferential spacing make peripheral surface grinding applicable for multiple edge cutting tools, such as reamers, milling cutters, as well as for construction parts of similar configurations. Figure 4-3.10 illustrates the example of a relatively simple type of indexing device which is successfully used on surface grinding machines for holding workpieces between centers. Other types of indexing devices, either with vertical or horizontal axes, may have notched dividing plates for rapid indexing, or face-geared discs which have their teeth mutually meshed when in operating position or temporarily disengaged while indexing. This latter system can provide extremely high circumferential spacing accuracy, in the order of 5 seconds of arc, with the capacity of retaining the indexing accuracy over an extended service life.

Courtesy of Brown & Sharpe Mfg. Co.

Fig. 4-3.10 Index centers for surface grinding machine, operated by manual rotation through a worm and dividing plate with six rows of holes, for producing the more common circular spacings.

Devices for Wheel Truing and Dressing on Peripheral Surface Grinders

The basic method of peripheral surface grinding for producing a flat surface consists in taking several traversing passes which are adjacent to each other in the crosswise direction. Most of these passes engage the entire width of the grinding wheel, a condition favorable for the uniform wear of the operating wheel surface. Furthermore, the overlapping positions of the adjacent passes help to eliminate the adverse effect of the more rapid breakdown of the wheel edges. These conditions contribute to reducing, although never entirely eliminating, the need for wheel dressing in such basic surface grinding operations.

In "plunge" type peripheral surface grinding, such as is used for slots and profiles, the lateral overlaps of consecutive passes are not present and thus cannot exert their equalizing effect. Consequently, plunge type surface grinding requires regular wheel truing, particularly when the contour of the ground surface, whether straight or profiled, must be produced to tight tolerances.

Another factor calling for regular wheel dressing as a means for maintaining a "sharpness," that is, the optimum cutting capacity of the wheel, can be the hardness of the bond which, in the case of high accuracy requirements, may be selected by giving precedence to size and form-holding capability, over assuring the self-dressing properties of the wheel.

Finally, in profile grinding, the contour of the grinding wheel must be produced and maintained by means of regularly applied truing; the required frequency of these operations is only slightly affected

by the effectiveness of the wheel's self-dressing characteristics.

It can be concluded that wheel dressing and truing have important functions in peripheral surface grinding and are key elements in extending the adaptability of the process, particularly into the field of profile grinding.

Table 4-3.1 presents a survey of wheel dressing and truing systems which are the most frequently used in peripheral surface grinding. Subsequently, typical representatives of several systems will be illustrated and discussed in greater detail.

When there is no particular technical reason for using a table-mounted wheel-truing device, the truing of the grinding wheel with a wheelhead mounted attachment offers the advantage of wheel dressing without removing the work, its holding fixture, or carrying out other time-consuming steps. A compact version of the wheelhead mounted type, also called over-the-wheel truing attachment, designed for power operation, is shown in Fig. 4-3.11.

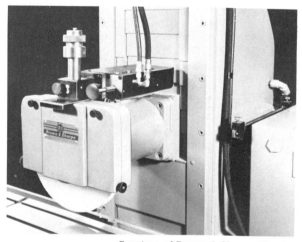

Courtesy of Brown & Sharpe Mfg. Co.

Fig. 4-3.11 Over-the-wheel truing attachment with power drive, operating without disturbing the workpiece beneath the wheel.

A similar device of different design, also with power drive of the diamond slide over the wheel is shown in Fig. 4-3.12. Generally, hydraulic power is used, which permits stepless variation of the truing speed and adjustable reversal point positions. The diamond infeed is manually adjusted with a micrometer screw.

Wheel edge forms of simple geometry, such as angles or radii can be produced by hand-operated table mounted attachments which may serve either of these basic forms, or are designed as a combination device, such as that shown in Fig. 4-3.13, which can

Table 4-3.1 Methods and Equipment of Wheel Truing in Peripheral Surface Grinding

Wheel Profile		Truing Device System	Discussion of Applications and Operation	Illustration Examples
Category	Configuration			
Plain	Straight wheel periphery	Table mounted (mounted either directly, or on a magnetic chuck installed on the table)	The wheel dressing tool, e.g., serrated steel discs for coarse dressing, or diamond nibs for fine truing and general use, are retained in a holder which can be mounted on the machine table either directly or on the surface of an installed magnetic chuck, and then using the machine movements for infeed and for traversing the tool across the wheel periphery. A simple device, but its use involves positional adjustments each time, and then the restitution of the operational positions prior to the resumption of grinding.	
		Wheel head mounted, manual	The wheel-head mounted truing device permits the truing of the wheel without disturbing the operational setup and is recommended for continuous operations on similar workpieces. The manual traversing of the wheel holder bracket requires some skill for assuring a uniform rate of movement.	
		Wheel head mounted, with power drive	Power operation, usually hydraulic, permits the sensitive adjustment and consistent maintenance of the optimum traversing speed and the selected infeed rate of a self-contained truing device, such as the wheel-mounted type. The operation of such devices is, in some cases, combined with automatic size compensation.	Figs. 4-3.11 and 4-3.12
Simple geometric shapes	Angles, circular arcs (radii)	Table mounted, single element type	For producing a narrow inclined surface by peripheral surface grinding, the face of the wheel is trued to the required angle. Table-mounted, manually operated truing devices may be designed to have a sine-bar type base for very precise angle control; others are adapted for radius truing also and have a swivel base for the adjustment of the truing angle.	Fig. 4-3.13

(Continued on next page.)

Table 4-3.1 (*Cont.*) Methods and Equipment of Wheel Truing in Peripheral Surface Grinding

Wheel Profile		Truing Device System	Discussion of Applications and Operation	Illustration Examples
Category	Configuration			
Simple geometric shapes	Angles, circular arcs(radii)	Table mounted, double element type	In applications which require the simultaneous grinding of two adjoining surfaces, each at a different incline angle, double angle truing devices are used which provide precise control of the angular relationship of the ground surfaces.	Fig. 4-3.14
		Wheel head mounted, adjustable to different angles	The benefits for repetitive operations of wheel-head mounted truing devices also apply to angular truing. Such wheel-head mounted truing devices are usually designed to permit rapid adjustment to different frequently used angles.	Fig. 4-3.15
Composite profiles	Blended profiles	Continuous radius and tangent truing devices	The generation of composite profiles in truing the wheel is needed mostly in toolroom operations where the flexibility of adjustment for radius length and angle of tangency are essential requirements, while manual operation and table mounting are acceptable design characteristics.	Fig. 4-3.16
Complex or special profiles	Repetitive or individually complex profiles, comprising several elements of regular or irregular contour.	Profile bars (templates) applied with direct transfer of the contour	Duplicating devices of relatively simple design use profile bars (templates), usually made of hardened steel, which are traced by a follower, that in turn is attached to the upper slide of the device carrying the diamond nib. As the base slide of the truing device is moved across the periphery of the grinding wheel, usually by hydraulic power, the follower, riding on the profile bar, will cause the upper slide to move radially in or out with respect to the wheel. The system is rarely applied to intricate profiles or to those with sharp transitions or steep angles.	
		Templates with profile transfer by pantograph	Recommended for intricate profiles which must be duplicated with precisely defined details. Used for toolroom work or for production, in limited quantities, considering the time consuming characteristics of pantograph duplication.	

Table 4-3.1 *(Cont.)* **Methods and Equipment of Wheel Truing in Peripheral Surface Grinding**

Wheel Profile		Truing Device System	Discussion of Applications and Operation	Illustration Example
Category	Configuration			
Complex or special profiles	Repetitive or individually complex profiles, comprising several elements of regular or irregular contour.	Profiled diamond blocks	Blocks studded with diamonds on the top surface which has the profile required on the wheel, are mounted on the machine table, conveniently in a position corresponding to the extension of the finished work profile. Some types of wheel-head mounted truing devices are also designed to accept profiled diamond blocks. Applied to repetitive work which justifies the initial high cost of such specially made tools.	
		Rotary diamond dressing rolls	Similar to the preceding, but the tool is a profiled roll with diamond impregnation around its entire periphery and rotated at substantial speed by an individual power drive. Performs the profile truing of the wheel very rapidly, with consistent accuracy which is maintained over an extensive service period. However, due to the high cost of the diamond-impregnated profiled roll, the application is limited to high-volume production.	
		Crush form dressing	A profiled roll made of hardened steel or cemented carbide is supported on a freely rotating axle and approached radially toward the grinding wheel which is run at a greatly reduced special speed. By applying force the profiled roll is made to penetrate gradually into the wheel surface, dislodging, under pressure, the interfering grains, thus transferring the inverse replica of the roll profile into the wheel, In some cases two rolls are used, one being retained as a master for the occasional truing of the wheel which then is used to reshape the worn working roll. A very productive method requiring special, but not particularly expensive, tooling.	Fig. 4-3.17

Fig. 4-3.12 Hydraulically actuated dressing device, attached to the wheel hood, with control valves for direction and speed. Infeed movement is controlled by a micrometer screw.

Fig. 4-3.13 Table mounted angle and radius truing attachment.

be used for truing both angles and radii, as separately applied alternative operations.

When the wheel edge must contain two angular sections which are precisely interrelated, thus forming a double bevel, special table mounted truing attachments, such as illustrated in Fig. 4-3.14 may be needed. This device has two slides, each tiltable around its own pivot axis, and individual cranks are also provided for manual operation.

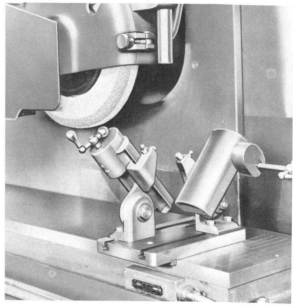

Fig. 4-3.14 Table mounted double bevel truing device.

For truing single angles, particularly in the frequently used angular positions of 30°, 45°, and 60°, some models of large grinding machines can be supplied with wheel-slide mounted, hydraulically actuated, special truing devices (see Fig. 4-3.15) which offer the advantage of quick truing of the wheel in any position of the work table.

A frequently used profile element is the circular arc, commonly referred to as a *radius*, executed with a specific radius length and adjoined by one or two tangent straight lines. The point of tangency is usually defined as a point on a circle having the desired radius and centered at a specified distance from the side of the wheel. For producing such composite wheel profiles with one continuous movement of the truing diamond, special continuous radius and tangent truing attachments are used, a typical example of which is shown in Fig. 4-3.16.

Courtesy of Mattison Machine Works

Fig. 4-3.15 Wheel-slide mounted single-angle truing device for 30°, 45°, and 60° angles.

Profiled workpieces comprising straight line elements in the direction normal to the profile plane, are particularly well adapted for grinding on peripheral surface grinders with reciprocating table movement.

Wheel truing, such as is employed on peripheral surface grinders in general, in distinction to the generating type of profile grinding process, is a fundamental operation of plunge type profile grinding. In plunge grinding, the ground profile is an inverse replica of the wheel profile, consequently, the latter has to be produced consistently with a definition and accuracy at least equal to that expected from the ground

surface. Because of the particular accuracy requirements and varying degrees of form intricacy which characterize the profiled workpieces ground on peripheral surface grinding machines, the applied methods and devices of profile wheel truing differ in many respects.

The more frequently used systems of profile truing on peripheral surface grinders are listed in Table 4-3.1. As an example, Fig. 4-3.17 illustrates a wheel crushing device installed on a horizontal spindle surface grinder. This particular attachment is motor driven for rotating the crush roll at a peripheral speed close to that of the grinding wheel. Motorized wheel crushing devices are preferred for use as attachments on general purpose surface grinding machines, which thus can be operated at regular wheel speed during the crush-dressing. In grinding machines which are built especially for crush dressing, a separate low wheel speed is provided for operating in engagement with a freely rotating crush roll.

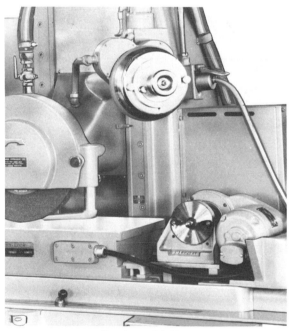

Courtesy of Gallmeyer & Livingston Co.

Fig. 4-3.17 Crush-dressing device installed on a toolroom type surface grinder; the device has motor drive to eliminate the need for specially low wheel-speed during the crush dressing of the grinding wheel.

Size Control in Peripheral Surface Grinding

Peripheral surface grinding is a method used predominantly for precision work requiring accurate

Courtesy of Brown & Sharpe Mfg. Co.

Fig. 4-3.16 Continuous radius and tangent truing device.

dimensions, consequently, the consistent control of the produced work size is an important process function. The size which must be controlled in all surface grinding operations, even when other dimensions are also critical, is the distance of the ground surface from a reference plane, commonly, the locating surface of the workpiece. That dimension is controlled by the downfeed of the grinding wheel, retained in the wheel head which is mounted on a slide that is movable along the vertical guideways of the upright column.

The basic elements of this vertical wheel head movement are a threaded spindle, similar to a leadscrew, and rotated through gears by a handwheel which has a scale ring graduated in small increments; a typical value for precision type grinders being 0.0002 inch (or 0.005 mm). Machines intended for high precision work may have additional fine feed knobs with graduations of 0.0001 inch (or 0.0025 mm). For quick positioning, such as is needed during the setup, grinders of more advanced design, are also equipped with rapid power movement of the wheel head.

The various systems of automatic downfeed have been discussed in the preceding chapter. These systems operate on the principle of attaining the finish size by means of a wheel advance over a set distance from a fixed starting position. The effect of wheel wear is generally neglected, except for the reduction of the wheel diameter in dressing; this is corrected either manually or by an automatic wheel dress compensating device.

For controlling the wheel downfeed by its actual movement, in distinction to controlling the downfeed input, surface grinding machines can be equipped with vertical wheel head position dial indicators, such as that shown in Fig. 4-3.11.

To assure the direct control of the workpiece thickness, by relying on the result of the grinding instead of the wheel head downfeed movement, surface grinding machines can be equipped with *in-process gaging instruments*. Generally such measuring instruments operate by means of a sensing finger contacting the work whenever it is brought into a position under the gage head by the reciprocating movement of the machine table. The dependable functioning of the gage sensing member is not affected by interruptions of contact with the work. The electrical signals originating from the sensing head are transmitted to a remotely located amplifier and appear on a dial graduated in fine increments, 0.000 050 inch (or 1 micrometer), showing the progress of the stock removal and the momentary work size in relation to a previously set finish dimension.

The operation of such electronic gages can be tied in with the control of the grinding machines which are equipped with automatic feed movement. Such controls can either cause the termination of the downfeed when the finish size is reached, or reduce the downfeed rate, when the finish size has been approached, to a certain distance, and then continue the downfeed at a slow rate, followed by a few spark-out passes.

Size control in conjunction with the transverse (cross feed) movement is generally not critical in peripheral surface grinding, when the crosswise advance serves positioning purposes only, such as for the initial setup, and for the successive presentation of adjacent work surface sections into the path of the grinding wheel. Transverse size control has, however, an important function in the grinding of slots, of horizontal surfaces with an adjoining shoulder, of features in parallel positions located at specific crosswise distances apart, and in other, similar types of work configurations. Such work is most common in toolmaking operations, and for that reason peripheral surface grinding machines of the toolroom type are usually designed with fine adjustment of the crosswise moving member, commonly in increments equal to those available for the fine adjustment of the downfeed movement.

Peripheral Surface Grinding – Process Practices and Special Adaptations

Operational Data for Peripheral Surface Grinding

Peripheral surface grinding, in the majority of its common applications, is a finishing method whose primary objective is the dimensional and form accuracy of the produced surface. Although in some operations, such as in abrasive machining, the amount of stock removed from the work surface may reach several cubic inches per minute (1 cu in. equals 16.39 cu cm), commonly, the rate of stock removal is in the 0.250 to 2 cubic inches /minute (about 4 to 33 cu cm/min.) order. Such performance is within the capacity of grinding machines equipped with spindle drive motors in the 5- to 12-hp range, these two values corresponding to the above mentioned limits of commonly used maximum stock removal rates. In toolroom applications using grinders which may have drive motors of 1 to 3 hp, the attainable stock removal rates are substantially lower, considering that the relationship betweeen the wheel drive power and the machine capacity for stock removal is not a linear one.

In traverse type peripheral surface grinding the rate of stock removal in cu in. per minute is the result obtained by multiplying three factors: the table speed per minute, the amount of the downfeed, and the amount of the cross feed applied during the same period; in plunge grinding, only the first two factors are considered.

While the actually applied operational values will have to be selected so that they will not produce a load in excess of the stock removing capacity of the grinding machine, the available power of the drive motor does not determine the applicable grinding data. These are controlled by several other factors, examples of which follow:

(a) The general size of the work surface and the amount of material left for the grinding operation, i.e., the stock allowance. These conditions may not permit operating the grinding machine at full capacity.

(b) The configuration of the workpiece may make it sensitive to deflections by the grinding pressure or could limit the applicable holding force.

(c) In finish grinding, rigorous specifications for size, flatness, surface finish, etc., generally require stock removal rates substantially smaller than those used for roughing.

(d) Profiled work surfaces, particularly when composed of intricate contour elements, are usually ground with reduced stock removal rates in order to retain the shape of the wheel between the applicable truing operations.

(e) Concern about the potential damaging of the work surface, such as causing heat cracks, deteriorating the surface integrity of highly stressed construction members in critical applications, etc., often requires the reduction of the stock removal rate to below the level established by other limiting factors. In the case of certain workpieces, materials and heat treat conditions using grinding data which are entirely acceptable for some common applications may have abusive effects here which must be avoided.

It may be concluded that while the available drive power is indicative regarding the maximum rate of stock removal, acting as a ceiling for the combined effect of the selected data, the actually determined values for work speed (table speed), infeed (down-feed), cross feed (transverse increments), and even for the peripheral speed of the grinding wheel, are controlled by several other, generally workpiece-related circumstances.

The following operational values, applicable to peripheral surface grinding operations carried out on medium size machines with drive motors in the above mentioned 5- to 12-hp range, reflect common practices and are intended for general information only:

(1) *The work speed* results from and is equal to the speed of the reciprocating table movement. Most generally applied table speeds are within the 50 to 100 ft per minute (about 15 to 30 m/min) range, using lower speeds of 40 fpm (about 12 m/min) for certain special alloys or materials (e.g., titanium), and occasionally also for obtaining a particularly fine finish on common work materials. Exception-ally low table speed is used in creep grinding, a method which will be discussed later in this chapter.

(2) *The downfeed* per pass has as basic values .001 inch (0.025 mm) for roughing and one-half of that amount for finishing. Much heavier downfeed rates, multiples of the above, may be used in some roughing operations, while for finish grinding to extreme dimensional tolerances, downfeed increments as fine as 0.000 04 inch (or 0.001 mm) may be needed.

The downfeed is applied (a) in plunge grinding at each or either end of the table travel, and (b) in transverse grinding at each reversal of the cross feed. In plunge type surface grinding where the entire width of the grinding wheel must penetrate into the work material to the full amount of the downfeed, the applicable rates are generally smaller than in transverse grinding. In the transverse grinding process a wheel, substantially wider than the actual cross feed, distributes over its face the developed cutting force, thus permitting higher downfeed rates.

The cross feed, which is used in traverse grinding only, is commonly selected to equal a specific part of the wheel width, such as 1/4 to 1/12. These fractions express the limits of the frequently used cross-feed rates, defined in relation to the width of the grinding wheel. The higher cross-feed rates, which may be, exceptionally, even in the order of 1/2 to 1/3 wheel width, are used for roughing, particularly of work materials having low hardness. For finishing, as well as for grinding high strength or sensitive materials, a slow cross feed, in the range of 1/8 to 1/12 of the wheel width, is applied.

The wheel speed, in terms of peripheral speed, is commonly in the range of 5,500 to 6,500 fpm (about 30 to 35 m/sec). High-speed grinding, operating at about twice that speed, is used on specially built surface grinding machines only, which are designed generally for abrasive machining and require appropriately manufactured grinding wheels. On the other hand, very low grinding wheel speeds, in the order of 3,000 to 4,000 fpm (about 15 to 20 m/sec) are needed for the surface grinding of various types of high alloy and high strength work materials, using the lower wheel speed as a means for minimizing the development of heat, to which certain types of materials and workpieces are particularly sensitive.

Grinding Wheel Selection

Most of the basic rules which govern the selection of grinding wheels for other methods of precision grinding apply to peripheral surface grinding also. In this respect it is well to remember that the length of the arc of contact between the wheel and the work which results in peripheral surface grinding is greater than that in cylindrical (OD) grinding and less than in internal (ID) grinding. This characteristic of peripheral surface grinding affects the desirable composition of the applied grinding wheels.

In the following, a few general aspects of grinding wheel selection will be discussed rather than tabulating a list of recommendations.

The abrasive material. For steel grinding, alumi-num oxide grains, the regular type for construction parts and general purpose work are used. However, the more friable types are selected for hardened steel and particularly for tool steel. High purity aluminum oxide abrasives, with some types including special metallic additives, provide a cooler cut and thus permit the increase of the infeed rate without burning the work material. Grinding wheels made of alumi-num oxide grains of special composition may also provide advantages in profile grinding by retaining their size and shape longer, due to their freer cutting action.

Silicon carbide of the regular type is used for grinding cast iron and nonferrous metals, and a special high purity type is used occasionally for the rough grinding of cemented carbides. Diamond, wheels, however, are preferable and are always used for cemented carbide finish grinding.

The grain size generally varies over a narrow range only. Numbers 36 to 46 are used for roughing, and for soft materials in general. Number 46 may also satisfy the requirements of varied operations in general purpose work. Finer grains, commonly Number 60 and exceptionally, Number 80, or finer, may be selected for the grinding and particularly for the finishing of hardened work materials. One degree coarser grain is generally used for larger wheels 16 inches or over (or about 400 mm), in view of the more extended arc of contact with the work.

The bond of the wheels used in peripheral surface grinding is, with few exceptions, the vitrified type. *The structure* of the wheels generally corresponds to the Density Number 8, using a more open structure for wheels operating with broad work contact.

The hardness of the wheels used in this method commonly varies from H to K. Harder bonds are used for soft materials, also for narrow or interrupted surfaces in hardened steel. Softer grades are preferred for general purpose work and for hardened steel in workpieces presenting broad contact with the wheel. In this latter case when fine grain size is selected, even softer bonds such as F, may be recommended. For larger diameter wheels, one grade softer bond may be needed in some operations than for the same workpiece when ground with a relatively small wheel.

Additional hints regarding the original, and perhaps corrective, selection of grinding wheels may be gathered from a discussion of the common faults and their probable causes in peripheral surface grinding in Table 4-4.1.

Dry and Wet Grinding

Peripheral surface grinding is often carried out dry, particularly in toolroom type operations. *Dry grinding*, by avoiding the need for guards and the visual obstruction caused by the coolant, permits a continuous observation of the workpiece while it is in the process of being ground. Due to the narrow area of contact in peripheral surface grinding carried out with shallow wheel penetration into the work, the heat developed in the process may not cause objectionable work damage, particularly when only light cuts are taken.

The major inconvenience of dry grinding is the development of dust, harmful to both the operator and the equipment. Modern surface grinding machines intended for dry grinding operations are, therefore, equipped with an air exhaust attachment (see Fig. 4-4.1). This contains a motor driven fan which is mounted on a separator tank, erected on the side of the grinder, usually within the area required by the full length travel of the longitudinal table movement. The exhaust nozzle is attached to the wheel guard and is connected with a flexible pipe through which the dust-laden air is drawn into a separator unit. This latter is usually comprised of two major elements: a spiral separator, in which the whirling air throws the heavier particles against

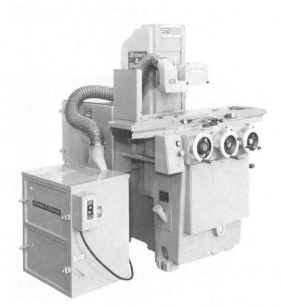

Fig. 4-4.1 Exhaust attachment for dry grinding arranged alongside a toolroom type surface grinder. The separator tank and fan are of compact design, fit under the reciprocating table and permit installation without additional floor space.

deflectors by centrifugal force, from where they pass into the dust chamber. The air, now containing fine particles only, enters an expansion chamber, where removable filter pads trap most of the remaining solids. The separating efficiency of such attachments is claimed to be 95 per cent, or better.

For surface grinding operations involving substantial stock removal, *wet grinding* is recommended. Heavier type surface grinding machines are supplied with accessories for wet grinding as standard equipment (see Figs. 4-2.7 to 4-2.9). For lighter or medium-size surface grinders the equipment for wet grinding can be provided as an optional attachment, such as is shown in Fig. 4-4.2.

A supply tank, adjacent to, but usually separated from the machine bed, is equipped with baffle plates for supporting the particle settling. The motor driven centrifugal pump is mounted on the tank and supplies the coolant for the wheel through flexible tubing which ends in a nozzle attached to the wheel guard. A regulating valve and a splash guard are also provided. The machine bed has troughs through which the coolant flows back into the tank; the machine table is surrounded by adjustable guards.

In shops operating a large number of similar grinding machines of both types, the exhaust for dry grinding and the coolant supply for wet grinding, can

Table 4-4.1 Examples of Common Faults in Surface Grinding and Their Potential Causes

General Area	Designation of the Faulty Performance	Discussion of Potential Causes and Remedies
Geometric accuracy of the ground workpiece	Flatness deficiencies	Failure to produce flatness of the required accuracy within the limits of the tested machine capability, is often due to one of the following causes: (a) stresses, particularly from heat treating, within the workpiece. This calls for stress relieving prior to surface grinding; (b) insufficiently flat mounting surface of the workpiece which then can be distorted by magnetic pull; avoiding magnetic retainment, e.g., by simply using lateral confinement, may relieve the trouble.
	Out-of parallelism	Out-of-parallelism which exceeds the limits of machine capability is generally due to either of two conditions: (1) the inadequate flatness of the reference surface on which the part is supported or (2) the inaccuracy of the work mounting. To correct the second condition the surface of the magnetic plate or any other holding fixture, must be free of burrs and parallel with the grinding plane; the interference of foreign matter such as grit or chips between the work and fixture surfaces must also be avoided.
	Out-of-squareness	The squareness produced in surface grinding is predicated on three basic conditions: (1) the flatness of the reference surface, (2) the flatness of the ground surface, and (3) the location of the part in a manner assuring the perpendicularity of the reference surface to the plane of the wheel action. It is usually the combined effect of inaccuracies in these respects which limits the attainable degree of squareness.
Conditions of the ground surface	Chatter marks	Chatter marks on the ground surface are generally due to one of the following conditions: (a) vibrations, and (b) reduced cutting capacity of the grinding wheel, such as from glazing (see below: Wheel glazing). Vibrations can be transmitted from outside sources, but more frequently originate within the machine from conditions such as wear in the wheel spindle or its bearings. An out-of-balance wheel is also a common cause of chatter. Sometimes chatter can be eliminated by reducing the cross-feed rate and/or reducing the table speed.
	Scratches on the work surface	Scratches can appear to be repetitive although not uniform, or may be located entirely irregularly; the appearance is often indicative of the probable causes. Marring of the ground surface can result from causes such as accumulation of sludge in the cooling system, loose grit floating between the wheel and the work, or improper wheel dressing. Damage to the ground surface can occur in a subsequent operation when this surface is used to retain the workpiece on the magnetic chuck.

Table 4-4.1 *(Cont.)* **Examples of Common Faults in Surface Grinding and Their Potential Causes**

General Area	Designation of the Faulty Performance	Discussion of Potential Causes and Remedies
Conditions of the ground surface	Burnished work surface	It appears as highly polished irregularly located patches and is indicative of reduced cutting capacity of the grinding wheel. Causes of such conditions may be several, such as too hard or too fine a wheel, delayed dressing or improperly selected grinding data. Improving the cutting capacity of the grinding wheel by changing its specifications, correcting the feed rate or increasing the frequency of wheel dressing, are examples of measures by which the harmful condition can be corrected.
	Burning of the ground surface	This condition is noticeable by the discoloration of the ground surface and, in hardened steel, also by the presence of very fine cracks, known as checkmarks. Mainly responsible are local heating of the surface and inadequate dissipation of the generated heat. Softer wheel, reduced contact area, lighter cut, coarser wheel dressing, increased traversing speed are some of the remedial measures, perhaps in combination with more effective coolant supply and distribution.
Location and retainment of the work during grinding	Inadequate locational control (e.g., for the grinding of grooves)	Lack of positional accuracy of related surfaces, such as parallel grooves produced in surface grinding, may be due to the limited precision of the crosswise machine table adjustment, making the use of a direct sensing position indicator advisable. The true running of the grinding wheel and the proper dressing of both the periphery and the side, preferably to a slight degree of concavity, are further examples of conditions contributing to well defined and accurately located ground features.
	Unstable retainment of the work	Workpieces with mounting surfaces which offer limited contact area or which do not have the proper flatness, may not be retained adequately by the magnetic plate and may slide under the effect of the grinding force. Using magnetic plates with pole pattern better suited for the work, or applying additional lateral support by using retainer blocks or rings, often prove to be effective remedies. Magnetic plates with adjustable force are also useful for grinding workpieces requiring different degrees of retaing pull.
Performance of the grinding wheel	Wheel glazing	A glazed wheel surface can readily be detected by the shiny appearance of the wheel surface and also by the slick feel when touched. Measures to avoid that harmful condition include: (a) use of better suited, softer and coarser grinding wheels or those with more open structure; (b) increased work traverse speed to accelerate wheel breakdown, and (c) wheel dressing with sharp diamond applied with faster traverse rate to produce a coarser wheel face.

(Continued on next page.)

Table 4-4.1 (*Cont.*) **Examples of Common Faults in Surface Grinding and Their Potential Causes**

General Area	Designation of the Faulty Performance	Discussion of Potential Causes and Remedies
Performance of the grinding wheel	Wheel loading	Excessive bond hardness, sometimes chosen to prolong wheel life, may prove harmful by causing wheel loading, thus reducing the effectiveness of the grinding process and possibly damaging the work surface. While the obvious solution is the use of the proper wheel specifications, intermediate corrective measures may include coarser dressing, increased work speed, and a reduced work contact area to accelerate the breakdown of the wheel surface.
	Excessive wheel wear	Besides shortening the service life of the wheel, a too-soft bond will also reduce the effectiveness of the form and size control required in the process. To avoid the harmful effects on work geometry and size, temporary measures until wheels of proper composition are available, may include reduced work speed, lighter cut, more frequent wheel dressing with size compensation, these actions involving, however, reduced operational efficiency.

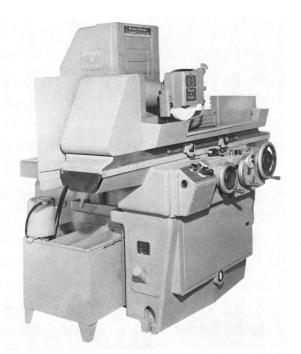

Courtesy of Brown & Sharpe Mfg. Co.

Fig. 4-4.2 Wet grinding attachment with floor tank, pump, hose, nozzle, and guards, installed on a small-size, toolroom type surface grinder.

be part of a centralized system, for assuring higher operational efficiency and easier maintenance.

Coolant supplied through the wheel is successfully applied in certain operations in which the direct introduction of the coolant into the work area by avoiding the deflecting effect of the air barrier around the operating wheel, and the continuous flushing of the operating wheel surface by high pressure coolant emerging from the hub toward the periphery, provide distinct advantages. The grinding of hardened tool steels which are sensitive to thermal cracks and of workpieces which are susceptible to distortion, are examples of preferred areas of application. Figure 4-4.3 illustrates the operating principles of such equipment which require coolant delivery at very high pressure and a particularly effective filtering system.

Mist cooling which represents a transition between dry and wet grinding, is sometimes used in peripheral surface grinding, particularly in toolroom applications. This system does not obstruct the observation of the part and still provides a more effective cooling than the simple air flow. Mist coolant arrangements usually operating with compressed air, dispense the coolant in the form of a fine mist, which on contact with the work, evaporates, thus exerting a cooling effect. Such equipment, which is usually rather compact, can be easily mounted on the surface grinding machine and is operated with air obtained from the shop line.

Common Faults and Their Probable Causes in Peripheral Surface Grinding

The reviewing of faults which commonly occur peripheral surface grinding together with a listing

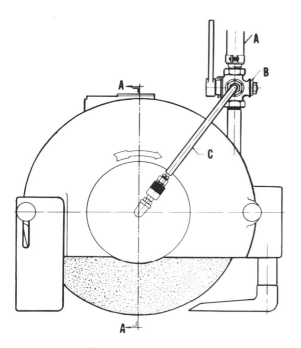

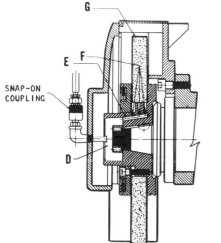

SNAP-ON
COUPLING

Courtesy of Gallmeyer & Livingston Co.

Fig. 4-4.3 Coolant-thru-the-wheel—operating principle. The coolant passing through hose A, valve B, and hose C, enters the central cavity D, of the wheel bushing. Dispersal duct E, and metering plugs F guide the coolant which, by centrifugal force, will traverse the pores of the wheel to its operating surface G.

the probable causes of unsatisfactory performance, may serve the double purpose of correction and prevention. Such a review is presented in Table 4-4.1. Obviously, it does not aim at a complete listing, nor are the listed probable causes considered the only ones that could be responsible for any specific fault.

Use of Peripheral Surface Grinders for Test Operations in Grinding Research

It is worthwhile noting that most of the tests serving research on grinding wheels and grinding technology, are carried out on peripheral surface grinders operating with a reciprocating machine table. Such tests may be directed at determining the comparative performance of grinding wheels, the grindability of different work materials, or examining the effects of variables on the grinding process. Peripheral surface grinding machines are chosen for such tests because of the characteristics of this method which are particularly well suited for carrying out grinding operations under precisely controlled and accountable conditions.

A brief review of these characteristics in their relation to test requirements will indicate the reasons for choosing peripheral surface grinding and reciprocating table travel in preference to all other grinding methods in order to assure precisely controlled and dependably reproducible parameters.

(1) A definite contact area between the wheel and the work, practically from the start to the end of the wheel engagement, only insignificantly affected by the reduction of the wheel diameter during grinding. (In face grinding, the contact area gradually increases from zero to the maximum following the start of the engagement, and then gradually decreases as the work moves out of the wheel area.)

(2) Constant work speed over the entire effective length of the table travel. (In rotary surface grinding, which operates also with the periphery of the wheel, the work speed decreases from the outer edge to the center of the work-holding table which commonly rotates with constant speed.).

(3) The feed rates and the work speed can be regulated in fine increments or steplessly over a wide range to suit the need for extensive and precise adjustment of these data, this being a frequent requirement in test operations.

(4) Simple and secure mounting of the test specimen on a magnetic chuck, which also assures well defined and precisely reproducible work location.

(5) Easily and precisely measurable amount of stock removed from the specimen and of the amount of wear on the straight grinding wheel.

(6) Convenient observation of the work during grinding, should such monitoring of the process and of the momentary state of the test be desired.

The listed characteristics of the method have been pointed out as beneficial for tests in grinding research. Several of these properties of the method are also of advantage in production type operations, particularly when the consistent control of the selected grinding data has a significant effect on the quality of the grinding.

Adaptation for Particular Purposes by Using Basic or Special Equipment

The fundamental principles of peripheral surface grinding, namely: (a) Using its periphery as an operating surface of the grinding wheel, either specifically or essentially; and (b) Causing the workpiece, which is mounted on a reciprocating machine table, to pass longitudinally along the grinding wheel in a straight line, are adaptable to several special operational objectives. The designation "special" refers, in this case, to uses which differ from the basic purpose of producing a flat work surface in a horizontal plane.

The preceding definition for the basic applications does not preclude certain variations from the plain condition resulting from an extended flat work surface parallel to the locating surface of the workpiece. The flat work surface may have defined and projecting boundaries such as the bottom of a slot. The flat surface may also be in a nonparallel position with respect to the locating surface, which then will have to be supported in an other than horizontal plane such as for grinding the sides of a prism.

Applications closely related to the basic may be set up with the aid of readily available auxiliary devices, such as sine plates, dividing heads, or compound action vises for work holding, and angle or radius truing devices for shaping the operating wheel face to regular forms other than straight and parallel with its spindle.

The use of peripheral surface grinding machines for such operations, as well as the required standard and optional accessories, have been mentioned in preceding chapters, together with the basic types of equipment and their more frequent applications.

In the following, the less common applications of the basic system involving adaptations based on special machines or attachments, will be explained by examining several characteristic methods and the means by which they are carried out.

Creep Traverse Grinding

This is a method of peripheral surface grinding developed for the grinding of deep profiles into solid work material. In creep traverse or, by an alternate designation, creep-feed grinding, (a) the form trued wheel is set to the full depth of the profile, and (b) a very slow longitudinal advance movement is substituted for the table reciprocation applied in regular form grinding with gradual infeed. These principles, as well as some of the resulting conditions are shown and explained in the diagram, Fig. 4-4.4.

In creep traverse grinding the wheel will, except for the start and the end of the traverse movement, stay in continuous engagement with the work along the entire contour of the profile. This is a condition substantially different from the regular downfeed type profile grinding, where the crest of the wheel will have to penetrate into the work material to the full depth of the contour, before the actual contact with the work extends over the entire operating face of the wheel.

Creep traverse grinding requires special equipment, sometimes available as an optional accessory of certain makes of peripheral surface grinding machines. The longitudinal table movement must have variable speed, starting at a very low value, but adjustable steplessly over a rather wide range, in order to suit the many different operational conditions. A typical range of a specially designed creep grinding machine extends from 0.4 inch to 80 inches (about 10 to 2000 mm) per minute. For that purpose, special machines are built with DC drive for the creep feed, and an AC motor with mechanical or hydraulic power transmission for the rapid table movement such as is needed for setup and for bridging gaps within the workpiece or between in-line mounted workpieces.

The table guideways must be designed to assure smooth movement even at the lowest speed; table ways of the antifriction type, such as those provided by rows of balls or rollers, are preferred for creep grinding.

Due to the extended length of contact between the wheel and the work, the dissipation of heat and the expulsion of chips are impeded, therefore, ample coolant supply at high discharge pressure is needed, and wheels with open structure are used. The filtering device of the coolant tank must be capable of separating the very fine chips which are typically produced in creep grinding.

In order to produce profiled parts efficiently by creep grinding, the wheel must be trued by a method of comparable effectiveness, such as by crush dressing or rotary diamond truing. For creep grinding in general, and particularly when crush dressing is used, variable speed wheel spindle drives are of advantage and are used on special purpose grinding machines.

The potential advantages of creep traverse grinding, when applied to the proper type of work, include the following:

(a) High rate of stock removal, because of grinding the whole profile simultaneously (A typical value is

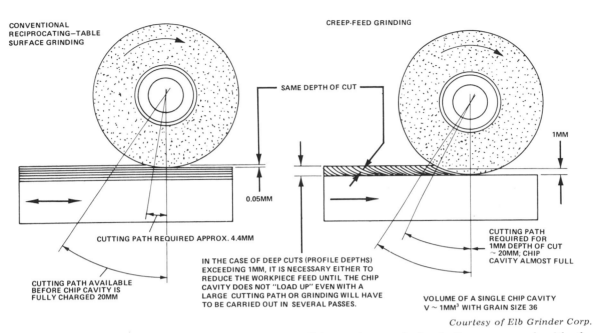

Courtesy of Elb Grinder Corp.

Fig. 4-4.4 Operational characteristics and related conditions of creep-feed grinding, compared with the conventional, reciprocating table type, peripheral surface grinding.

about one cubic inch [about 16.4 cu cm] per minute).

(b) The avoidance of pre-machining, such as by shaping or milling.

(c) Good surface finish, produced by the fine grit wheels which are generally used in this method.

(d) Relatively small specific wheel wear, an important factor in assuring the high form accuracy of the ground profile.

Profile Grinding on Peripheral Surface Grinding Machines

Peripheral surface grinding machines operating with a traversing table are frequently used for grinding profiles with straight line elements. The cross-sectional contour of such profiled workpieces may contain sections which are straight, curved with regular circular or with irregular form, as well as with any combination of such constituent elements. The differences in the character of the profile affect, however, the required or preferential method of wheel truing to such a degree that the applicable method of wheel truing may become a governing factor in the design of surface grinding machines intended for profile grinding operations.

On peripheral surface grinding machines the grinding of the profiled work surface is carried out by plunge grinding, without a transverse movement, consequently the width of the applicable grinding wheel is a limiting factor in the selection of workpieces which can be profile ground by this method.

The length of the profile in the direction of the constituent straight line elements would, however, only in very rare cases cause a limitation, given the commonly great length of the table travel of peripheral surface grinders in relation to the width of the applicable wheels. Actually, the available table travel which exceeds the length of the workpiece may be utilized for mounting several identical workpieces in-line, or for operational functions connected with wheel truing or workpiece gaging.

Way Grinding Machines

Machine members whose operation involves a traverse movement along a precisely straight line path, are generally supported and guided by machine ways. Consequently, in addition to supporting the traversing member, the guideways must also provide surfaces in other than the support plane. These serve for guiding those surface elements of the moving machine member which confine its movement to a specific track, thus avoiding lateral digressions from the basic straight line of the theoretical path.

Machine tools are typical, although not exclusive examples of equipment whose proper operation is dependent on guideways which must be adequately designed and precisely executed.

The operation of machine tools usually involves forces which act on the traversing member. The direct and the resultant actions of these forces only very rarely coincide with the direction of the force of gravity, consequently, the guideways must be designed to support both the operational and the gravitational forces. That requirement is met by a combination of guideway surfaces, each in a plane essentially normal to the expected direction of the acting force.

Finally, in various types of machines the ways must guide different members which may have overlapping positions. That operational condition calls for multiple guideway surfaces such as, for example, the machine bed ways of a lathe, which support and guide independently of each other, the carriage and the tailstock.

The functional purpose of the machine ways requires, besides excellent flatness, straightness and smoothness of the individual surface elements, also mutual parallelism and angular relationship at a high degree of accuracy.

The quality requirements of the individual surfaces can be met by properly built basic surface grinding machines which operate with reciprocating table movement. However, for assuring the very exacting specifications regarding positional interrelations of the participating surfaces, such as are common in modern machine tools, it is also mandatory that all the way surfaces shall be finished in the same setup of the workpiece, which may be, for example, a complete machine tool bed.

Considering that the surface sections to be ground may be on the top, at the sides, or even on the reverse side of the element which contains the top; furthermore, that such surfaces may be at different angles, sometimes even in recessed locations, it is obvious that the regular types of peripheral surface grinding machines are not adapted for such operations.

The operational requirements thus arising call for special types of surface grinding machines. That need has been supported also by the fact that way grinding offers considerable economic advantages over the earlier, exclusively used manual finishing method by hand scraping. Such manual operations are dependent on special skill, and may also present manufacturing problems, particularly in the case of very extensive and complex surfaces.

The special way grinding machines which have been developed and are now in general use in industry, particularly in machine tool building, are comparable in two major respects:

(1) The general dimensioning, characterized by the large mounting surface of the table and the high

Courtesy of Mattison Machine Works

Fig. 4-4.5 Openside way grinder with the horizontal and the tiltable vertical-spindle wheel heads arranged side-by-side in individual slides of the upright column. A. General view of the grinder. B. Close-up of the grinding of a V-way with beveled edge straight wheel.

clearance above the mounting surface, resulting in substantially heavy type machine tools

(2) The use of two or more grinding heads, one of them with a fixed horizontal spindle position and the other, often designated as the "universal head" because of its many uses, inclinable in both directions in a plane normal to the table traverse.

The general design of the basic types of way grinders follows either of the following two principles:

(1) The open-side design in which the horizontal spindle wheel head and the universal wheel head are arranged side-by-side, along the traversing table, each head mounted on its own outrigger slide on the upright column of the machine (see Fig. 4-4.5).

(2) The bridge type design, very similar in its construction to a planer, with the two heads guided in the ways of a cross rail. That member is supported and displaced vertically on two columns which straddle the machine bed (see Fig. 4-4.6).

Courtesy of Waldrich Coburg - Cosa Corp.

Fig. 4-4.6 Heavy guideway grinding machine in bridge type design, with independent horizontal and universal grinding heads mounted on separate slides, which are arranged on a vertically displaceable cross rail.

Another system of way grinding machines, also in bridge type construction, comprises two independently operated grinding heads: a horizontal spindle head for peripheral surface grinding, and a universal grinding head which can be swiveled to a position agreeing with the angle of inclination of the machine way to be ground. Examples of this system, representing a medium size and a large size way grinder are shown in Figs. 4-4.7 and 4-4.8.

For the small volume or occasional production of machine tool beds or similar workpieces with several parallel surfaces in different orientations and, perhaps, at various levels, the procurement of a special way grinding machine may not be warranted economically. From a technical standpoint, however, the grinding of such interrelated surfaces should be carried out in a single setup, for assuring the accurately parallel directions of these functionally complementary ways.

The special surface grinding machine shown in Fig. 4-4.9, designated as "universal" by its manufacturer, may be the type of equipment which satisfies

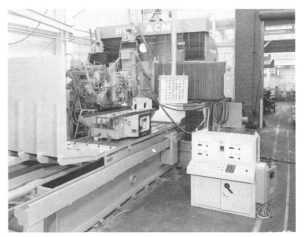

Courtesy of The Hill Acme Company

Fig. 4-4.7 Medium-size way grinding machine in bridge type construction, with a horizontal spindle type and a universal wheel head.

the technical requirements of the described operations, at moderate investment. As seen from the photo, and more clearly displayed in the diagram, Fig. 4-4.10, this type of special surface grinder has two wheel heads, each installed on individual arms which are supported in their own vertical ways on two opposite sides of a common mounting column.

Courtesy of The Hill Acme Company

Fig. 4-4.8 Large-size bridge type grinding machine with two independently operated wheel heads, one of which has a horizontal spindle while the other is swivelable for grinding inclined work surfaces.

This column is rotatable and by swinging it around at 180 degrees, alternate wheel heads can be brought into operating position over the machine table. After having brought the proper wheel head into the operating position the rotational movement of the column is locked in a positive manner, assuring a positioning accuracy stated by the manufacturer as ± 0.00008 inch (± 2 micrometers).

One of the wheel heads has a horizontal spindle for mounting 18 inch (450 mm) diameter straight

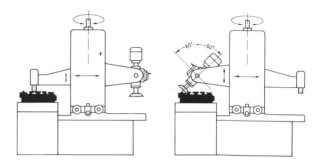

Courtesy of Elb Grinders Corp.

Fig. 4-4.10 Diagrams indicating the major adjustments and feed movements of the versatile surface-grinding machine in Fig. 4-4.9, showing the central column of rotatable design for bringing alternate wheel heads into operating position. The individual arms of the two wheel heads have independent vertical movements, while the transverse positioning and feed movements are provided by the slide on which the common vertical column rests. (*Left*) Horizontal spindle head with straight wheel grinding a flat surface. (*Right*) Tilted universal head with conical cup wheel grinding a V-way.

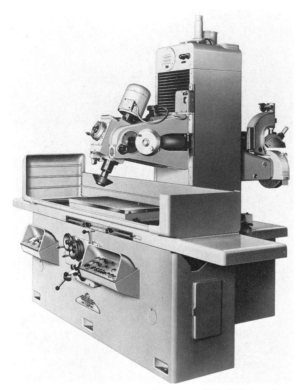

Courtesy of Elb Grinders Corp.

Fig. 4-4.9 Versatile surface grinding machine with two alternately applicable wheel heads, one with horizontal spindle and the other, a universal type, with tiltable vertical spindle. This machine is designed for the grinding of directionally parallel surfaces at different angles of inclination, such as machine tool ways, which can be finished in a single mounting of the work on the grinder table.

grinding wheels. The other arm carries the vertical spindle type universal wheel head, which can be tilted in both directions over a swing angle of 60 degrees. This spindle will accept a straight or beveled cup wheel. Power actuated rapid and adjustable feed traverse in crosswise direction of the column, as

well as rapid and variable downfeed movement of the individual wheel heads, are provided, in addition to the coarse and fine manual positioning adjustments.

This versatile type of surface grinding machine is built in different sizes with table lengths ranging from 59 to 256 inches (1500 to 6500 mm).

The capability of the design with grinding heads arranged side-by-side (see Fig. 4-4.5) is indicated in Fig. 4-4.11, showing diagrams of way surfaces which may be ground (A) with the horizontal spindle wheel head, and (B) with the universal head having a 55-degree tilt capacity in either direction.

The actually applicable methods of grinding the machine tool ways vary, of course, according to the design of the way grinding machine used for carrying out the operations. Another variety of methods is illustrated by the diagrams in Fig. 4-4.12, showing examples of processes, applied on the way grinder

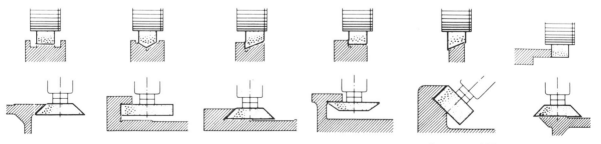

Courtesy of Mattison Machine Works

Fig. 4-4.11 Diagrams of different machine tool way-configurations and locations, which may be ground with: (*Top Row*) The horizontal wheel head; and (*Bottom Row*) The tiltable vertical wheel head of way grinders in the system illustrated in Fig. 4-4.5.

illustrated in Fig. 4-4.5, for the grinding of typical machine tool way surfaces. In this case characteristic operations performed with the two side-by-side mounted wheel heads, one with a horizontal spindle and the other of the universal type, are shown in a manner permitting a comparison of the respective capabilities.

The wheel head with horizontal spindle has a very large diameter spindle for mounting heavy wheels and, depending on the machine size, has drive motors with power ratings ranging from 15 to 50 hp. This head is equipped wth a swivelable wheel-dressing device, permitting the wheel to be dressed with

straight or beveled profile, for the grinding of either flat or inclined way surfaces located at the top of the workpiece. While these surfaces, which comprise the largest area of machine tool ways finished by grinding, could also be ground with the universal head, it is obvious that the use of the more powerful horizontal spindle head, with its large diameter and wide wheels, is preferred as the more productive method.

The universal wheel head is installed on a rotating base, has feed movements both in vertical and horizontal directions and is designed to accept wheels of different dimensions and shapes, including cup wheels. In order to provide the proper peripheral speed for wheels of widely different diameters, the spindle of the universal wheel head is driven by a variable speed DC motor of either 10 or 15 hp. The universal wheel head is equipped with a versatile dressing device for truing wheel surfaces of different shapes and in orientations which result from the selected tilt setting of the wheel head.

The general dimensions of way grinders built as standard sizes, vary over a wide range. Typical, but not necessarily extreme dimensions for table widths range from 24 to 140 inches (about 600 to 3,500 mm), with table lengths from 8 to 32 feet (about 2400 to 9,600 mm). In view of the types of workpieces for which these grinding machines are designed, such as machine tool bed and uprights, as examples of the bulkier workpieces, the useful heights of way

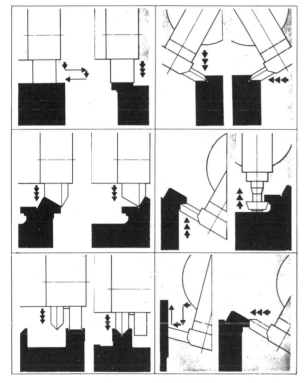

Courtesy of Waldrich Coburg - Cosa Corp.

Fig. 4-4.12 Examples of machine tool way-grinding operations which can be carried out with the use of (*Left Column*) the horizontal spindle type wheel head; (*Right Column*) the universal wheel head. (*Top Left*) Grinding flat surfaces with straight wheel; (*Center*) Grinding inclined surfaces with bevel-edge straight wheel; and (*Bottom*) Grinding flat and V-shaped surfaces in a single setup with side-by-side mounted straight- and double-bevel-edge wheels. (*Top Right*) Grinding the two faces of a dovetail-shaped way, using double-bevel-edge wheel mounted on a tilted spindle, and by applying vertical and horizontal feed, respectively; (*Center*) Two alternate methods for grinding the retainer way surfaces, one with open access, the other in recessed location; and (*Bottom*) Alternate methods of grinding vertical surfaces by reciprocating motions (wide surface) and by plunge grinding (narrow surface).

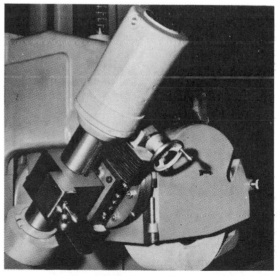

Courtesy of Mattison Machine Works

Fig. 4-4.13 Auxiliary vertical spindle installed as optional attachment on a regular type peripheral surface grinder, for the grinding of guideway surfaces which are not accessible to the wheel on the horizontal spindle head.

grinders vary over a typical range of 20 to 120 inches (about 500 to 3,000 mm).

Auxiliary vertical spindles. It is a general practice to carry out the grinding of guideways on workpieces which can be accommodated on general-purpose surface grinding machines when these are adapted to way grinding, by means of a vertical or swivelable wheel-head attachment. Such attachments, an example of which is shown in Fig. 4-4.13, are supplied by a few manufacturers of surface grinding machines.

Diverse Types of Special Purpose Surface Grinding Machines

In addition to the methods and equipment which have been discussed as examples of the special applications of the basic system, peripheral surface grinding is used for other types of operations also, but which require machines or accessories in a particular, application-related design. A few more examples will be reviewed briefly with the purpose of indicating the wide scope of potential uses for the common fundamental principles.

Electrolytic Surface Grinders (Fig. 4-4.14). Electrolytic surface grinding is successfully applied for workpieces which, by the configuration of the areas to be

Courtesy of Thompson Grinder, Div. of Waterbury-Farrel

Fig. 4-4.15 Spline grinder, a special type of peripheral surface grinding machine, designed for automatic operation, including work indexing, wheel dressing, and size control. A. General view. B. Close-up of a spline shaft being ground.

when burr-free finishing, or the avoidance of grinding stresses is critical.

Peripheral surface grinders built expressly, or equipped by retro-fitting for electrolytic grinding, must have insulated wheel spindles, hoses, and nozzles for the distribution, guards and return lines for the retainment, and tanks with a special pump for the circulation of the electrolyte, in addition to the electrical power unit. Depending on the size of the grinding machine and on the expected performance of the process, the output of the DC power unit is selected from a generally available range of 50 to 500 amperes.

Spline and Slot Grinders (Fig. 4-4.15) operate with automatic indexing of the workpiece carried out

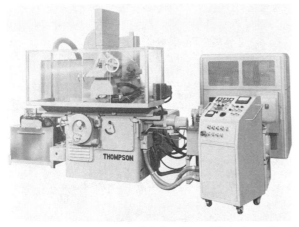

Courtesy of Thompson Grinder, Div. of Waterbury-Farrel

Fig. 4-4.14 Electrolytic peripheral surface grinder, equipped with special guards, electrolyte circulation system and electrical power unit.

finished, call for peripheral surface grinding but the work material presents difficulties in grinding by regular methods. Cemented carbides and various types of special alloys, sometimes referred to as "exotic," are examples of such work materials which can be ground much more efficiently when the grinding action is supported by electrolytic dissolution. Electrolytic surface grinding is also advised

Courtesy of Thompson Grinder, Div. of Waterbury-Farrel

Fig. 4-4.16 Flat broach grinder, a specially equipped traversing table type surface grinder with wheel head designed to mount both dish and cup wheels. Available for manual or automatic operation, also for grinding round broaches which require the rotation of the workpiece in addition to the incremental traverse movement.

synchronously with the reciprocating movement of the machine table. Such machines, built with longer tables for spline grinding and with shorter table movement for grinding slots, are usually designed for continuous production purposes, with automatic cycle, including wheel truing and size compensation.

Broach Grinders (Fig. 4-4.16) are, in their general appearance (resulting from the basic design of the machine body and of the work table), a variety of peripheral surface grinding machines. However, the essential grinding function of these machines, whether designed for flat or round broaches, is carried out with relatively small grinding wheels in shapes generally different from the basic straight wheels, and the principal movements are adapted to the fundamental objective of tool sharpening.

Rotary Table Peripheral Surface Grinding Machines

General Characteristics of the Process and Equipment

Rotary surface grinding is the commonly used designation of a grinding method in which the periphery of the grinding wheel is in contact with the workpiece mounted on a rotating table, while a reciprocating movement—radially, with respect to the circular table—extends the contact over the entire work surface. (See Fig. 4-5.1.)

Functionally, this system differs from common peripheral grinding in that the workpiece travel path is rotational in distinction to the more commonly used translational path, which results from the reciprocating movement of the table. The rotational path of the workpiece movement, and the radial direction of the relative wheel traverse movement, create grinding conditions significantly different from those in regular traverse grinding, and offer process characteristics which are beneficial to specific operational objectives.

Several potential advantages associated with rotary surface grinding are:

(1) Development of a circular grinding pattern, essentially concentric with the center of the workpiece (when the work is mounted with its center closely coincident with that of the rotating machine table).

(2) The possibility of grinding wide-angle conical surfaces, convex or concave, executed with consistently high angular accuracy. (See Fig. 4-5.2.)

(3) The continuous engagement of the wheel with the work from start to finish of the operation, a condition which can contribute to increased performance and improved work quality.

(4) Adapted to the efficient grinding of annular surfaces by adjusting the reciprocating stroke length to the radial width of the work surface, thus avoiding interruptions in the grinding contact or interference with protruding elements inside or outside the ring-shaped work surface.

There are also certain drawbacks in the system, limiting its application to operations in which the attainable advantages govern selection of the method. Examples of such limitations of the method are the following:

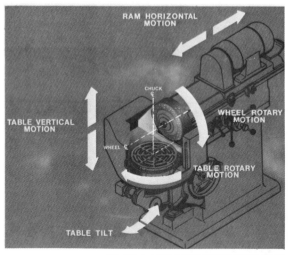

Courtesy of Sundstrand Machine Tool, Div. of Sundstrand Corp.

Fig. 4-5.1 Diagram showing the principles of rotary table, peripheral surface grinding.

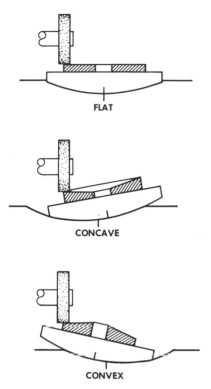

FLAT

CONCAVE

CONVEX

Courtesy of Cincinnati Milacron, Inc., Heald Machine Div.

Fig. 4-5.2 Principles of producing flat or conical (concave or convex) work surfaces by tilting the table of rotary surface grinding machines.

(a) The effective work speed, resulting from the constant rotational speed of the machine table, varies continuously with the radial transverse position of the wheel contact, the surface speed of the work being the maximum at the periphery of the table, decreasing gradually toward its center. The effect of this drawback can, however, be greatly reduced in modern machines with steplessly regulated speeds of the table rotation and of the reciprocating movement.

(b) The system offers the greatest advantages in the grinding of single workpieces of round form and is also effectively applied for multiple smaller parts which can be arranged in an essentially ring-shaped area of the work table. Application to workpieces of other shapes may be warranted by specific operational reasons which outweigh considerations based purely on grinding efficiency.

The Reciprocating Machine Member

Rotary table peripheral surface grinding machines are built according to two different design concepts with regard to the selection of the machine member which carries out the reciprocating movement:

(1) The reciprocating ram type (see Fig. 4-5.17).
(2) The reciprocating table type (see Figs. 4-5.15, 4-5.16, and 4-5.18).

Both systems find application in machines built by reputable manufacturers, some of whom adhere in the design of their entire line of rotary surface grinders to one or the other of these two basic design systems. Manufacturers in whose products both systems find application, usually design their models to be applied to varied operations, such as for toolroom work, in the ram type version. The production type machines in the larger sizes, are designed with a rigid column supporting a vertically moving wheel head slide. In this latter system the rotary table is mounted on a slide or saddle which carries out the transverse movement.

The speed of the reciprocating machine member is usually adjustable steplessly over a range expressed by the ratio of about 1 to 15 and having for typical extreme values 1 and 15 ft/min (about 0.3 and 4.5 m/min). Most types also have a rapid speed of the reciprocating member for the approach and the return movements.

The Feed Movements

Depending on the basic concept of design with respect to the reciprocating member, the feed movement is carried out in either of two ways:

(1) The ram type machines support the work table on an extension, or knee, of the main body, and it is by raising the table toward the plane of the ram-held wheel head that the feed movement is accomplished. The feed can be applied manually by a handwheel with fine incremental graduations or, in production type operations, automatically at different rates, extending over a typical range of 0.0001 to 0.0015 inch (or about 0.0025 to 0.038 mm), per reciprocating movement of the ram.

(2) The slide-supported table type machines have the wheel head mounted on a vertically adjustable slide of the upright column, with typical feed ranges similar to those of the alternate design.

Automatic Grinding Cycles

Rotary surface grinders are frequently operated in automatic cycle, limiting the work of the operator to a few activities only, such as the loading of the workpiece, the starting of the grinding cycle and, after termination of the automatic cycle, the unloading.

There are several varieties of automatic cycling in rotary surface grinding, however most of these consist of the following major phases:

(1) Start, which sets the table in rotation and initiates the rapid approach of the traversing member to the point of the wheel/work contact. That point can be set by mechanical trip dogs actuating electrical switches, or it can be detected by a load sensor.

(2) The reciprocating strokes start at the speed which has been selected for the grinding.

(3) At each end of the stroke, inside and outside of the work, or at the end of each double stroke, the feed takes place by the set incremental amount.

(4) Production type operations are often cycled with two different feed rates: coarse for the first part of the grinding and fine for the finishing strokes.

(5) A timed spark-out phase may follow, consisting of a few reciprocating strokes without feed advance.

(6) At the end of the operation the traversing machine member, the table slide, or the ram, moves the work or the wheel out of contact and then a rapid return of the wheel head or of the work table to the start level is carried out. At that point the automatic cycle is completed and table rotation stops.

Automatic wheel dressing, optionally combined with size compensation, may also be a part of the automatic cycle.

In operations which do not have to accomplish an accurate size control, the feed cycle is controlled by a timer, instead of applying a specific number of incremental feed movements.

For operations which require very rigorous size control, automatic cycling may be applied only up to a point close to the finish size, then a change must be made to manual control, which may be guided by the measurement of the actual work size.

Continuously Variable Speeds for the Table Rotation and the Reciprocating Movement

One of the limitations of the common models of rotary table peripheral surface grinders is the dependence of the effective work speed on the radial position of the wheel with respect to the center of the machine table. As the grinding wheel approaches the center of the table, during its stroke, the peripheral speed of the contacted work surface decreases, whereas during the return stroke toward the table periphery, the effective surface speed increases. Consequently, the table speed must be established at an rpm value which, in the extreme outside operating position of the wheel, will still not result in an excessive surface speed of the work.

In order to avoid the ineffectually low stock removal rate which would result from the reduced surface speed of the workpiece toward the center, various methods are used on different makes of rotary table surface grinders, each intended to partially or fully correct the inherent limitations of the system:

(a) Increasing the rotational speed, revolutions per minute, of the machine table as the position of the wheel approaches the center of the table. When the traverse speed remains the same, the wheel track spacings will then vary, and while an extremely unfavorable work surface speed is avoided, the overall effectiveness of the process is only slightly improved. (See Fig. 4-5.3A.)

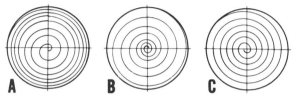

Fig. 4-5.3 Diagrams illustrating the pattern of the wheel path in rotary surface grinding operating with: A. Varying rotational speed of the table while the traverse speed remains constant; B. Increasing or decreasing the traverse speed without changing the table rotational speed; C. Varying both the rotational speed of the table and the traverse speed in synchrony with the positional changes of the grinding wheel.

(b) When the speed of the traverse movement is varied according to the momentary position of the wheel with respect to the center of the table, increasing on the inward and decreasing on the outward stroke, the rate of stock removal will tend to be closer to uniform. However, the conditions of the grinding will vary due to the changes in the surface speed of the work. The width of the grinding wheel used is also a limiting factor because a minimum overlap of the adjacent contact paths must be assured. (See Fig. 4-5.3B.)

(c) The most effective speed control can be accomplished by varying both the rotational speed of the machine table and the traverse speed, in synchronism, according to the momentary position of the grinding wheel in relation to the center of the machine table. (See Fig. 4-5.3C.)

Typical Applications—Examples and Analysis

Rotary table peripheral surface grinding has its preferential applications in operations where the inherent characteristics of this grinding system are particularly beneficial to the major process objectives.

Such objectives may include high productivity, excellent accuracy of size, flatness or finish, the development of a particular grinding pattern, easily accomplished work loading and holding, and possibly some others as well.

To provide a guide with regard to the conditions warranting the selection of rotary table peripheral surface grinding, a few operations will be discussed as examples of typical applications of this method.

Grouped under titles indicating the workpiece characteristics, the pertinent advantages of rotary surface grinding will be pointed out briefly and complemented with illustrations of typical operations.

Round, thin parts with flat surfaces (see Fig. 4-5.4), are preferably ground with the periphery of the grinding wheel, for assuring better heat dissipation and thus avoiding warpage, burns, or other heat-related damages of the workpiece. Rotary grinding offers the advantage of continuous engagement of the wheel with the work, thus avoiding thickness variations near the edges of the ground part. Further advantage of rotary surface grinding for parts exposed to frictional load, such as clutch plates, or for parts which must have dependable sealing properties, such as valve plates, is the essentially concentric pattern of the grinding.

Courtesy of Abwood Machine Tools Ltd.
Bentley Industrial Corp.

Fig. 4-5.4 Grinding to a high degree of flatness, the surface of a thin disc-shaped workpiece mounted on a rotating magnetic chuck.

Round, thin parts with hollow surfaces (see Fig. 4-5.5). Typical examples are circular saws and circular knives which need to be relieved from the operating periphery toward the center in order to reduce the friction of the engaged tool on the nonfunctional areas of its surface. Such tools must be ground on their surfaces with a gradually increasing concavity toward the center, often in a manner leaving a circular central section with parallel surfaces. That geometry is easily and accurately produced on rotary surface grinders, by tilting the table to an incline corresponding to the angle of the required concavity.

Courtesy of Cincinnati Milacron, Inc., Heald Machine Div.

Fig. 4-5.5 Grinding the faces of a circular saw to a concave shape; the work is mounted on a magnetic chuck and located by a centering hub.

Small parts assembled in a ring shaped area in a continuous row (Fig. 4-5.6), *or in sectional groups* (Fig. 4-5.7). The grinding wheel engagement along a continuous path, without the need for reversals involving some overtravel, can be well utilized when surface grinding small parts which are arranged in a ring shaped pattern, thus presenting an aggregate grinding surface which is comparable to a single ring shaped part. In the case of very small parts, such as shown in Fig. 4-5.7, the method may be considered as efficient even when the parts are grouped along noncontinuous sections of a circular area, thus avoiding the need to install a chuck with annular poles for occasional work, in exchange for the more

Courtesy of Cincinnati Milacron, Inc., Heald Machine Div.

Fig. 4-5.6 Grinding the faces of a batch of gears mounted on an annular section of the magnetic chuck and located by a retainer ring.

Fig. 4-5.7 Mounting groups of very small parts (bearing rings) on the arcuate sections of a radial pole chuck which is equipped with a retainer ring.

generally used magnetic chuck with radial poles. The use of a retainer ring (Fig. 4-5.6) is generally recommended for assuring additional part retention without the need for very high magnetic pull, and also as a means of facilitating the loading of the parts in a proper arrangement.

Surface grinding medium-size workpieces in groups. For grinding the surfaces of medium-size parts which are processed in lots, grinding methods operating with the face of the wheel are often preferred because of higher productivity in respect to stock removal

Fig. 4-5.8 Substantial number of small parts are surface ground while held in random arrangement on a magnetic chuck with retainer ring.

rates. However, in frequent cases, the characteristic advantages of peripheral surface grinding, such as cooler cut and the commonly higher flatness accuracy, may outweigh the benefits derived from higher productivity, particularly when the required stock removal is comparatively small. For grinding small or medium-size parts in groups on rotary table peripheral surface grinders, the applicable system of loading will depend on the general size and configuration of the individual parts. Accordingly, either loading in random order (see Fig. 4-5.8) or in a particular pattern (see Fig. 4-5.9) may be the preferred method. For such applications it is recommended not to fill the chuck surface completely but to leave the innermost area empty. This will avoid grinding in the least effectively functioning portion of the chuck if it is rotating at a uniform number of revolutions per minute.

Fig. 4-5.9   Medium-size parts of special shape (rotary knives) are held in an appropriate positional pattern on the magnetic chuck.

Producing a concave tool surface   (Fig. 4-5.10). A concave face shape may have the role of a rake angle in particular types of cutting tools, of which gear shaper cutters are an important example. While such operations can also be carried out on other types of equipment, such as universal tool grinders, the inherent capability of rotary surface grinders for concave grinding, in combination with the stable work support and powerful grinding wheel drive of these machines makes them advantageous, particularly for larger size tools and for production runs.

Grinding surfaces in recessed location (Fig. 4-5.11). When a flat surface, which is in a recessed location, has to be ground without contacting the farther

Courtesy of Cincinnati Milacron, Inc., Heald Machine Div.

Fig. 4-5.10 Grinding for sharpening the hollow face of a gear cutter held in a special fixture mounted on the magnetic chuck. The angle and the centered position of the conical work face are ground with high accuracy.

extending elements which surround the work surface of a substantially round part, rotary surface grinding may be preferable to alternative methods, such as those operating with the face of a small wheel mounted at the end of a vertical grinding spindle. Grinding spindles for such uses are generally designed for less powerful drives than the regular wheel spindles of rotary surface grinders. However, the application of rotary surface grinders for recessed surfaces is necessarily limited to parts whose configuration and dimensions provide sufficient clearance for straight wheels which operate with their periphery.

Grinding annular shoulders around an extending hub (Fig. 4-5.12). This is a typical workpiece con-

Courtesy of Abwood Machine Tools Ltd.
Bentley Industrial Corp.

Fig. 4-5.12 Grinding to the shoulder of a flange mounted in a precisely centered position on the face of a magnetic chuck.

figuration for which rotary surface grinding is positively the most suitable method to use. While other types of grinding machines, such as universal cylindrical or universal internal grinders in special set-ups can be adopted for these operations, the applied method will be a duplication of rotary surface grinding by using the system without benefiting from the advantages of mounting the work on the horizontal surface of a magnetic chuck.

Grinding the flat surfaces of parts with irregular shape held in multiple fixtures (Fig. 4-5.13). The need for fixture mounting of such parts does not exclude the successful use of rotary surface grinding.

Courtesy of Cincinnati Milacron, Inc., Heald Machine Div.

Fig. 4-5.11 Grinding a ring shaped surface in recessed position. The workpiece is the impeller shell of a torque converter.

Courtesy of Cincinnati Milacron, Inc., Heald Machine Div.

Fig. 4-5.13 Holding several components for surface grinding in a special fixture with manual clamping.

While the method of part holding results from the configuration and the locating requirements of the part and applies to any system by which the flat work surface is being produced, the advantages of the continuous grinding path of rotary grinding, in combination with the light grinding forces needed and the convenient access to the successive work positions during loading and unloading by an indexing type rotation of the table, are factors favoring the use of a rotary surface grinding machine.

Surface grinding with direct size control (Fig. 4-5.14). The most positive way of controlling the work dimension produced by a metalworking operation is by the direct in-process measurement of the affected dimension, such as the work thickness in surface grinding. The indications of an appropriately selected, installed, and set-up gage will provide a dependable guide to the operator regarding the amount of stock which still needs to be removed and will signal the position where the finish size has been reached. That finish size may be specified as a fixed value or can be a dimension varying from part to part, such as in match-grinding where the mating component of the workpiece is used to provide the applicable reference dimension. Rotary table peripheral surface

grinders equipped with fine incremental adjustment of the infeed are well suited for operations with direct size control. In applications where the grinder is occasionally used for convex or concave grinding, the resetting of the table swivel to its zero position may be a critical adjustment prior to the controlled size grinding and it is carried out preferably with the aid of a zero setting gage, generally available as an optional accessory.

Rotary Surface Grinding Machines—Representative Models

The principles of rotary table peripheral surface grinding are effectively applicable to a wide range of workpieces which differ with respect to size, quality requirements, and production quantities. Accordingly, rotary surface grinding machines are manufactured in a variety of models, each designed primarily to meet the objectives of a particular group of products.

Figure 4-5.15 illustrates one of the smaller types of rotary surface grinding machines; these are available both as manually operated models with table reciprocation by hand wheel, and also with additional

Courtesy of Cincinnati Milacron, Inc., Heald Machine Div.

Fig. 4-5.15 The operating and control area of a small-size, rotary-table peripheral surface grinder.

Courtesy of Cincinnati Milacron, Inc., Heald Machine Div.

Fig. 4-5.14 Match grinding the surface of an angular contact ball bearing ring by precisely controlling the thickness with the aid of an electronic in-process gage.

power movement of the table reciprocation. The former version is intended for toolroom type operations, while the automatic reciprocation makes the machine applicable, within its work-size capacity, for production type operations on workpieces requiring the degree of accuracy which is generally assured by rotary surface grinding. It should be mentioned that rotary table surface grinders in the smaller size ranges, such as the illustrated model, are commonly manufactured in the reciprocating ram design with stationary position of the axis of the rotating table, which can be raised or lowered for vertical adjustment. In this particular model, also representing a proven design, these relative movements are reversed, the supporting slide of the table carrying out the traverse movement, while the vertical adjustment is accomplished by moving the wheel head along the guideways of the vertical column.

For assuring the high level of performance accuracy the table ways of the illustrated model are of the hardened box type, riding on hardened steel rollers which are ground within 0.000 010 inch (about 0.000 25 mm) for size, and assembled as a continuous complement, alternating with undersize rollers which act as separators. These machines are supplied with 6-inch (about 150-mm) or, optionally, with 8-inch (about 200-mm)-diameter magnetic chucks, have 4-inch (about 100-mm) maximum grinding stroke, use 7-inch (about 175-mm)-diameter grinding wheels, for whose drive motors with either 1½, 3, or 5 hp can be provided.

Rotary surface grinders in a design similar to the preceding example, are also manufactured in larger sizes (see Fig. 4-5.16) with typical maximum chuck diameters of about 16 and 30 inches (about 400 and 750 mm) with grinding wheels of 9 or 12 inches (about 225 or 300 mm) in diameter and having a 15 hp motor drive. These heavier machines have the table slide supported on plain guideways, one flat and one V-shaped, widely spaced and arranged to straddle the vertical plane of the grinding wheel action. One of the noteworthy features of this design is the wide tilt range of the table, listed as table swivel, permitting the grinding of a conical surface to 17 degrees convex and 10 degrees concave. The chuck speeds can be varied steplessly and, depending on the model (characterized by its chuck diameter), have a range of 50 to 225 and 30 to 80 rpm, respectively. The table speeds for grinding can be varied from 0 to 15 fpm about 0 to 4.5 m/min), and a higher speed of 20 fpm (about 6 m/min) is provided for the runout used for work loading.

Rotary surface grinders in this heavier size category are used frequently for production purposes, which commonly require operation by automatic feed. The power feed of these models operates at each reversal of the table slide movement and is adjustable from 0.0001 to 0.0005 inch (about 0.0025 to 0.0125 mm) in increments of 0.0001 inch (about 0.0025 mm). As an optional accessory, feed by a hydraulic cylinder, designated as an impulse feed box, is offered; this can be operated either with constant feed or with incremental feed at reversals only.

These machines can also be supplied for automatic cycle, requiring only manual starting by the table lever, following which both reciprocation and feed will operate automatically to finish work size, the cycle also controlling the runout of the table and the retraction of the wheel slide.

Another type of grinding machine is shown in Fig. 4-5.17, representing a design concept which is widely used for small and intermediate models of

Courtesy of Cincinnati Milacron, Inc., Heald Machine Div.

Fig. 4-5.16 Rotary-table peripheral surface grinder with reciprocating table movement, designed for general purpose production work.

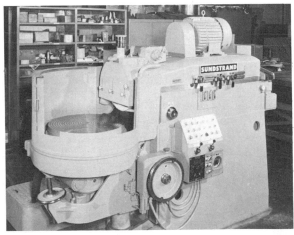

Courtesy of Sundstrand Machine Tool,
Div. of Sundstrand Corp.

Fig. 4-5.17 Rotary-table peripheral surface grinder with reciprocating ram and vertically adjustable machine table.

rotary table peripheral surface grinders. This particular type is manufactured in models with chuck diameters of 16, 20, 24, and 30 inches (about 400, 500, 600, and 750 mm). The basic design characteristic of this type is the reciprocating ram carrying the wheel head and moving in a constant plane, while the adjustment of the work level with respect to the wheel, including the application of the feed during the operation, as well as the retraction at the end, are accomplished by raising or lowering the work table, also designated as the "chuck holding slide."

The vertically adjustable design of the machine table does not interfere with the steplessly regulated drive for the rotation of the chuck, which is provided by a hydraulic motor offering a speed ratio of about 1 to 4 between the lowest and the highest number of revolutions per minute. Other types of rotary grinders designed with reciprocating ram, may have even wider ranges for the chuck speed, such as in a ratio of about 1 to 10, thus providing for an extremely flexible adaptation to a variety of operational requirements. The actual speeds have rpm values related to the largest chuck diameter of a particular model, generally not exceeding a maximum peripheral chuck speed of about 1,000 fpm (about 300 m/min), but for many models, only about one-half of this value. Of course, the actual work speed is always less than the peripheral speed of the chuck.

To indicate the applicational potentials of the system, it should be pointed out that rotary table peripheral surface grinders are produced by some specialized machine tool builders in exceptionally large sizes, too, such as with about a 50-inch (1250-mm)-diameter chuck mounted on a table with reciprocating movement, which is similar, but significantly heavier, than those used in the conventional models of this system.

For rotary surface grinders with a very large chuck diameter some manufacturers prefer to have the rotating table supported on a solid base, and to incorporate the reciprocating movement into the slide which supports the wheel-head column. Rotary surface grinders of this latter system are built in several sizes, a typical range comprising grinders with 32- to 60-inch (800- to 1500-mm) table diameter, such as the machine shown in Fig. 4-5.18. The illustrated largest model of this series has a 29-inch (735-mm) travel for the wheel-head column, which can be operated as rapid traverse for setup and positioning, or with steplessly variable feed traverse with stroke lengths from 0.04 to 2.4 inches (1 to 60 mm). The vertical movement of the wheel head has rapid motion as well as automatic feed.

Courtesy of Elb Grinders Corp.

Fig. 4-5.18 Large-size, rotary-table surface grinding machine with stationary axis type 60-inch (1500-mm)-diameter rotating table and wheel-head column mounted on a traversing slide. This system can be supplied with tiltable table for grinding beveled edges on ring-shaped workpieces.

Supporting Functions and Auxiliary Equipment

The following discussion will briefly review the three major supporting functions, (a) work holding, (b) wheel dressing, and (c) work gaging, subsequently listing examples of auxiliary equipment, generally supplied as optional accessories.

(a) *Work holding on rotary surface grinders* is, in the majority of applications, by means of a circular magnetic plate, covering the entire work table surface and commonly designated as the "chuck." Two basic types of magnetic chucks are used with regard to the general pattern of the poles, namely those with concentric poles and those with radial poles. Concentric, or ring shaped poles are more generally used because they are adaptable to a wide variety of work sizes, including small parts loaded in groups. Radial poles, sometimes also named "star" poles, may provide a greater holding force for a single piece mounting in a concentric position. Other, less frequently used pole arrangements include the fine pitch concentric poles and a combination of radial (inside) and concentric (outside) poles. Special single-ring shaped poles may be used in toolroom work in combination with easily exchangeable top plates which are supplied with different radial pole patterns. Magnetic chucks may have center holes to accept mechanical fixtures or locating members in a central position. Retainer rings, mounted on the periphery of the magnetic chucks are generally advisable for multiple part loading. When special mechanical fixtures are used for holding multiple parts of configurations not adaptable for magnetic holding, face plates with tapped holes for the mounting bolts may be preferred over the regular magnetic chucks.

(b) *Wheel dressing methods and equipment.* The dressing of the grinding wheel in rotary surface grinding is, with rare exceptions, limited to dressing the periphery of the straight wheel. The rotary grinding method is not adaptable to form grinding, therefore, the several varieties of wheel truing devices which are frequently used in the traversing type of peripheral surface grinding have no applications in this method.

The design and the operation of wheel dressing devices on rotary surface grinders differ for the ram type and the reciprocating table type machines.

The ram type machines have the wheel dresser device built on top of the wheel head (see Fig. 4-5.17), and are generally operated by an individual hydraulic traversing slide, using a micrometer screw for adjusting the feed.

In the reciprocating table type machines there is no need for a separate drive of the wheel dresser device. The dresser diamond is held in a retractable mount on the table slide outside the periphery of the chuck and has an elevating mechanism for adjusting the diamond feed (see Fig. 4-5.19). For dressing the

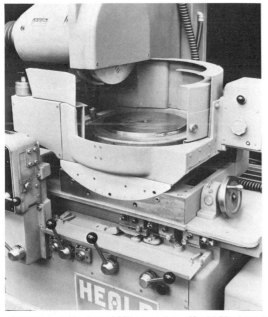

Courtesy of Cincinnati Milacron, Inc., Heald Machine Div.
Fig. 4-5.19 Wheel dresser diamond of a reciprocating table type rotary surface grinder, mounted on the table slide adjacent to the chuck (left), and adjusted by an externally located control knob (top, far left).

wheel the length of the table stroke is extended to cause the diamond to pass along the periphery of the wheel, while the traversing speed is adjusted to the rate needed for wheel dressing. Mounting the diamond dresser on the reciprocating table slide offers additional advantages in certain operations, such as (1) incorporating the wheel dressing into the automatic operational cycle, i.e., the wheel dressing taking place when the slide runs out at the end of the operation; (2) using the wheel dressing as a means of size control by setting the height of the diamond at the level of the finish work size.

(c) *Work size gaging* in rotary table surface grinding is simplified by the fact that generally, a single dimension only—the thickness of the part with respect to the work surface—has to be controlled. The emphasis in size control is on the infeed by vertical adjustment, either by raising the work or lowering the wheel head, depending on the basic design of the machine. For this reason grinding machines of this category are built with sensitive vertical adjustment devices, mechanical or hydraulic, permitting positional control with incremental changes which, on several models are as fine as 0.0001 inch (about 0.0025 mm).

A circumstance which could interfere with holding the set size is the dressing of the wheel, unless it is adequately compensated. As just discussed, that compensation is simplified in the reciprocating table type machines, while the ram type rotary surface grinders have wheel dress compensators for the feed hand wheel, to which the amount of diamond advance can be transferred.

It is feasible, and in special cases, warranted (see Fig. 4-5.14) to use direct sensing gaging devices on rotary table surface grinders, particularly when the reference dimension for the sizing of the workpiece has a varying value.

Optional Accessories

Optional accessories for rotary table peripheral surface grinders are discussed in the following by listing a few examples to indicate the means by which the adaptability of the method can be expanded, satisfying manufacturing requirements which are generally beyond the capabilities of the basic equipment, in a technologically adequate and economically advantageous manner.

Work-holding devices, in addition to the earlier mentioned magnetic chucks with different pole patterns, and auxiliary top plates, may include face plates and various types of fixtures designed to accommodate workpieces of particular configurations.

Electrical control elements for making the operation of the magnetic chuck better adapted to the requirement of particular workpieces such as very thin parts which are sensitive to distortion under the effect of excessive magnetic force. Equipment for these purposes may include rheostats to provide adjustment of the chuck holding power, demagnetizer unit for minimizing the demagnetizing time after grinding, and a rectifier in case an adequate DC supply is not available for operating the chuck.

Angular setting device. The angular setting of the inclinable machine table is usually guided by the graduated segment of the table base (see Fig. 4-5.19). For operations which require a very accurate incline position, some machine models can be equipped with special angular setting devices which operate on the sine bar principle, using a combination of micrometer screw and gage bars of different lengths for the accurate adjustment of the distance between the reference points.

Positive stops for limiting the traverse movement in a direct acting and dependably accurate manner when grinding a surface close to a centrally located shoulder.

Special Types of Rotary Table Peripheral Surface Grinders

The fundamental principles of rotary table peripheral surface grinding are also applicable to other positional arrangements of the operative machine members besides the common system; namely, the horizontal table surface and the horizontal grinding spindle.

Two less common applications of the peripheral grinding of workpieces held on a rotating table will be pointed out in the following:

The vertical rotary table peripheral surface grinder has the work table with the magnetic clutch supported on a horizontal spindle, comparable to the design principle of a facing lathe. The reciprocation of the grinding-wheel head is accomplished by the horizontal movement of a slide which has a path parallel with the table surface. The guideways of the traversing wheel head slide are at a 30-degree incline with respect to the vertical, a design feature considered to counteract effectively the resultant of the forces originating from the weight of the slide and from the grinding pressure. The vertical arrangement of the work table is claimed to offer a more convenient handling of workpieces with large diameters and small thickness, such as brake and clutch discs, circular saws and knives, etc.

Rotary table peripheral surface grinders with vertical table surfaces are built with chuck diameters from 12 to 40 inches (about 300 to 1,000 mm) and with infinitely variable table speeds, adjustable manually or automatically.

Rotary table peripheral surface grinders with inclinable wheel heads. Such machines with a single, or two independent wheel heads (see Fig. 4-5.20) are built in a variety of sizes, incorporating various optional accessories, with design details adapted to particular work requirements. The example illustrates the versatile adaptations of the basic operational principle of rotary table surface grinding.

Courtesy of Campbell Grinder Co.

Fig. 4-5.20 A representative model of specially designed rotary-table surface grinders with inclinable wheel heads.

Rotary Grinding Fixtures

For tool and occasional production work when the stock removal capacity is not considered as a critical requirement, rotary grinding fixtures, such as that shown in Fig. 4-5.21, can be used as a substitute for rotary surface grinding machines. These are essentially precision made rotating tables with bearings designed for continuous rotation and motor drive. The illustrated model, a typical representative of this kind of fixture, operates with speeds variable from 20 to 60 rpm, and has a 16 5/8-inch (about 270-mm) table diameter. Other models with table diameters ranging from 6 to 15 inches (about 150 to 380 mm) are also available.

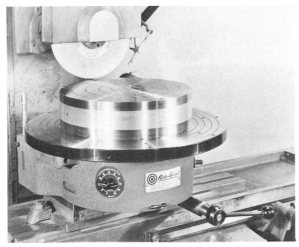

Courtesy of M & M Precision Systems, Inc.

Fig. 4-5.21 Motorized rotating table mounted on a reciprocating type peripheral surface grinder as a supplementary rotary grinding fixture.

Rotary grinding fixtures are mounted on regular peripheral surface grinders with reciprocating table traverse or on universal tool grinders, using the downfeed movement of the base machine to produce the vertical adjustment and the wheel feed. When transverse reciprocation is needed, because the radial width of the work area is greater than the width of the grinding wheel, that is produced by the manually actuated crosswise table movement of the base machine.

A more versatile design of rotary grinding fixture is shown in Fig. 4-5.22, installed on a conventional surface grinder. This model has a 90-degree tilt axis and is equipped with a variable speed motor for operating the table with a speed of from 0 to 100

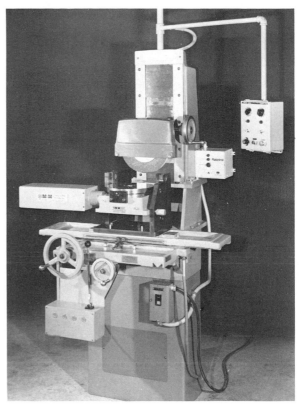

Courtesy of M & M Precision Systems, Inc.

Fig. 4-5.22 Tiltable axis type rotating table with power drive installed on a conventional surface grinder, for which cross-feed control for automatic rotary face grinding is also provided.

rpm. A cross-feed control for the machine table is also supplied as a supplemental accessory.

Rotary Table Face Grinding

General Method Appraisal

As the designation of the category indicates, this method of surface grinding has two distinctive characteristics:

(1) The workpiece is mounted on a rotating work table

(2) The face of the grinding wheel is used as its operating surface.

The designation in the title is not a currently used term, although it is considered descriptive of the method and equipment to be discussed in this chapter. Some grinding machine manufacturers call them rotary surface grinders, vertical surface grinders, or just surface grinders. In common shop language the machine is often referred to as a "Blanchard" grinder, with reference to the company believed to be the originator of the method, still a prominent, although not the sole supplier of this category of surface grinding machines.

In principle, as well as in many actual applications, rotary-table face grinding is the most productive method of surface grinding, particularly when considering the rate of stock removal as the significant property.

The comparatively high rate of stock removal, a potential implemented by adequately dimensioned equipment, results from two dominant characteristics of the method:

(a) The work moves continuously and at a constant speed along the contact area of the grinding wheel.

(b) The entire radial width of the circular work table on which a single workpiece or a group of workpieces may be mounted, is covered by the operating face of the wheel, consequently, a full turn of the work table will expose the entire work surface to the action of the grinding wheel.

In comparison with other methods of surface machining, rotary face grinding offers the advantage of combining substantial stock-removal rates with the capacity of producing surfaces with very good dimensional accuracy, flatness, and surface finish, all carried out in a single setup, applying effective roughing followed by tightly controlled finishing, in successive phases of the same operation, with the work retained in its original mounting position.

Comparing rotary face grinding with peripheral surface grinding, whether linear or rotary, the transverse movement normal to the work traverse or rotation is eliminated, a condition which greatly reduces the operational time needed for applying the wheel contact over the entire worked surface. However, for selecting the appropriate method for particular workpieces and operations, the following applicational limitations of surface grinding with the face of the wheel should be considered:

(1) Many types of workpieces and work materials are not adaptable for grinding with the face of the wheel which, when mounted on a spindle normal to the work table, produces an area contact, in distinction to the nominal line contact of the wheel periphery. A grinding process with wide area contact may be too harsh in its effect due to the application of more power and the generation of more concentrated heat on the worked surface than in a process carried out with a narrow contact area.

(2) In face grinding, particularly when applied for the purpose of increased stock removal rates, the created grinding forces may be excessive for delicate

parts and will require higher work-holding power obtained by increased magnetic pull, or pressure, in the case of mechanical fixtures.

(3) The configuration of the workpiece may be such that the surface to be ground is not parallel with the holding surface, or may have elements protruding over the level of the area being ground.

The preceding listing is, of course, not complete. It does not discuss, e.g., the differences in the resultant grinding patterns, the attainable flatness and size accuracy, the possible effects of face grinding on the leading edges of certain workpieces, etc. It does indicate, however, a few major aspects in which, under specific conditions, surface grinding with the face of the wheel, may not represent the best suited method, notwithstanding its generally superior productivity.

Considerations other than purely technological ones may also control the choice of method and equipment, such as by taking into account the following circumstances:

(a) The limited potential utilization of a high-productivity grinding machine, whose application areas may be restricted to certain categories of workpieces.

(b) The handling of the work is facilitated by bringing the chuck into a position clear of the grinding wheel by means of the traverse movement of its supporting carriage. Nevertheless the time required for loading and unloading (particularly for parts held in fixtures or mounted with auxiliary support elements) as compared with the pure grinding time, may be such as to substantially lessen the benefits attained by increased stock-removal rates. Limitations from this aspect can, in certain cases, be partially eliminated by applying continuous work loading and discharge with the aid of special equipment, (See Fig. 4-6.10.)

While the applicational conditions for which rotary table face grinding is not adaptable, or does not represent the best suited method, have been discussed to some length, this should not detract from the fact that in a wide range of uses this method provides the optimum process for developing flat work surfaces, often in comparison not only to other surface grinding methods, but also to other metalworking processes such as milling, planing, etc. These positive aspects of rotary face grinding will be examined in the following paragraphs.

Applications of Rotary Face Grinding Machines

Examples of typical applications, each selected to represent a particular kind of workpiece and opera-

tional condition, will now be presented. The purpose of this discussion is to point out the areas in which the method is commonly used and offers advantages, sometimes quite substantial, over alternate systems and processes. Of course, the examples shown are not considered as a complete listing, but should be indicative of the method's potential and thus stimulate the examination of rotary table face grinding as a possibly advantageous alternative process for working flat surfaces.

In order to facilitate the appraisal of various applications in which rotary table face grinding can offer significant advantages, Table 4-6.1 surveys various characteristic aspects of each of the selected examples. Because the operations discussed are intended to illustrate principles, no particular considerations have been given to the actual dimensions of the workpieces, grinding data, and the required models of rotary table face grinding machines.

(1) Large workpieces mounted directly on the magnetic chuck, either uniformly distributed over the chuck area or, in the case of a single part, in the approximate center of the magnetic chuck (see Fig. 4-6.1). A central mounting, even at the cost of accomplishing partial chuck surface utilization only, may be applied when, for a particular reason, a concentric grinding pattern of the ground surface is desired. Variations in the location of the work to suit particular purposes and workpiece configurations can be easily accomplished on the magnetic chuck which represents the work-holding system generally used in this grinding method. The figure also illustrates the relatively large height capacity which can be provided by the extended column type rotary face grinders. Six sides of the workpiece shown can be ground in consecutive order on the same machine.

Courtesy of Cone-Blanchard Machine Co.

Fig. 4-6.1 Single workpiece mounted in the center of the magnetic chuck of a rotary table, face-grinding machine.

Table 4-6.1. Examples of Typical Rotary Face Grinding Operations

Workpiece Type	Work Mounting Method	Beneficial Process Characteristics	Examples
Large workpieces with plain flat surfaces	Direct mounting on magnetic chuck or with auxiliary support elements when needed.	Mounting of large workpieces directly on the magnetic chuck without the need for mechanical clamping is one of the major advantages of rotary face grinding over other methods of metalworking.	Fig. 4-6.1
Small and medium-size workpieces with irregular contours	Direct mounting on magnetic chuck in an arrangement permitting best work area utilization.	Round chuck surface without corners, the most effectively coverable work area and generally the best suited for accepting a large number of parts with irregular contours, when arranged in a properly selected pattern.	Fig. 4-6.2
Small workpieces with round or other regular contours	Random placement with subsequent shifting into adjacent positions on the chuck surface area optionally bounded by a retainer ring.	Small parts densely covering the effective grinding area, which may be physically bounded for easy loading and secure work retainment, provides uniform wheel engagement in continuous operation, a condition conducive to efficient process performance.	Fig. 4-6.3
Workpieces with portions extending over the plane of the support surface	The parts are arranged to have the protruding portions overhanging the border of the chuck, or auxiliary support elements may be used for raising the part over the chuck surface.	Parts which offer a flat mounting surface of sufficient area and location for assuring stable work support, even when limited portions of the surface extend over the plane of support, can be mounted on a magnetic chuck by adequate arrangement or by auxiliary supports providing clearance for the extending portions.	Fig. 4-6.4
Tall workpieces supported and ground on the small faces	Held by magnetic chuck on the small locating surface but supported laterally by auxiliary elements which are retained magnetically also.	When the relatively small area of the mounting surface is insufficient for assuring the magnetic retention needed to resist the grinding force, auxiliary elements—generally plain steel blocks—placed closely around the workpiece will often adequately supplement the direct magnetic action.	Fig. 4-6.5
Parts of awkward shape without a dependable locating surface	Special purpose supports may be provided by adjustable elements which, in turn, are retained magnetically.	A flat surface may have to be produced as a starting operation to provide, e.g., a locating surface for subsequent machining. To bring and then to hold the surface to be ground into the plane of wheel action, an irregularly shaped heavy workpiece may only need to be lined up by properly adjusted support elements.	Fig. 4-6.6

Table 4-6.1. (*Cont.*) **Examples of Typical Rotary Face Grinding Operations**

Workpiece Type	Work Mounting Method	Beneficial Process Characteristics	Examples
Parts with ground surfaces in different, yet parallel, planes processed individually or together	Either mounted directly on the magnetic chuck in an adequately arranged pattern or with use of mechanical fixtures which positively assure the position of the workpieces.	In rotary table surface grinding the grinding wheel action can extend over the entire chuck surface or over an annular area only, which may have any radial width inward from the chuck periphery. By the proper mounting of the parts selective grinding at different levels can be accomplished in consecutive steps, starting near the center.	Fig. 4-6.7
Concave or convex round parts with shallow incline angles	Generally mounted on the magnetic chuck, directly when the bottom is flat, otherwise with shims to compensate for deviations from the plane of the chuck face.	Rotary face grinding may be given preference over rotary peripheral grinding because of higher productivity, although the application of face grinding is limited to rather shallow angles. A well-centered mounting of the single workpiece is an important requirement in this method.	Fig. 4-6.8
Parts requiring fixture clamping because of their general shape and the position of the locating surfaces	Work-holding fixtures with locating surfaces adapted to the workpiece, and mechanical clamping by manual or power action.	Rotary face grinding, particularly in its center column version, is successfully applied to grinding, in a single operation, flat surfaces on multiple parts which have their locating surfaces in positions other than opposite the ground surface. The work-holding fixtures are usually installed around the periphery of the rotating table.	Fig. 4-6.9
Small parts adapted for automatic loading and discharge	Direct magnetic retention on a rotating chuck with selective magnetic action limited to sectors outside the loading and unloading areas	The rotary face grinding machine can be equipped with accessories and supplementary devices to operate continuously in an automatic process which is based essentially on a sectored chuck with selective magnetization, permitting the introduction and discharge of the work while the rotating chuck is passing over the nonmagnetic sectors.	Fig. 4-6.10
Nonferrous metal parts with shapes adapted to surface grinding parallel to the mounting surface	Indirect magnetic holding can often be accomplished with auxiliary elements made of steel which either restrain lateral movement or actually exert hold-down force on the work.	The direction of the grinding force in face grinding tends to hold the workpiece against the locating support opposite the surface being ground, permitting the omission of hold-down retainment when the work is secured against lateral shift. Special auxiliary elements combine confinement with some degree of hold-down force when exposed to magnetic action.	Fig. 4-6.11

(*Continued on next page.*)

Table 4-6.1. (*Cont.*) **Examples of Typical Rotary Face Grinding Operations**

Workpiece Type	Work Mounting Method	Beneficial Process Characteristics	Examples
Nonmetallic parts in shapes suitable for supporting on a flat surface	Cementing or gluing small flat and thin parts to the surface of a steel mounting plate which is then transferred onto and retained by a magnetic chuck.	Face grinding with its large area of contact is often the most efficient method for developing or improving flat surfaces on thin nonmetallic parts. With proper mounting the method is applicable even to parts which, due to fragility, need full support on the side opposite the surface being ground.	Fig. 4-6.12

(2) Small and medium size parts, even with irregular contour configurations can be mounted with efficient chuck surface utilization when a pattern is applied in loading the parts (see Fig. 4-6.2). Filling the chuck surface is not only beneficial, with respect to a higher production rate of the operation, but will also present a more continuous surface area to the grinding wheel, thus providing better control of the wheel performance.

Courtesy of Cone-Blanchard Machine Co.

Fig. 4-6.2 Large number of small workpieces arranged in a regular pattern on the surface of a rotating chuck.

(3) Parts with circular contour permit the effective filling of the chuck area, even when a random arrangement is used, thereby accelerating the chuck loading process (see Fig. 4-6.3). An essentially complete coverage of the chuck surface contributes in producing an excellent flatness and finish of the ground part surfaces, when that is required for functional reasons, such as in the case of the illustrated valve plates. Of course, other significant factors of the process, such as the wheel composition, work speed, feed rates, etc., must also be appropriately selected.

(4) Parts with protruding elements can frequently be adopted for surface grinding on a rotary face grinding machine. The method of work mounting must be selected to conform to the general dimensions

Courtesy of Cone-Blanchard Machine Co.

Fig. 4-6.3 Small round workpieces mounted in random arrangement over the entire surface of a rotary chuck.

and the configuration of the part, taking also into account whether the protruding element is on the side of the surface being ground or on the opposite side of the part. In the former case the parts can often be arranged in a radial pattern with the extending boss near the center of the chuck, which is then left outside the area of wheel contact when moving the table into the operating position. When the protruding boss is on the underside of the mounted part, then one solution is the use of auxiliary steel rings on which the part is supported and held magnetically. The rings, having the proper thickness, provide the free space for the overhanging boss. Another solution consists in arranging the parts near the periphery of the chuck, with the protruding elements hanging down outside the chuck as shown in Fig. 4-6.4. All these work-mounting methods are predicated on the presence on the part of a flat surface, parallel with and substantially opposite to that being ground. Otherwise, the use of a mechanical holding fixture is needed.

(5) Slender workpieces of regular shape which offer only a relatively small support area, can still be mounted directly on the magnetic chuck by combining the directly applied magnetic action with a

Courtesy of Cone-Blanchard Machine Co.

Fig. 4-6.4 Parts ground on the flat surface portion, with the projecting boss on the underside hanging down outside the rotating chuck.

mechanical lateral support provided by steel blocks which are retained by the magnetic chuck. These blocks, which must have a somewhat lesser height than the workpiece, confine the position of the adjacent parts to a closed area, thus preventing lateral shift or "tipping," under the influence of the grinding force (see Fig. 4-6.5). The application of this mounting system is limited to parts whose shape assures positive confinement by mutual contact.

Courtesy of Cone-Blanchard Machine Co.

Fig. 4-6.5 Bushings or round parts of similar shape with a high length-to-diameter ratio require lateral support to complete the directly acting retention of the magnetic chuck.

(6) Occasionally single, or a very limited number of parts of irregular shape must be ground flat on one side which is not opposed by a flat and parallel locating surface at all, or with an area which is insufficient for adequate magnetic retention. For this type of rare application, particularly when the general dimensions of the part are beyond the capacity of other available equipment, such as that of a

reciprocating type peripheral surface grinder, it is possible to carry out the surface grinding on a rotary table face grinder by adding temporary support elements to the direct action of the magnetic chuck, or improvising a work-holding assembly by using standard clamping and mounting devices held on the magnetic chuck, as illustrated in the example shown in Fig. 4-6.6.

Courtesy of Cone-Blanchard Machine Co.

Fig. 4-6.6 Workpiece of awkward shape is held for surface grinding by the combination of a vise and jacks which, in turn, are retained magnetically.

(7) Workpieces with discrete surface areas which are located at different levels can be ground in the same setup by grinding first the lower surface and leaving the higher one outside the contact area of the wheel, which in this phase of the operation is confined to the outside zone of the chuck surface. Thereafter, the wheel head is raised to conform to the higher surface level, the table is advanced to extend the wheel action over the entire chuck, and the work areas located near the center then will be ground. For better utilization of the chuck capacity, and depending on the general size of the workpieces, two parts (see Fig. 4-6.7), or even more, arranged in a radial pattern can be mounted for the concurrent grinding of first all the lower, and then of all the higher, surfaces.

(8) For grinding circular surfaces which have a limited amount, up to about 1 degree, concentric incline, convex or concave, the column of most types of rotary table face grinding machines can be tilted,

Courtesy of Cone-Blanchard Machine Co.

Fig. 4-6.7 Flat parallel bosses of different height are ground in successive phases of the same operation.

commonly by inserting gage plugs under the appropriate support points. While the range of tilt is substantially smaller than in rotary table peripheral grinding machines, it is adequate for such parts as, e.g., circular saws which have to be relieved slightly only, from the periphery toward the center (see Fig. 4-6.8). Generally, the dishing of the sides of circular saws does not extend to the center, where a flat circular hub is left for clamping the tool in operation. That condition is produced by mounting the saw blade in a well centered position and then limiting the wheel engagement to an area which does not include the central portion of the chuck. For grinding the second side of the saw blade, opposite to that on which the concave surface has already been

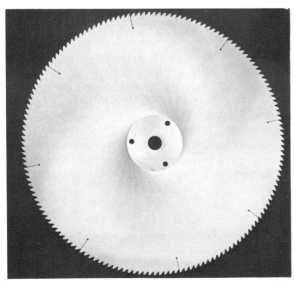

Courtesy of Cone-Blanchard Machine Co.

Fig. 4-6.8 For grinding the concave surfaces of circular saws in a well centered position, the work is located and retained on the magnetic chuck and the column of the grinding machine is tilted by an amount corresponding to the radial incline of the surface.

developed, a ring shim is used as an auxiliary supporting member to compensate for the amount of dishing.

(9) Work holding by mechanical clamping is the proper solution when the parts to be ground do not have adequate surfaces for magnetic retention, or the reference elements to which the ground surface must be related, are not contained in a flat area which is suitable for supporting the part. Work-holding fixtures for small and medium size parts which will be ground on a rotary table face grinder are commonly arranged side-by-side around the periphery of the rotating machine table (see Fig. 4-6.9). This arrangement of the fixtures has several advantages, including the best space utilization permitting the installation of the largest number of fixtures, and the comparable ease with which the fixtures can be serviced, generally involving the manual mounting and dismounting of the workpieces. Depending on the design of the fixtures, the clamping is done by hand, individually for each unit, or power clamping is provided to act simultaneously on all positions after the fixtures have been loaded. Although the individual handling of the parts is time consuming, often substantial savings can be accomplished by rotary face grinding in comparison to milling, e.g., which requires similar fixturing of the parts, often with more powerful clamping. The actual stock removal rate may be higher by grinding, particularly when the work material is very hard, the tool maintenance—limited in grinding to occasional wheel dressing—is much simpler and the ground surface generally satisfies the finish part specifications without the need for a subsequent finishing operation.

Courtesy of Cone-Blanchard Machine Co.

Fig. 4-6.9 Grinding the cap assembly surfaces of large connecting rods in a peripherally arranged fixture on a rotary table face grinder.

(10) Automatic handling, including loading and discharge, of small- and medium-size parts with parallel flat surfaces, such as the small ball-bearing races shown in Fig. 4-6.10, permits continuous operation of the rotary table face grinder and results in very high production rates. The parts can be brought to the rotary table and moved into its annular operating area which is bounded by retainer

Courtesy of Cone-Blanchard Machine Co.

Fig. 4-6.10 Rotary-table face grinding in continuous operations by use of a fixtured chuck in conjunction with automatic part-loading and discharge.

Courtesy of Cone-Blanchard Machine Co.

Fig. 4-6.11 Brass rings mounted between abutting steel blocks whose magnetic retention assures the adequately secured position of the nonmagnetic workpieces during the surface grinding operation.

rings, by a vibratory trough as shown in the illustration, by a vibratory bowl, or a rotating disc, the latter preferred for larger diameter parts. After passing under the grinding head, or heads as discussed in the following chapter, the rotating chuck carries the parts ground on one side to the discharge area where a deflector plate will cause the work to leave the chuck and slide down on an inclined channel, the starting section of which is surrounded by a demagnetizer coil. An appropriate device may turn over the parts and retain them in that orientation, to be introduced in a subsequent operation into the same grinder, after its grinding head has been lowered to conform to the finish thickness of the workpieces. For operations of this type, where the opposite and parallel sides of the part have to be ground flat, the double disc grinder, discussed in Chapter 4-8, has the obvious advantage, when compared with the rotary-table face grinder, of grinding concurrently two opposite sides of the part. However, for particular workpiece dimensions, material allowances, production requirements, or lot sizes, the use of rotary-table face grinders, often in the double head design, still may be considered preferable over alternate methods, including the double disc grinders.

(11) Nonferrous metal parts, as well as those made of nonmagnetic stainless steel, when processed by grinding on a rotary-table face grinder, require particular work-holding methods. If the configuration of the part being ground permits the location on an opposite parallel surface, then mounting on a magnetic chuck by applying indirect clamping is generally feasible and represents the simplest method of holding such workpieces. For indirect clamping, steel blocks, thinner than the work, are placed adjacent to the nonmetallic workpiece (see Fig. 4-6.11). After turning on the electric magnet the blocks will be

solidly held, preventing any lateral movement of the confined workpieces. The retainment exerted by the blocks can be made more effective by using blocks with serrations or points on the contacting surfaces.

(12) The surface grinding of nonmetallic parts to a high degree of size and flatness accuracy, as well as excellent surface finish, is frequently carried out successfully on rotary-table surface grinding machines. Typical nonmetallic materials to which this grinding method is applied include glass, quartz, plastics, ceramics, carbon, etc. When mounting nonmetallic workpieces for surface grinding, magnetic retention cannot be used and, therefore, other work-holding methods have to be devised. These include fixtures with spring pressure into which the parts are loaded individually, separator plates with nests for each part, or the workpieces are embedded into some cementing substance, such as plaster, melted wax (see Fig. 4-6.12), sulfur, etc. Intermediate plates are often used for attaching the parts and then these loaded plates

Courtesy of Cone-Blanchard Machine Co.

Fig. 4-6.12 Silicon wafers mounted for surface grinding on a retainer plate by using melted wax.

are transferred to the table or chuck of the grinder. Although the mounting of the parts is not as simple as in the case of direct magnetic retention, when processing workpieces with critical requirements, the advantages of rotary-table face grinding may outweigh by far the costs of the applicable fixturing and the time spent on mounting the parts.

Finally, a few points are listed which *apply in general* to most cases of part mounting on the magnetic chuck or rotary-table surface grinders:

(1) For assuring the maximum magnetic pull the workpiece should be placed over at least two adjacent poles of the magnet. If that is not possible because of the small size of the parts or the wide spacing between the successive opposite poles of the magnet, then supplementary mechanical support by blocking elements, such as retainer rings, may be needed. Another approach might be the use of concentrically placed support rings, which could serve as a bridge for the magnetic flux.

(2) Unbalanced parts may have to be supported under the overhanging portions.

(3) Projecting elements on the workpiece:

(a) when on the worked side (upperside) of the part, may be located in the center portion of the chuck which then should stay outside of the area of grinding wheel contact;

(b) when on the locating side (underside) of the part, then sufficient free space must be assured for the projecting portion by raising the part on support blocks and letting the projection overhang the periphery of the chuck.

(4) Parts held with light magnetic force only, due to small size or thickness, are prone to be moved by the grinding action of the wheel and should be secured by adjacent blocks. For placing these blocks the direction of the wheel action must be taken into account, the ingoing edge of the cylindrical wheel pointing toward the center, and the outgoing edge toward the periphery of the chuck. The need for a two-directional blocking of the part's lateral movement can be avoided by a slight tilt of the wheel to an incline in the order of a few minutes of arc, which would reduce the shifting action of the wheel to a single direction. The resulting reduced area of contact will also contribute to a more efficient cutting action. Such a wheel setting would affect the flatness of the resulting work surface only to a small degree, a condition generally acceptable for rough grinding and noncritical parts. The appearance of the grinding pattern will be altered, the absence of the typical cross-hatch pattern being indicative of a less-than-perfect flatness.

The diagram in Fig. 4-6.13 illustrates the relationship between the diameter of the cylinder wheel, the amount of tilt, and the maximum depth of the concave path which that wheel will produce at the deepest point of its penetration into workpieces of different widths. That depth of concavity will decrease with larger wheel diameter, narrower contact path, and lesser amount of tilt. For precision type work, however, not even a small amount of concavity of the nominally flat surface can be tolerated. Therefore, the procedure is often applied to carry out the rough grinding, requiring substantial stock removal with the more efficiently cutting tilted wheel and then the wheel head is brought back into its vertical position for the finish grinding. That wheel head adjustment is a simple and quick operation on modern rotary surface grinders.

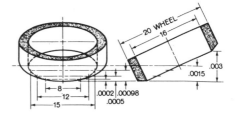

Courtesy of Mattison Machine Works

Fig. 4-6.13 Diagram showing, in highly exaggerated proportions, the relationship between cylinder wheel tilt, diameter, width of cut, and the maximum depth of the resulting concavity on the ground surface.

The Grinding Wheels and Operation of Rotary Table Face Grinders

Essentially, two types of wheels are used in face grinding: the cylinder wheels and the segment wheels, also designated as segmental wheels (see Fig. 4-6.14).

Cylinder wheels are, in their basic shapes, plain hollow cylinders with wall thickness equal to about 1/10 of the outside cylinder diameter. Cylinder wheels are commonly made in the diameter range of 10 to 20 inches (about 250 to 500 mm), although variations from that median range are not infrequent. The length of the cylinder wheels is usually in the range of 4 to 6 inches (about 100 to 150 mm). Cylinder wheels are supplied either as plain straight cylinders or with inserted mounting nuts.

For mounting the plain cylinder wheels either of two methods is used: cementing (generally using sulfur), or clamping, by a special wheel holder. For the latter method of mounting, which permits a faster wheel change, cylinder wheels with a resilient gasket

Fig. 4-6.14 Grinding wheel types used on rotary-table face grinders: (top) Cylinder Grinding Wheel; (bottom) Segmental Grinding Wheel—one of several designs.

material, applied as a strip around the top, must be used.

Cylinder wheels of the inserted mounting nut type have several equally spaced nuts permanently molded into and flush with the top face; these are mounted by bolting to a face plate adapter. The inserted nut design provides a better utilization of the wheel's abrasive material and is preferred for automatic grinders.

Cylinder wheels are commonly selected for general operations. They are definitely indicated for grinding small parts and for producing fine finishes, because the wheel contact with the work is continuous and the wheel pressure is distributed over the maximum area. For safety purposes cylinder wheels are banded, using either wire or glass tape. The latter, which grinds away with the wheel, does not require cutting as the wheel wear progresses, and is recommended particularly for automatic operations.

Segment wheels consist of a number of essentially arc shaped abrasive segments, held firmly by mechanical clamping in an appropriate holder, designated as the segment chuck. The equally spaced segments mounted into the chuck are arranged to cover a ring

shaped area which, however, is not continuous due to the open spaces between the adjacent segments and the configuration of the individual segments.

The use of the segmental design for the large diameter face grinding wheels is a technical necessity, because it is not possible to manufacture safely operating solid cylinders in such large sizes. Using segment wheels which provide an interrupted operating surface also offers functional advantages, particularly in grinding large continuous work surfaces; the open space between the consecutive segments provides clearance for the discharge of chips and for the penetration of coolant. For that reason, even in wheel diameter ranges in which solid cylinder wheels are also available, the use of segments is preferred for the rough grinding of broad work surfaces, particularly when substantial stock removal rates are to be achieved.

The diameters of the segment chucks are adapted to the capacity of the surface grinding machine on which they are used. The range of the commonly manufactured segment chuck diameters is 20 to 80 inches about 500 to 2000 mm), exceptionally, even larger, and the chucks are designed to hold, 6 to 12 segments, depending upon their diameters. The height of the segments varies by chuck system and diameter, usually within a range of 6 to 10 inches (about 150 to 250 mm).

The configuration of the individual segments is determined by the clamping method used in the chuck which, in turn, varies with the manufacturer. Table 4-6.2 provides a review of the segment chuck types of American manufacturers, with diagrams showing the method of clamping and the general shapes of the segments. The listing, having only the purpose of general information, is not complete and does not show foreign made chucks. Some of the segment clamping methods are adaptable to a wide range of chuck diameters; others are intended for a more limited diameter span.

The segments, while positively locked laterally, are adjustable vertically in order to set the section extending over the clamps to the specified distance, generally not exceeding 2 inches (about 50 mm). When the segments wear down, the projecting section can be readjusted, thus using up most of the abrasive material. Many chucks can hold the segments securely when only one inch (about 25 mm) is clamped, however, in this respect the manufacturer's instructions must be followed to assure safe operation.

In *the selection of grinding wheels* for rotary-table face grinding the recommendations of the machine and wheel manufacturers, which are based on exten-

Table 4-6.2. Abrasive Segment Types and Chuck Mounting Systems*
(Metric Equivalents are Approximate)

Type Designation (Manufacturer and Style)	Diagram	Range of Dimensions		
		Chuck Diameter	Segment Height	Pieces per Set
BLANCHARD Style "BL"		11 to 42 in. (280 to 1060 mm)	5 to 8 in. (125 to 200 mm)	4 to 10
CAPITAL Style "CP"		14 in. (355 mm)	5 in. (125 mm)	8
CORTLAND Style "CD"		6 to 120 in. (150 to 3050 mm)	3½ to 6 in. (90 to 150 mm)	4 to 30
FMR Style "AA1"		18 to 88 in. (460 to 2250 mm)	6 in. (150 mm)	4 to 19
FMR Style "AA2"		10 to 16 in. (255 to 410 mm)	5½ in. (140 mm)	3 to 5
HANCHETT Style "HA"		16 to 60 in. (410 to 1525 mm)	6 to 8 in. (150 to 200 mm)	8 to 20

Table 4-6.2. (*Cont.*) **Abrasive Segment Types and Chuck Mounting Systems**
(Metric Equivalents are Approximate)

Type Designation (Manufacturer and Style)	Diagram	Range of Dimensions		
		Chuck Diameter	Segment Height	Pieces per Set
HANCHETT Style "HL"		14 to 16 m (355 to 410 mm)	5 in. (125 mm)	8
HANCHETT Style "HM"		14 to 32 in. (355 to 810 mm)	6 to 7 in. (150 to 175 mm)	6 to 12
HARRIS-SEYBOLD Style "SE"		12 in. (305 mm)	3 in. 75 mm)	6
HILL ACME Style "HI"		23 to 32 in. (580 to 810 mm)	7¼ in. (185 mm)	8 to 11
ITT ABRASIVE Style "PE"		11 to 12 in. (280 to 305 mm)	4 to 6 in. (100 to 150 mm)	6
ITT ABRASIVE Style "PE" (for larger sizes)		18 to 22 in. (455 to 560 mm)	5 to 6 in. (125 to 150 mm)	8 to 10

(*Continued on next page.*)

Table 4-6.2 *(Cont.)* **Abrasive Segment Types and Chuck Mounting Systems**
(Metric Equivalents are Approximate)

Type Designation (Manufacturer and Style)	Diagram	Range of Dimensions		
		Chuck Diameter	Segment Height	Pieces per Set
ITT ABRASIVE Style "STG"		8 to 30 in. (200 to 760 mm)	4 to 6 in. (100 to 150 mm)	4 to 12
NORTON Style "NO"		14 to 42 in. (355 to 1070 mm)	6 in. (150 mm)	6 to 20
ROGERS Style "RO"		12 to 20 in. (305 to 510 mm)	4 to 4½ in. (100 to 115 mm)	7 to 8

*Excerpted from the publication: *Specifications of Segments Used In Chucks,*
the Grinding Wheel Institute, Cleveland, Ohio.

sive experience, may provide the best guidance for concrete applications. A general review of such recommendations will show them indicative of a few characteristics of the method:

(1) The grain material, mostly semifriable aluminum oxide or, for cast iron, austenitic stainless and nonferrous metal, silicon carbide, are defined by the work material, the same as for most other grinding methods.

(2) The bond material is for roughing and semifinishing operations generally vitrified, while for obtaining higher grade finish, a resinoid bond is recommended.

(3) The required grain size is typically coarser than for comparable operations carried out with the periphery of the wheel. For example, surface finish in the order of 25 microinches (about 0.6 micrometers) can be obtained with grain size 30 on soft steel, and with grain size 46 on hardened steel, while for producing a surface finish in the order of 6 to 8 microinches (about 0.15 to 0.2 micrometers), a wheel with grain size of 100 is needed.

(4) The grade, expressing the "hardness" of the wheel is, in general, softer than in comparable operations using other grinding methods. That almost

generally applicable difference is due to the relatively large area of contact between the face of the wheel and the work surface. Depending on the general configuration of the workpiece and the closeness of such workpieces to each other on the chuck, the effective work area will vary. The relative density of the worked surface with respect to the overall area of the machine table is expressed by designating the work area or the width of the work surface as narrow, medium, and broad, as illustrated in the diagram in Fig. 4-6.15. Because the narrower work surface, with a larger ratio between edge length and surface area, tends to act as a wheel dresser, it requires harder wheels than the medium, or the broad work surface.

Courtesy of Cone-Blanchard Machine Co.

Fig. 4-6.15 Symbols representing the relative density of the work area which results from workpieces having different widths.

As a close approximation, although not a generally valid rule, the difference in the applicable wheel hardness is one grade between each area type, such as, e.g., it is recommended to use I grade for narrow, H grade for medium, and G grade for broad work surface when grinding hardened tool steel.

Maintaining the proper operating conditions of the wheel face by assuring its free cutting properties is of particular importance for the effective performance of the face grinding process.

One of the main causes of deteriorating performance is the glazing of the wheel surface, which reduces its free cutting properties and may actually cause burns on the ground work surface. That condition can originate from various causes, which may be present individually or by several acting concurrently. The more frequently occurring causes of wheel glazing are pointed out in the following:

(a) The conditions of the wheel and of the coolant:
    (1) The composition of the wheel, which may have a too-hard grade or too-fine grain
    (2) The condition of the dressed wheel surface which may be too smooth or the frequency of wheel dressing which may be insufficient.
    (3) Improper type or inadequate amount of coolant being used.

(b) The selection of the grinding data or the development of the work area:
    (1) Too-slow chuck speed makes the wheel act harder; thus it becomes more prone to glazing.
    (2) Low infeed rates, sometimes necessitated by insufficient power, detract from the self-dressing action of the wheel.

(3) Continuous work surface may be inherent in the configuration of the workpiece, however, when the chuck is loaded with a large number of smaller parts, thus leaving wider spaces between the adjacent pieces, it can accelerate wheel breakdown and thereby prevent glazing.

For *dressing*, the face-grinding wheel, serrated, cutter type dressers are commonly used. These have several cutters mounted on a common spindle, often in an arrangement having the lands of the adjacent cutters staggered to produce a more open and sharper wheel face. The dresser head is held in an arm supported on a vertical shaft so that it can be swung across the face of the wheel by means of a hand lever within easy reach of the machine operator.

Measurement of the work size in rotary face grinding is generally carried out by one of the following systems:

(1) In manual control (a) by the direct measurement of the work thickness following an initial pass and then adjusting the downfeed stop for the required stock removal; (b) using a "witness" part, called a "size block," mounted, together with a batch of unground workpieces, on the chuck. The top surface of the block, which is made to finish size, is painted or copper plated, thus when touched by the wheel it will indicate for visual observation that the downfeed has reached the finished worksize.

(2) In automatic operations (a) the depth of the downfeed is set to a predetermined level and a device is used to compensate for wheel size changes due to dress-off; or (b) direct sensing gages, with probes in contact with the work, which produce signals for controlling the automatic downfeed.

Surface Grinding Machine Operating with the Face of the Wheel

Producing a flat work surface by grinding with a ring shaped area of the face of the wheel is a system most frequently applied in rotary face grinding. Various aspects of that method have been discussed in the preceding chapter based on the application of rotary-table face-grinding machines.

The vertical spindle, rotary-table face-grinding machine, although the most commonly used, is not the only type of equipment by which face grinding is applied. While the basic characteristics of the face grinding processes are similar, regardless of the type of machine used for its application, many operational aspects are different according to the design system of the equipment which is actually used for carrying out the face grinding process.

In the following discussion various types of face grinding machines used for producing flat work-surfaces will be reviewed. Separate paragraphs will examine the significant characteristics of regularly manufactured design systems of surface grinding machines which operate using the face of the wheel.

Table 4-7.1 serves to survey the types of surface grinders to be discussed, and points out several of the applicable distinguishing characteristics of each category. In this context the term "type" refers to the fundamental design concept of the machine, with the understanding that such concepts could be common to models of widely different sizes and capacities.

Vertical Spindle, Rotary-Table Surface-Grinding Machines

Machines assigned to this general category are manufactured in a very wide range of sizes. Generally applied, but not necessarily extreme limits of the chuck diameter, a dimension commonly characterizing the individual machine models, are 16 to 160 inches (about 400 to 4000 mm).

Typically, surface grinding machines in this design system are intended for production work, generally involving substantial stock removal, consequently, even the smallest standard models have wheel spindle drive motors of 15 hp, the corresponding value for the largest models being 250 hp, optionally, even more. A metalworking process which can utilize that amount of drive power must be carried out on appropriately dimensioned equipment and, again quoting typical values as a means of visualizing the range, the net weights of such machines are in the order of 3 tons to 150 tons.

For the concise review of grinding machines of this category a few typical models will be discussed briefly, each of the selected models being considered as representative of one of the following arbitrarily established groups:

Medium size rotary table face grinders are intended for general purpose production work, combining the system's high flexibility for various adaptations, as discussed in the preceding chapter under the heading *Applications*, with the favorable stock removal and other characteristics of the method. These latter include inherently good accuracy in producing flat work surfaces, which may be further improved by following the rough grinding by finish grinding in the same operation, each applied with appropriately selected grinding data.

Figure 4-7.1 illustrates a representative model of this size group; the grinder shown has a 36-inch (about 915-mm) chuck diameter. Machines of similar

Table 4-7.1 Systems of Face Grinding Machines

Spindle Position	Table Movement			Diagram	Discussion of Characteristics and Applications
Vertical	Rotating	Table position	Stationary		The most commonly used face grinding system with the widest range of potential application and adaptable to a large variety of work configurations. The work is generally held on a large magnetic chuck, although it is used also with machine table and mechanical work-holding devices.
			Indexing		A special model developed for small workpieces for which rotating table type surface grinding is the most suitable method, however, consecutive operational steps are needed to complete the process.
			Alternating		Designed for workpieces requiring rotary face grinding but also involving time-consuming mounting in special fixtures. By alternating two rotary tables into mounting and grinding positions the actual grinding process can be almost continuous.
	Traversing — reciprocating				Workpieces with general configurations resulting in a predominantly longitudinal shape are more efficiently processed in a traversing movement; such parts would require very large diameter rotary tables with poor utilization of the table surface passing through the grinding zone.
	Special system — stationary table and oscillating wheel head				Represents a partial modification of the reciprocating face grinding system, functionally similar but accomplished by a reversal of the traversing members.

(Continued on next page.)

Table 4-7.1 (*Cont.*) **Systems of Face Grinding Machines**

Spindle Position	Table Movement	Diagram	Discussion of Characteristics and Applications
Horizontal	Traversing — reciprocating		Preferred for long, yet thin workpieces, such as machine knives which are easier to handle and less prone to deformation when mounted on a vertical table surface.
	Stationary — Traveling wheel head		For applications similar to the preceding type. Rarely used system, which is of advantage for very large workpieces whose reciprocation on a traversing table would involve equipment requiring very extensive floor space.

design are also manufactured with chucks which are substantially smaller, e.g., 16 inches (about 400 mm) or larger, e.g., 42 inches (about 1070 mm). While the smallest models use only cylinder wheels of 11-inch (about 280-mm) diameter, the larger models operate either with cylinder wheels of 20-inch (about 510-mm) diameter or with segment chucks having the same size.

Heavy rotary surface grinders, Fig. 4-7.2, for the surface grinding of large and heavy workpieces, singly or several at a time, in dimensions which may be

Courtesy of Cone-Blanchard Machine Co.

Fig. 4-7.2 Heavy type, rotary-table face grinder for large workpieces, operating with segmental grinding wheels.

characterized by the capacity data of the selected typical model: on a chuck of 84-inch (about 2135-mm) diameter, workpieces 24 inches (about 610 mm) in height can be mounted, with new segments installed in the 42-inch (about 1070-mm)-diameter segment chuck.

Similar models are also manufactured with 96- and 100-inch (about 2430- and 2540-mm) chuck diameters, using segment wheels of 48- and 54-inch (about 1220- and 1370-mm) diameter. Spindle motor outputs range from 75 to 125 hp for the smallest, and

Courtesy of Cone-Blanchard Machine Co.

Fig. 4-7.1 Medium-size, rotary-table face grinder for general purpose production work.

100 to 150 hp for the larger models, according to user's requirements.

The effective utilization of such highly powered wheel drives results in very substantial stock removal rates, making these machines adaptable to *abrasive machining*, a term which expresses the economically justified substitution of grinding for a comparable operation by a more traditional metalworking process, such as milling or planing.

Extra heavy rotary-table face grinding machines are designed for the grinding of flat surfaces on very large workpieces, such as machine bases, welded structural parts, large plates, etc., up to a weight of about 100 tons. Figure 4-7.3 provides a general idea of the overall dimensions of this type of very large grinding machine which is built with table diameters of 120, 144, and 160 inches (about 3, 3.5, and 4 m). The largest of these models is normally equipped with an 80-inch (about a 2-m) diameter segment chuck, has a 250-hp spindle drive motor and weighs about 165 tons, and is used for such uncommon operations as machining the faces of steam turbine casings.

It is characteristic of the rotary-table face-grinding method that even with such mammoth size machines, designed and dimensioned for very substantial stock removal, good accuracy and finish of the ground surface are still realistic objectives. These particular machines are also built with quick-acting wheel head tilt adjustment, permitting the use of the slightly inclined spindle position for efficient roughing and the quickly established precisely perpendicular spindle position for accurate finishing, as two consecutive phases of the same operation.

Multiple-Head Face Grinders with Rotary Work Table

Automatic processing in a continuous operation of small and medium-size components, is also desirable for the grinding of flat surfaces. Most of the conventional systems of operation which were discussed in the preceding examination of surface grinding are, however, noncontinuous processes. These produce the required work surface gradually, applying an incremental infeed of the wheel against the workpiece which passes repeatedly across the operating wheel surface, until the final work size has been reached. Thereafter, the work, a single piece or batch, is removed, new work is loaded and the same operational sequence is repeated. In the previously reviewed surface grinding methods it is in exceptional cases only, such as in the rarely used creep grinding, that repetitive wheel and work contact can be dispensed with. For reasons related to the mechanics of work handling, continuous processing in surface grinding is generally predicated on single pass contact between the automatic work loading and discharge.

Rotary face grinding, a process having the capacity of substantial stock removal per pass, has the potential for continuous work processing. However, a single pass carried out with a wheel selected to assure the removal of the total grinding allowance, will have limitations with respect to the attainable size control and surface finish. A second, finishing pass in a subsequent operation, possibly carried out on a similar machine, is feasible and sometimes actually applied, but it involves two pieces of equipment, additional tooling, more floor space and material handling equipment.

Courtesy of Mattison Machine Works

Fig. 4-7.3 Extra heavy, rotary-table face-grinding machines, representing the adaptability of the system for grinding flat surfaces efficiently and accurately even on exceptionally large workpieces.

Installing two grinding heads, side-by-side around the periphery of the rotary work table, the work passing consecutively beneath the adjacent heads, combines functionally, the operation of two successively used grinding machines. *Double-head rotary face-grinding machines*, the first head with a coarse wheel for roughing, and the second head with a finer grain wheel for finishing, each wheel set to the appropriate depth of penetration, are successfully used for the automatic surface grinding of continuously processed small-size workpieces (see Fig. 4-7.4). Such machines are frequently equipped with automatic size control, for both the roughing and the finishing stations, and are serviced by automatic work-handling devices. The work-handling comprises, in the typical case shown in the illustration, a power driven rotary feed table introducing the parts on a sectional rotary magnetic chuck which holds the parts only while passing under the wheel, without interfering with the free loading and discharge, the latter preceded by demagnetization on the chuck. This leaves only a minimum of remanent magnetism in the workpiece which will subsequently be eliminated by the part passing through a demagnetizer coil, also shown in the illustration.

Fig. 4-7.4 Double-head, rotary face-grinding machine, successively roughing and finishing workpieces which are carried continuously across the wheel faces by the rotary magnetic chuck.

Depending on the amount of stock to be removed from the work surface and on other product-related conditions, more than two grinding steps may be needed, such as, e.g., a roughing, two semi-finishing, and a finishing pass. Multiple-head, rotary-table face-grinding machines are also built with three, four, or five grinding wheel heads, for complying with the requirements of specific workpieces the continuous

process type surface grinding of which requires several successive passes.

Rotary-table face-grinding machines in the *center column design* represent another concept for applying surface grinding as a continuous operation for workpieces which (a) require several grinding passes, and (b) must be held in a mechanical fixture, the configuration or the locating surfaces of the part not permitting the use of direct magnetic retention on a chuck. In this design the rotary table has the shape of a ring surrounding the center column, whose vertical guideways support several grinding heads, arranged around the inside border of the rotating table (see Fig. 4-7.5).

Fig. 4-7.5 Center column type, rotary surface grinder with several heads mounted centrally inside the ring-shaped work table. Adapted for single-pass automatic grinding of workpieces which require several grinding steps.

This arrangement offers a longer work-table periphery, with easy access to the work, a condition particularly desirable for the usual operation of these machines which involves the loading of the workpieces into fixtures mounted on the rotating table; the clamping is actuated manually or automatically. On rotary surface grinders of this design it is even possible to grind, in a single operation, work surfaces with different locations and dimensions, by the appropriate adjustment of the pertinent grinding heads beneath which the rotating table will move the properly positioned and clamped workpieces.

Rotary Face Grinders with Changing Table Positions

In the conventional types of rotary-table face-grinding machines the position of the work table, as defined by its axis of rotation, is fixed in relation to the grinding wheel spindle. Both the introduction of the workpiece into the wheel contact area, as well as the traversing of the work across the wheel face, are accomplished by the rotation of the work table or chuck, around a fixed axis. The work table of these standard machines is rotating continuously while the machine is in operation, consequently, the loading of the work has to be carried out either (a) by stopping the operation including the rotation of the chuck or (b) by loading and unloading the work while the table is in rotation.

Such a loading "on the fly" is possible with a sectional magnetic chuck on which the parts have to be transferred simply by shoving and removed by the action of a mechanical deflector. When individual mechanical part holding is applied then the rotational speed of the table must be adjusted to allow a sufficient time span for the fixtures during their travel over the loading sector.

The above outlined limiting conditions, however, are not always acceptable, particularly for parts whose surface grinding requires faster rotational speed, or whose loading and unloading involves more time than is afforded by a continuously rotating work table.

The following are examples of special rotary-table face-grinding machines in designs permitting work loading in a stationary position while still maintaining a substantially continuous production process.

Indexing Multiple-Chuck Face-Grinding Machines

The grinding machine shown in Fig. 4-7.6 has a circular indexing table containing four groups of two individually rotating work spindles, each of which is equipped with a work-holding device, such as a chuck or an expanding arbor. The table has four indexing positions, 90 degrees apart; one of the positions is for loading and unloading, while the other three are available for grinding operations. The work spindles are rotating in all operational positions, but are braked to a stop in the loading position. Individually controlled, vertical spindle grinding heads, each comprising two adjacent wheel spindles, are located opposite the operational positions of the indexing machine table. Two, as shown in the illustrations, or three of these wheel heads can be installed, permitting four or six concurrent operations for each indexing cycle of the machine. The adjacent work spindles of the groups can hold identical workpieces, and present the same part of their surface for grinding. Other alternatives are two different parts held in adjacent work positions, or identical parts but with different portions of their surfaces presented to the wheels, which are selected and adjusted according to the requirements of the operation.

The consecutive operational stations, two or three —depending on process requirements—are used for roughing and finishing and, possibly, honing. Each grinding head has its individual feed control mechanism with interchangeable cams which control the approach, infeed, spark-out, and retraction of the spindle to its high position, where the automatic wheel dressing and wear compensation takes place.

Rotary Duplex Surface Grinder

This designation applies to a particular type of rotary-table face grinder (Fig. 4-7.7) which has two individually driven rotary tables or chucks, arranged side-by-side on a traversing carriage. During the grind-

Courtesy of The New Britain Machine Co.

Fig. 4-7.6 Vertical indexing type surface-grinding machine with main table equipped with four pairs of individually driven work spindles brought sequentially into work station under the double spindle grinding heads.

Courtesy of Mattison Machine Works

Fig. 4-7.7 Duplex rotary surface grinder with two individually rotating chucks mounted on a traverse indexing carriage.

ing operation one of the rotary tables is under the grinding wheel, while the other table is in the free access loading position. When the grinding operation is finished, a longitudinal indexing movement brings the newly loaded table into the operational position and moves the previously operating table into a position away from the wheel, for unloading and re-loading. Consequently, these auxiliary operations can take place while the grinding is in progress and thus are essentially continuous, except for the few seconds required for the indexing movement.

For workpieces, whose loading and unloading time is approximately the same as the required actual grinding time, the duplex design can assure, with one operator, a production rate which is approximately equal to that accomplished by two rotary face grin-ders, generally attended by two operators.

Face Grinding with Reciprocating Work Movement

For workpieces with general dimensions exceeding the capacity of the chuck surface of a rotary grinder having the proper wheel size, or when the configura-tion of the work is better suited for the grinding path of a traversing table, the reciprocating type surface grinder may provide the best suited grinding method. When the required rate of stock removal or other operational conditions call for grinding with the face of the wheel, the appropriate system combines a recip-rocating table movement with a vertical grinding spindle on which a cylinder or segment wheel is mounted.

Vertical spindle, reciprocating-table type face-grinding machines, although used to a much smaller extent than the rotary-table type, are produced as regular models by several manufacturers. Figure 4-7.8 illustrates one of the heavy models, which is manu-factured in table lengths of 36 to 240 inches (about 0.9 to 6 m) and table widths from 13 to 36 inches (about 330 to 915 mm).

Courtesy of Mattison Machine Works

Fig. 4-7.8 Reciprocating table, vertical-spindle face grinder. Photo shows one of the largest models of a system which is also applied to smaller machines.

Another type of reciprocating face-grinding ma-chine, used much less frequently and only for work-pieces of exceptionally large general dimensions both in length and width, are the planar type surface grinding machines. These machines, also called planos, are manufactured with either peripheral or face type grinding heads, also as a dual type with both systems of heads. Typical range for the table length is 120 to 384 inches (about 3.05 to 9.75 m) and for the table width, 50 to 96 inches (about 1.27 to 2.44 m). These types of heavy machines are needed particularly when the width of the work sur-face exceeds the pertinent capacity of the regular, open side, vertical-spindle surface-grinding machines.

Precision Type Face Grinding Machines

The principles of grinding a flat work surface with the face of a wheel which is mounted on a spindle perpendicular to the plane of the work movement offers several favorable process characteristics. One of these is the inherent capacity for producing a flat surface of excellent geometric accuracy. Due to the self-dressing property of the rotating wheel face in contact with a work surface which is moving in a plane normal to the wheel's axis of rotation, the flat-ness of the ground surface can consistently be assured. An essentially uniform cross-hatch pattern of the ground surface is an obvious proof of the high grade flatness of the ground surface.

In appraising the capabilities of face grinding in general, the emphasis is primarily on the work size capacity and the grinding performance, that is, prop-erties supported by the dimensions, the massive design, and the substantial drive power of these ma-chines. Grinding quality with regard to size control and particularly flatness accuracy, are commonly mentioned as favorable characteristics of the ground surface, but usually they are not considered the prime objectives of the process.

There are, however, face grinding machines which are designed and built for the primary purpose of producing surfaces of excellent flatness, dimensional accuracy, and finish, by utilizing the inherent cap-abilities of the face-grinding method. Figure 4-7.9 illustrates such a precision type face-grinding machine with traversing work table. Similar machines are also built with rotating work tables.

The illustrated model mounts cup wheels 2 to 10 inches (about 50 to 250 mm) in diameter, and the dimensions of the commonly used magnetic chuck are 18 × 7 inches (about 450 × 175 mm). The machine table is mounted on precision type pre-loaded bearing rollers guided in a flat and in a

Courtesy of George Mueller - Austin Industrial Corp

Fig. 4-7.9 Precision type, vertical-spindle face-grinding machine with reciprocating table.

V-shaped way. The speeds of the hydraulic table movement are infinitely variable to 400 inches per minute (about 10 m/min). For applying the "creep-grinding" or "deep-cut grinding" method, often of advantage for the grinding of cemented carbides or nonmetallic materials, fine table speeds, infinitely variable within a range of 0.48 to 19.20 inches per minute (about 12 to 490 mm/min) are also available. The downfeed by means of a stepping motor, can be set for increments of 0.000 040 to 0.004 inch (.001 to .1 mm) for general work, and from 40 to 360 microinches (.001 to .009 mm) for fine finishing. The machine is designed to operate also with diamond wheels.

These precision type surface grinders can be equipped for operating in an automatic cycle comprising roughing, fine grinding, spark-out, and retraction; the initiation of these subsequent steps being controlled by a direct sensing electronic gage.

Oscillating Wheel Type Face Grinders

Another, recently developed variety of face grinding machine is functionally comparable to a reciprocating-table face grinder, however, the actual movements of the machine members differ in two major respects from the conventional: (a) the traversing path between the wheel and the work is along a circular arc instead of a straight line; (b) the relative movement is brought about by the oscillation of the wheel-head column about a fixed axis, instead of the usual reciprocating table. Figure 4-7.10 illustrates this uncommon type of face-grinding machine, representing a system with advantages within a limited size range, such as a reduced cost in relation to comparable capacity machines of conventional design.

Courtesy of Universal Vise & Tool Co.

Fig. 4-7.10 Oscillating wheel type face-grinder with the wheel head reciprocating over an arc-shaped, fixed work table.

Face-Grinding Machines with Horizontal Wheel Spindles

These machines represent the type to which, in certain usage, the designation "face grinder" is exclusively applied. Horizontal-spindle face grinders have a rather limited field of application and are generally used for workpieces, such as machine blades, which, due to their small thickness, are more easily handled and less prone to deflection when mounted on a vertical table. Another field of preferred application for these machines is the grinding of very large parts, often of awkward shape, such as those requiring surface grinding in railroad shops and shipyards. Machines of this general type are built in two major design systems:

(1) Traveling table type face grinders (see Fig. 4-7.11), which represent the more common version and are manufactured with a wide range of capacities, up to about 240 inches (about 6 m) length.

(2) Traveling wheel-head type face grinders, usually made in the greater lengths within the above mentioned range, are preferred when savings in floor space achieved by this system are considered a deciding factor.

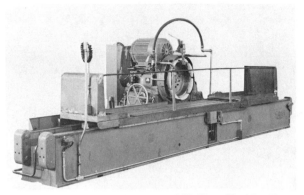

Courtesy of Mattison Machine Works

Fig. 4-7.11 Traveling table type horizontal-spindle face grinder for grinding flat surfaces on heavy parts of large length.

Disc Grinding

Disc grinding is the designation of a particular type of face grinding which uses the entire wheel face or a substantial portion of it as the operating surface. This is one of the major differences from the common methods of face grinding, the latter using as an operating surface the ring-shaped face of a solid or segmental cylindrical wheel.

The wheels used in disc grinding have a limited thickness in relation to the diameter, thus resulting in a disc shape, hence the designation of the method. The disc shaped wheels are supported on their entire back surface by a strong plate. Such a solid mounting is needed to compensate for the inherently limited strength of a large diameter, but relatively thin wheel which is exposed to grinding forces acting in a direction normal to its face. The whole assembly, comprising the disc and the backing plate, is then attached by bolts to the spindle flange.

Carrying out the grinding operation with the work in contact with a wide wheel face creates particular conditions which are advantageous in several respects, but also present various limitations. The preferential application areas of disc grinding are those in which the attainable benefits predominate and the system's drawbacks may not even affect the particular process.

Grinding flat work surfaces with disc wheels is a method applicable to several types of operations and equipment; these will be discussed in the subsequent paragraphs of this chapter. The most widely used application of the method, because of its obvious advantages for many types of workpieces and processes, is double-disc grinding operating with two opposed discs in simultaneous contact with opposite sides of the workpiece. Therefore, in the following brief review of the generally experienced advantages and limitations of the method, the appraisal is based primarily on the use of double-disc grinding machines.

Potential Advantages of Disc Grinding

(a) Full use of the entire or a substantial part of the wheel surface, resulting in uniform wheel wear and favorable abrasive utilization.

(b) Uninterrupted travel of the workpiece across the wheel surface can frequently be applied, thereby the method is adaptable to the continuous processing of large quantities of identical parts.

(c) In its double-disc version, it permits the simultaneous grinding of two opposite surfaces even on thin workpieces, by requiring only lateral support of the work being ground.

(d) The method is also adaptable to a high degree of process automation, including work loading and discharge.

(e) It produces work surfaces with very good flatness, parallelism and finish, as well as uniform size, particularly when automatically controlled.

Potential Limitations of Disc Grinding

(a) Due to the lack of free chip disposal and coolant access, particularly when grinding broad work surfaces, the depth of cut is generally less than in the common types of face grinding.

(b) Asymmetrical parts with substantially different ground areas on opposite sides do not have such a positive support as when held on a rigid machine table.

(c) The disc grinds the entire exposed work surface; the process generally does not permit selective grinding and is not directly applicable to parts with protruding elements.

The Systems and Equipment of Disc Grinding

Disc-grinding machines are manufactured in different design systems which are adapted to particular operational objectives and specifications.

The objectives of the grinding operations which are carried out with disc wheels can generally be associated with either of the following groups, listed in the order of increasing requirements:

(1) To produce an essentially flat surface on one side of the workpiece. The first may be followed by one or several similar operations for producing additional flat surfaces on the same part.

(2) To grind a flat surface parallel with and at a specific distance from a locating surface which is on the opposite side of the part, or to grind a flat surface in a particular position with respect to some other reference element on the workpiece.

(3) To grind two mutually parallel flat surfaces at a specific distance apart and in a position generally perpendicular to a third surface serving as the locating surface during the grinding of the workpiece.

For meeting the first objective (1), the part surface to be ground must be presented to the disc face and held by pressing it with an adequate force against the disc. As long as the orientation of the part does not change, the general shape of the wide contact abrasive disc will produce a surface whose flatness is commonly better than achieved by any other method applied in such a loosely controlled process.

Disc grinding machines used for such operations are either of the horizontal- or vertical-spindle design. The horizontal-spindle grinders may be of the single-end or the double-end type and are similar to pedestal grinders, except for the wheel shape which, instead of being a straight wheel cutting with its periphery, it is a disc cutting with its face. The design of the work support members is adapted to the operating characteristics of the wheel.

The vertical-spindle, open-end disc grinders with the disc surface in the horizontal plane, permit a generally more convenient work holding, provide a better control of the work orientation, and all or part of the required holding force may be produced by the weight of the workpiece. Such machines optionally can be equipped with a hand-operated swinging work table for controlling the distance of the ground flat surface from the opposite side of the part.

Although these grinding machines operate with abrasive discs, the objectives of the operations and the applied equipment differ from those considered as applicable to precision grinding.

The second objective (2), of grinding a single flat surface in controlled dimensional and positional relation to an existing one, commonly parallel with that to be ground requires a single spindle grinder only, although equipped with work-holding members which assure proper directional and size control of the workpiece.

Single-spindle disc grinding machines, not a very frequently used variety, are similar in design to the double-spindle type, except for the number of wheel spindles. Because of the need for dependably locating the workpiece on the side opposite to that exposed to the grinding action, these machines are commonly built in the vertical upper spindle design, that is, with a vertical grinding spindle installed above the horizontal work table. This system is also used for parts which are held in special clamping fixtures and are ground on a single surface only.

The third objective (3), of grinding simultaneously the two parallel surfaces of the workpiece can most effectively be accomplished with the double-spindle disc grinders. These operate by passing the work between two mutually opposite operating surfaces of two abrasive discs, mounted on independent but directionally and positionally aligned spindles. By adjusting the proper distance between the essentially parallel discs, which are set or dressed to a shallow entrance angle, the parts passing through that gap will be ground with a good parallelism of their opposite surfaces and to a uniform size. Furthermore, disc grinding with its wide area of contact tends to produce a good flatness and surface finish.

While this capability of double-disc grinding is basically inherent in the system, the actually achieved workpiece conditions will, of course, depend on many equipment and process factors, several of which will be discussed later.

Double disc grinding offers the highest potential for utilizing the beneficial properties of disc grinding. This is the system in which disc grinding is finding its widest applications in industry; therefore, the following discussion of disc grinding as a general method will primarily apply to the use of double disc grinders.

Design Systems of Double Disc Grinding Machines

These grinding machines which operate with the opposite faces of two rotating abrasive discs, supported independently on separate but essentially aligned spindles, are manufactured in two design systems: the vertical and the horizontal spindle type. Each of these basic design systems may be generally justified by particular operational advantages with respect to different types of workpieces. In some other cases the relative advantages are not distinctly in favor of one particular type. In such borderline cases the trend in this country is toward the horizontal spindle design, while some foreign manufacturers seem to apply the vertical spindle system more frequently. Following are a few relative advantages which may assist in appraising the suitability of each system for a particular set of operational conditions.

The vertical spindle design:

(1) Permits a more convenient hand loading of the workpieces into the feed wheel or carrier.

(2) After the feed wheel has carried the parts through the grinding area, the work can be loaded by gravity into a chute.

(3) Thin parts or any parts having a small thickness in relation to the area being ground, can be supported outside the actual grinding area in a more stable manner when the work feeding is in a horizontal position.

(4) Workpieces, such as coil springs, which do not offer a dependable locating surface perpendicular to that being ground, are more securely located during the feeding when resting on one of the flat surfaces which are to be ground.

The horizontal spindle design:

(1) Offers several alternatives for the feeding and retention of the work during grinding (see Table 4-8.1)

(2) Will more positively locate the workpiece on a surface normal to that being ground, because it is aided by gravity

(3) Provides the natural position for feeding round parts by gravity, as they are rolling or sliding down along the rails of a chute

(4) The weight of the part is not a factor which would disturb the application of a balanced grinding force.

(5) Coolant application and discharge can be controlled more uniformly, and gravity aids the effective flushing away of chips and loose grit.

Work Feeding in Disc Grinding

One of the major advantages of disc type surface grinding is the general adaptability of the method to fully automatic operation. The automatic processing on machine tools is predicated on continuously operating work feeding, therefore, the means and functions of work feeding systems used in disc grinding deserve particular attention.

Material feeding for continuous operations may be considered to have two phases: (a) transporting the parts to the machine, generally in a properly oriented position and at a rate assuring uninterrupted supply; and (b) the actual feeding into the machine, a function often termed "loading," which comprises the transfer of supplied parts into the operating area of the machine, also presenting the work to machining and then discharging it at the end of the operation.

Phase (a) of this process is for continuously operating disc grinding machines very similar to that used on other machine tools with comparable output, such as centerless thrufeed grinding machines. Such equipment consists of vibratory feeders, elevators, etc., with chutes and other types of guide tracks, adapted to the size, configuration, and other conditions of the workpiece. This phase of material feeding, because of its features in common with those used in other processes, will not be discussed in detail. The following examination applies to phase (b) of work feeding, the loading of the work, which has several particular characteristics in its application to disc grinding.

To facilitate the appraisal of the different work feeding systems used in disc grinding, particularly in the operation of the double horizontal-spindle disc-grinding machines, a survey of the most frequently used systems is presented in Table 4-8.1.

While the reviewed systems are applicable to operations in which two opposite sides of the workpiece must be presented simultaneously to discs which face each other, feeding devices of similar design are used also for single surface disc grinding.

In the subsequent paragraph on disc grinding applications the feeding devices serving the operations which are discussed as examples, will also be pointed out. This procedure is followed in view of the basic and controlling function which work feeding and loading have in disc type surface-grinding operations.

Typical Applications of Double Disc Surface Grinding Machines

The following examples are presented in the order of the listing of work feeding devices in Table 4-8.1.

Continuous feeding of round parts which are successively inserted into holes, serving as work-holding nests near the periphery of a rotating carrier disc. — The holes have bushings with bore diameters providing a slip fit for the outside of the ground workpieces, which thus are retained during surface grinding in a position resulting in good squareness control. The insertion of the parts into the bushed holes is usually assisted by mechanical devices and an air blast is often applied to discharge the work when the carrier emerges from the grinding area. The method is capable of high production rates, even in the order of 60 to 300 parts per minute, depending on the diameters of the workpieces.

Continuously feeding parts of irregular shape into specially designed nests, which are arranged in an annular pattern within, yet close to the periphery of the carrier disc (see Fig. 4-8.1). Retainment in work-holding nests whose form complies with the outside

Table 4-8.1 Work Feeding in Disc Grinding Systems, Methods and Devices of Work Loading

Work Loading System	Work Loading Operation	Designation of the Method	Diagram	Applicational Characteristics	Examples
Continuous	The parts are loaded at uniform rate in positions either close or adjacent to each other.	ROTARY — with disc shaped carrier plate		Nests around or near the periphery of the carrier plate, usually outside-notches or bushed holes, will accept parts from a feeder and carry them in an arcuate path across the disc wheel face, discharging the parts after they are ground.	Figs. 4-8.1 4-8.2
		FEED-THRU — with supporting and confining rails		The most productive method with respect to the linear speed at which the work is passing through the disc wheels. Applicable to round parts which can rotate and transmit the inserting force, thus pushing the load continuously through the disc contact area.	Fig. 4-8.3
Intermittent	The parts are introduced singly or in groups into the grinding zone, while other work holding positions of the loading fixture are disengaged for loading and unloading	INDEXING —with carrier plate containing a few, commonly three, work holding members		Used for large parts requiring individual loading which is carried out while the grinding of the preceding part is in progress. Third position of the carrier is used for work discharge.	Fig. 4-8.4
		SWIVEL-ING — a swivel arm with work holding fixtures at opposite ends		Parts can be mounted individually or in batches into the holding fixture at one end of the swivel arm while a similar fixture at the other end presents the work to the action of the disc shaped grinding wheels.	Fig. 4-8.5

(Continued on next page.)

Table 4-8.1 (*Cont.*) **Work Feeding in Disc Grinding Systems, Methods and Devices of Work Loading**

Work Loading		Designation of the Method	Diagram	Applicational Characteristics	Examples
System	Operation				
Discrete	The loading device contains a single work holding fixture which is either in the handling or in the grinding position.	GUN TYPE — with the work-holding fixture sliding along straight guides between the loading and the grinding positions		For the individual handling of parts at lower production rates, with manual or power movement. While held in the grinding zone the part is reciprocated between the contacting disc wheel faces.	Fig. 4-8.6
		SWING (OSCIL-LATING) TYPE — work-holder mounted on a swinging arm which also carries out an oscillating motion		Similar in purpose to the gun type loader, with the swinging and oscillating arm carrying out its movements around a pivot, instead of a straight traverse between guides.	Fig. 4-8.7

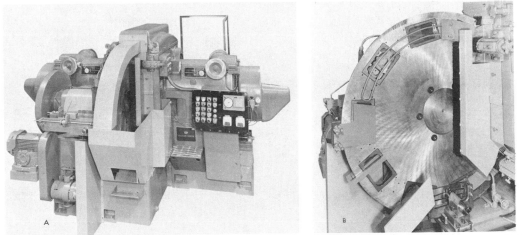

Courtesy of Bendix Industrial Tools Div.

Fig. 4-8.1 Double horizontal-spindle disc grinder with continuously rotating carrier disc with work-holding nests of special shape into which the workpieces are loaded automatically. A. General view of the grinding machine; B. Close-up of the specially designed carrier disc.

contour of the part, assures dependable positioning of the workpiece even while subjected to considerable grinding force during its passing between the abrasive discs.

Continuous feeding of parts into notches around the periphery of the carrier disc. — This system is used for longer parts, and even those of asymmetrical shape, such as automotive valves whose head and stem ends must be ground flat, mutually parallel and perpendicular to the stem axis. The notches are contained in a single, or in two parallel, plates of the carrier, with forms and dimensions corresponding to that portion of the part for which the notches are to provide a well-fitting seat. Similar carriers with V-block type notches are used to receive and to retain the round portions of parts of various shapes such as universal joint spiders (see Fig. 4-8.2) whose opposite trunnion ends are ground parallel, first normal to one axis, then the part may be indexed by 90 degrees and introduced for a second time to be ground perpendicular to the other axis. The workpieces are loaded into the carrier in a centered position and, during the grinding, are retained in the seats of the carrier by means of a linked chain.

Rotating carrier discs also can have clamping fixtures around their periphery, to retain parts which must be located on certain spots of their surface. The clamping may be operated manually or by power.

Feed-thru disc grinding offers the highest productivity for the parallel face grinding of round or, less frequently, for otherwise shaped thin parts, the operation being carried out on double-disc surface grinders tooled and equipped for such work (see Fig. 4-8.3)— The tooling comprises guide rails for the support and confinement of the workpieces during their travel between the abrasive discs, and a pair of driven belts or rolls, for forcing the parts through the gap between the discs. Such machines must also be equipped with feeding devices for transferring the parts from a hopper or elevator to the loading belts. It is also practicable, by this method, to grind parts with nonuniform face diameters by selecting the composition and adjusting the rotational speed of the grinding discs to compensate for the differences in the work surface areas, thus assuring essentially uniform abrasive wear. Thin and long workpieces may be transferred to the loading area on a belt in a direction perpendicular to the feed-thru movements, and then picked up, one at a time, by a chain which introduces the parts, end-by-end, on the feed-thru guide rails. Similar feeding methods are also used for the feed-thru disc grinding of piston rings and other thin round parts.

Intermittent work loading approaches the effectiveness of continuous grinding by reducing idle time to an insignificant portion of the grinding cycle—This is accomplished by providing separate loading positions

Courtesy of Gardner Machine Co.

Fig. 4-8.2 Work area of a double disc grinder equipped for the grinding of universal joint spiders held in peripheral notches of the rotary carrier. Also shown is an automatic indexing device for the reorientation of the parts following the first pass.

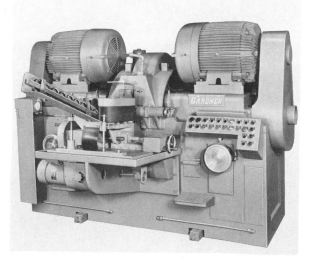

Courtesy of Gardner Machine Co.

Fig. 4-8.3 Double horizontal-spindle disc grinder equipped for the feed-thru grinding of bearing rings which roll down an inclined rail, and are introduced into the grinding area by the action of a pair of feeding belts.

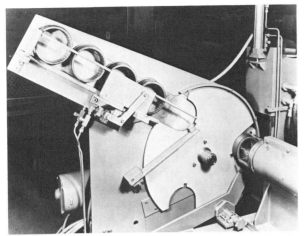

Courtesy of Gardner Machine Co.

Fig. 4-8.4 Work area of a double disc grinder with indexing carrier plate, operating in combination with a rotating work-holding arbor which is mounted on a swing arm.

of execution of intermittently operating loading devices are shown. Figure 4-8.4 illustrates an indexing carrier plate with three positions, a device which is widely used for operations requiring intermittent work loading. In this particular application the workpiece, a large, tapered roller-bearing inner ring, is picked up from the carrier plate by a rotating arbor which then introduces the part into the grinding area; here the abrasive disc advances in an infeed movement for gradual stock removal from both end surfaces of the rotating workpiece. The operation is carried out in an automatic cycle which also comprises spark-out and wheel wear compensation. In Fig. 4-8.5, a swivel type work-holding fixture is

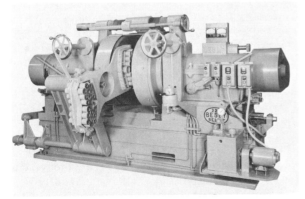

Courtesy of Bendix Industrial Tools Div.

Fig. 4-8.5 Large size, horizontal-spindle, double disc grinder equipped with an intermittently operating swing arm type work-holding device, designed to accept a batch of workpieces.

shown. This fixture has two positions, one for work loading and unloading and the other for the actual grinding, which again is carried out gradually, as the result of the slow infeed movement of the abrasive discs.

Separate (discrete) feeding is needed for large workpieces, sometimes of awkward shape, and is also acceptable, even economically justified for small quantities.—Generally, the workpiece is inserted into the work-holding portion of the fixture while in the loading position. The fixture then transfers the workpiece into the grinding area and keeps it there while being moved between the faces of the discs until the grinding is completed. The reciprocating or oscillating movements involved are either in a straight line or an arc, depending on the design of the feeding device. The *gun type feeder* shown in Fig. 4-8.6, transfers the part in a straight line movement from the loading into the grinding position and will keep it there in a reciprocating motion. The *swing type feeder* has an arm moving around a pivot, first in a swing movement transferring the part into the grinding area, then oscillating it between the abrasive discs while the grinding is in progress (see Fig. 4-8.7).

Courtesy of Gardner Machine Co.

Fig. 4-8.6 Double disc grinding machine with power operated gun type fixture for loading parts larger than acceptable in automatic processing.

Discrete feeding is sometimes applied for quality reasons as the to and fro movement, whether reciprocating or oscillating, is conducive to developing excellent flatness and surface finish.

The preceding examples, although typical with respect to the majority of disc grinding operations, do not constitute a complete listing. There is a large number of workpieces which, because of size, general configuration, material allowances, the position of

Courtesy of Bendix Industrial Tools Div.

Fig. 4-8.7 The work loading area of a double, horizontal-spindle, wet disc grinder with oscillating mechanism for grinding the parallel faces of workpieces, such as power steering spools. The oscillating motion contributes to generating a fine finish and the sequentially controlled dressing with automatic compensation maintains close size tolerances.

the locating surfaces, etc., may present individual problems. These can often be resolved either by minor modifications of the listed basic systems, or by the development of new approaches for permitting the successful application of the disc-grinding method.

The Disc Grinding Machines

In reviewing the disc grinding machines, a rather broad category of functionally related equipment, the *double horizontal-spindle disc grinders* are discussed first because of their very wide application in industry. Although these machines are built in a wide range of sizes with respect to the diameters of the abrasive discs, the drive power and the machine weight, the capacity of workpiece width is rather similar for all machine sizes. Typical values are for feed-thru grinding, 2½ inches (about 60 mm) and for carrier held parts, maximum 4 inches (about 100 mm). These capacity data characterize the appropriate field of application of double-disc grinders, which are primarily adapted for relatively thin parts.

Considering a popular range of double-disc grinders, a typical model of which is shown in Fig. 4-8.3, the diameters of the applicable abrasive discs range from 12 to 42 inches (about 300 to 1070 mm). The diameter of the abrasive disc is one factor for the acceptable part diameter, the other factor being the applied feed system. Feed-thru has the lowest diameter capacity limit, reciprocating feed, a higher capacity, and the rotary or indexing carriers have the maximum, although the differences between these

capacities may only be in the order of about 20 per cent for the consecutively listed types.

The drive power for each of the independently driven wheel motors varies from 5 to 40 hp, however, double-disc grinders of modern design may have drive motors up to about 100 hp for similar disc diameters.

Several design details of the double-disc grinders vary according to the applicable feed systems which, therefore, are not interchangeable. This condition, however, is generally not regarded as a serious limitation since double-disc grinders are predominantly used for a specific kind of production work. In such applications the dimensional and other details of the workpieces may vary, thus requiring different tooling, but the suitable feeding system will remain the same.

Finally, as a frame of reference, the weights of typical representatives of double-disc grinding machines vary from 3 to 8 tons, however, some models of the modern high powered machines weigh about 12 tons.

Besides these informative data the different models of double-disc grinding machines vary in many respects, including the regular and optional equipment which different manufacturers offer and users require. Consequently, the degree of automation, as well as the extent of setup and operational controls can often be adapted to the actual requirements of the intended use.

Vertical-spindle disc grinding machines are used in the single, usually upper spindle design or, more frequently, as *double vertical-spindle disc grinding machines*. A rather common application, although not an exclusive one, is the grinding of coil spring ends, usually loading the springs manually into the holes or other shapes of nests in the rotating carrier disc. The horizontal position of the carrier has several advantages for such applications: it permits easy loading, the parts can be discharged by gravity after leaving the grinding zone, falling through an opening in the guide plate of the carrier, into an exit chute. Finally, the vertical spindle design is adaptable to the tandem arrangement of two pairs of grinding spindles, the workpiece being carried successively through two double end grinding zones; one used for roughing, and the other for finishing the part in a single operation (see Fig. 4-8.8).

The preceding very brief review could only point out a few characteristics of the most widely used types of disc grinding machines. Neither the listing of the types, nor the description of the individual models is complete, and there are a large number of applications in which, by using specially designed equipment, disc grinding proves to be the most advantageous method for producing flat work surfaces.

Courtesy of Bendix Industrial Tools Div.

Fig. 4-8.8 Double vertical-spindle, tandem, dry disc grinder specially designed for grinding the end of coil springs. The two pairs of spindles, one for rough and the other for finish grinding, are placed at a sufficient distance apart for providing ample cooling space between the successive phases of the surface grinding, (*Left*) General view of the double vertical-spindle, tandem disc grinder. (*Right*) Close-up of the multiple station carrier disc (feed wheel) with interchangeable spring holder blocks.

The Capabilities of Double-Disc Grinding

The method selected for a particular operation is, from a manufacturing engineering aspect, determined first by the adequacy and second by the capabilities of the method.

The types of workpieces and surface configurations for which disc grinding, particularly with two opposite abrasive discs, is the adequate method, have already been pointed out in the preceding discussions. However, it must also be appraised how fast and how well the method and its equipment are capable of carrying out the process, thus supplying data for:

(1) Determining whether the required product quality and production rates can be satisfied

(2) Comparing it with other methods which are equally adaptable for performing the operation

(3) Establishing the limits of the method's capabilities as a guide for developing engineering specifications.

In the following, some of the most frequently considered capability aspects of double-disc grinding will be examined, indicating values which are close to the upper limits for consistent performance at the present state of technological development.

Uniformity of size. In many types of engineering components the thickness of the part with respect to the distance between two parallel flat surfaces, does not have a primary functional role. For that reason the size-holding capability of disc grinding, which for operations with automatic size control is in the order of 0.001 inch (0.025 mm), will generally be considered satisfactory.

Flatness is a basic requirement in surface grinding. In a direct, functional respect the flatness of a surface must often assure an uninterrupted bearing or sealing area and stable support. Indirectly, the flat surface is the basis for other, often critical geometric conditions, such as parallelism, squareness, and also uniformity of size. Flatness in the order of 60 microinches (0.0015 mm), exceptionally, even better, may be accomplished in disc grinding on parts of 2-inch (about 50-mm) diameter, or smaller, with proportional values for larger sizes.

Parallelism of the surfaces on opposite sides of the workpiece is a condition for the assurance of which double-disc grinding has a uniquely outstanding capability, particularly when also considering the pertinent production rates. Parallelism in the order of 0.0001 inch per inch (0.0025 mm/25 mm) is often feasible, although predicated on the excellent control of the operating equipment and work preparation.

Squareness, expressing the perpendicularity of the ground surface to that serving as the locating surface during the grinding process, is dependent on the geometric condition of the latter. Predicated on an excellent locating surface and on part configuration permitting stable positioning for the grinding, squareness in the order of 0.0003 inch per inch (0.0075 mm/25 mm) may be consistently achieved in precision disc grinding.

Surface finish can be consistently held to 4 micro-inches in AA (0.1 micrometer R_a) when required for functional reasons. Such a fine finish, however, is only exceptionally needed for flat surfaces of industrial parts, and generally, finish values in the range of 16 to 32 microinches in AA (0.4 to 0.8 micrometer R_a) are specified. Such surface conditions commonly result in disc grinding processes whose data have been established with the purpose of good flatness and parallelism control.

Production rates in double disc grinding with continuous feeding may be indicated by the following typical values which, however, are in the territory of top performance:

(a) with rotary carrier discs having two rows of bushed holes to accept cylindrical rollers of 0.325 inch (about 8 mm) diameter—300 pieces per minute;

(b) in feed-through grinding of bearing rings with 2-inch (about 50-mm) diameter—a linear rate of 400 inches (about 10 m) per minute.

Stock removal per side in double disc grinding:

(1) in finishing operations with rigorous control of several parameters—0.004 to 0.010 inch (about 0.1 to 0.25 mm);

(2) in general type surface grinding, about twice the above amount; on high-production machines, even more.

The preceding data should serve the purpose of general evaluation, but do not represent the limits of the method's capabilities. Special processes and equipment are being continuously developed, consequently it is possible to obtain by disc grinding, in particular cases, extremely high surface qualities, such as indicated by the following values: flatness and parallelism $50/70\mu$ inches (0.00125/0.00180 mm), uniformity of size 0.0002 inch (0.005 mm), finish 6 to 9 micrometers in AA(0.015/0.022 micrometers R_a). Of course, such extreme values can only be achieved with special equipment, limited (about 0.001 to 0.0015 inch or about 0.025 to 0.038 mm) stock removal and by applying uncommon process data, such as controlled and varied speed for the work traverse between the abrasive discs.

The Abrasive Disc Wheels of Disc Grinding

The disc grinding method, using the entire wheel face or a substantial portion of it as the operating surface, requires grinding wheels which, in general shape and composition, differ from those used in other methods of grinding.

The general shape of these wheels, indicated also by their designation, is essentially a flat disc, with very small thickness in relation to the diameter. Even

for large diameters the thickness of the disc wheels remains essentially the same, and is commonly specified within the limits of 1 to 3 inches (about 25 to 75 mm).

The abrasive discs are made either as solid or as segmental wheels. Solid discs are used commonly in the diameter range of 10 to 48 inches (about 250 to 1200 mm), while for larger sizes, in the overlapping diameter range of 36 to 84 inches (about 900 to 2100 mm), segmental discs are required. The solid-disc wheels may have the general shape of an uninterrupted disc, such as that needed for feed-thru grinding, or may be made in the shape of a ring, with a hole in the center. The diameter of that empty space in the center may vary, depending on the radial width of the ring-shaped wheel face area needed to provide full contact to the work surface while it is being carried across the wheel face (such as along an arc-shaped path by a rotary carrier.)

Figure 4-8.9 shows a few typical abrasive grinding discs and a segment, each example representing different wheel face configurations. The disc wheel faces may be smooth, corrugated, or a combination of these two pattern types. The disc wheels can be used with uninterrupted face or with radial grooves, the latter for the purpose of improved coolant distribution.

The abrasive grain material of the disc wheels is generally the same as that used for other wheel shapes when grinding the same types of materials: Aluminum oxide for steel, silicon carbide for cast iron, most nonferrous metals and nonmetallic materials. Sometimes these two abrasive materials are combined to obtain balanced wheel properties tailored to a specific set of operational conditions.

Courtesy of Gardner Machine Co.

Fig. 4-8.9 Typical abrasive disc wheels, solid and segmental, full and with center hole made with different face configurations. (The abrasive ring shown at far left, is used on face grinding machines.)

The grain sizes selected for disc grinding may commonly be coarser than those used for producing an equal finish by peripheral grinding.

The grades to be used are generally softer than those needed for similar work materials in peripheral grinding. However, conditions of the process, the size of the workpiece, the method and rate of work and wheel feeding, the amount of stock removal, the requirements of the ground surface, will all affect the choice of the best suited wheel grade. In some cases the specialized manufacturers recommend combination grades, different for the outer and the inner zones of the wheel.

The structure of the disc wheels varies from dense to open, using, e.g., dense for heavy stock removal to bring more abrasive grains in contact with the work, and open structure for hardened steel, to assure a cooler cut.

The bond type which is the most frequently used for abrasive discs is the resinoid, because it can withstand higher speeds and greater shocks than the more rigid vitrified bond. Resinoid bond also possesses a resilience which is conducive to good surface finish and can be operated in both wet and dry grinding. Specialized abrasive disc manufacturers will use different types of resinoid bond material, according to the operational conditions for which the wheel is required. Oxychloride bond is used only for discs to be operated dry, such as in grinding the ends of coil springs.

The disc wheels must be mounted on a backing plate; these are known under different designations, "steel disc wheel" being the most commonly used. There are four different mounting methods used by manufacturers of abrasive discs:

(1) The inserted nut type, one form of which is shown in Fig. 4-8.10. This type has nuts molded into the disc material either directly or, for additional safety, into a strong, woven steel-wire net, which is embedded into the abrasive structure at the back of the disc (see Fig. 4-8.11).

(2) The inserted washer type which requires cored holes through the abrasive to provide access to the screws which are inserted from the front into the countersunk holes of the embedded washers.

(3) The plate-mounted type has a separate mounting plate vulcanized or cemented to the abrasive disc. This mounting plate will then be attached with disc mounting screws to the steel disc wheel.

(4) The projecting stud or stud-mounted type, which is manufactured in different designs, such as (a) with the stud anchored permanently into the abrasive disc, (b) with an intermediate mounting plate having tapped holes for the studs, or (c) with

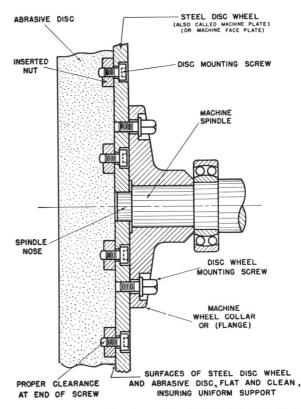

Fig. 4-8.10 Inserted nut type disc wheel holding; diagram shows one form of execution with the terms designating the major elements.

nuts, molded into the wheel, for retaining the ends of the projecting studs.

The steel disc wheel to which the abrasive disc has been attached by either of the above listed methods, is then mounted with bolts on the machine wheel collar or flange which is keyed on the grinding wheel spindle of the grinder.

Fig. 4-8.11 Abrasive disc wheel with wire net embedded into the abrasive structure to assure firmer hold for the inserted mounting nuts.

The dressing of the abrasive discs is not a frequently required operation, due to the substantially uniform wheel wear which is one of the characteristics of this method. For dressing the wheel face the commonly used tool is the diamond nib, two of which are held in a forked arm for double disc grinders. That arm carries out a swing movement which is usually power actuated for assuring uniform traversing speed of the diamond at a preset rate. For the dressing of the wheel head, slides must be moved into the dress-off position, and there advanced by the amount needed for the dressing passes, finally returned into the grinding position, also taking into account the size changes of the wheels caused by the dressing. To assure a dependable control of these slide adjustments, disc grinders are usually equipped with slide position gages, commonly dial indicators or, on some more advanced models, digital readout type movement counters.

Index

361